THE SCOPE OF
MILITARY GEOGRAPHY
ACROSS THE SPECTRUM FROM PEACETIME TO WAR

THE SCOPE OF
MILITARY GEOGRAPHY
ACROSS THE SPECTRUM FROM PEACETIME TO WAR

Eugene J. Palka

Francis A. Galgano

McGraw-Hill Primis
Custom Publishing

New York   St. Louis   San Francisco   Auckland   Bogotá
Caracas   Lisbon   London   Madrid   Mexico   Milan   Montreal
New Delhi   Paris   San Juan   Singapore   Sydney   Tokyo   Toronto

McGraw-Hill Higher Education
A Division of The McGraw-Hill Companies

The Scope of Military Geography
Across the Spectrum from Peacetime to War

Copyright © 2000 by The McGraw-Hill Companies, Inc. All rights reserved. Printed in the United States of America. Except as permitted under the United States Copyright Act of 1976, no part of this publication may be reproduced or distributed in any form or by any means, or stored in a data base retrieval system, without prior written permission of the publisher.

McGraw-Hill's Primis Custom Publishing consists of products that are produced from camera-ready copy. Peer review, class testing, and accuracy are primarily the responsibility of the author(s).

3 4 5 6 7 8 9 0 GDP GDP 0 9 8 7 6 5 4 3 2 1

ISBN 0-07-248480-2

Editor: M. A. Hollander
Cover Design: Dr. Jon C. Malinowski
Printer/Binder: Greyden Press

Contents

Contributing Authors   ix

Preface   xi

WAR FIGHTING

THE WARTIME CONTEXT   1

1. WORLD WAR II IN THE ALEUTIAN ISLANDS:
   Physical Geographic Challenges in the Battle for Attu   9

2. STREAMS AND THE CIVIL WAR MILITARY LANDSCAPE:
   The Battle for Island No. 10, Mississippi River, 1862   31

3. MILITARY GEOGRAPHY OF THE AMERICAN CIVIL WAR:
   The Blue Ridge and Valley & Ridge Provinces   53

4. AMPHIBIOUS WARFARE:
   Operation Chromite: Turning Movement at Inchon   75

5. PELELIU:
   Geographic Complications of Operation Stalemate   111

6. PROTECTING THE FORCE:
   Medical Geography and the Buna Campaign, WWII   133

MILITARY OPERATIONS OTHER THAN WAR

THE MOOTW CONTEXT   161

7. A DECADE OF INSTABILITY AND UNCERTAINTY:
   Mission Diversity in the MOOTW Environment   167

8. THE HUMAN LANDSCAPE AND WARFARE:
   The Military Geography of the War in Bosnia — 197

9. THE GEOGRAPHY OF HAZARD ANALYSIS:
   Disaster Management and the Military — 219

10. STRATEGIC MOBILITY IN THE 21ST CENTURY:
    Projecting National Power in a MOOTW Environment — 233

11. INSURGENCIES AND COUNTER-INSURGENCIES:
    A Geographical Perspective — 263

12. THE CASE OF BOSNIA:
    Military and Political Geography in MOOTW — 291

PEACETIME

THE PEACETIME CONTEXT — 323

13. MANNING THE FORCE:
    Geographical Perspectives on Recruiting — 327

14. TRAINING THE FORCE:
    Bioclimatological Considerations for Military Basic Training — 347

15. TRAINING A GLOBAL FORCE:
    Sustaining Army Land for Readiness in the 21st Century — 359

16. U.S. ARMY TRAINING AND TESTING LANDS:
    An Ecoregional Framework for Assessment — 373

17. MILITARY LAND:
    Multi-Disciplinary Land Use Planning — 393

18. INTEGRATED MILITARY LAND MANAGEMENT:
    The Warfighter's Terrain Perspective — 411

Conclusion   437

Glossary of Common Military Terms   439

About the Editors and Authors   455

Contributing Authors

Eugene J. Palka, Department of Geography and Environmental Engineering, United States Military Academy, West Point, New York 10996

Francis A. Galgano Jr., Department of Geography and Environmental Engineering, United States Military Academy, West Point, New York 10996

Jon C. Malinowski, Department of Geography and Environmental Engineering, United States Military Academy, West Point, New York 10996

Ewan W. Anderson, Centre for Middle Eastern and Islamic Studies, University of Durham, Durham, England

Julian V. Minghi, Department of Geography, University of South Carolina, Columbia, South Carolina 29208

Richard W. Dixon, Department of Geography, Southwest Texas State University, San Marcos, Texas 78666

William W. Doe, III, Center for Ecological Management of Military Lands, Colorado State University, Fort Collins, Colorado 80523-1500

Mark W. Corson, Department of Geology/Geography, Northwest Missouri State University, Marysville, Missouri 64469

Robert B. Shaw, Director, Center for Ecological Management of Military Lands, Colorado State University, Fort Collins, Colorado 80523-1500

Robert G. Bailey, USDA-USFS Inventorying & Monitoring Institute, Fort Collins, Colorado 80524

Kurt A. Schroeder, Department of Social Science/Geography, Plymouth State College, Plymouth, NH 03264-1595

David S. Jones, Research Associate Center for Ecological Management of Military Lands, Colorado State University, Fort Collins, Colorado 80523-1500

Thomas E. Macia, Project Officer, Integrated Training Area Management (ITAM), Headquarters, Department of the Army, Office of the Deputy Chief of Staff for Operations, Washington, D.C. 22332

Joseph P. Henderson, Department of Geography and Environmental Engineering, United States Military Academy, West Point, New York 10996

Andrew W. Lohman, Department of Geography and Environmental Engineering, United States Military Academy, West Point, New York 10996

Wiley C. Thompson, Department of Geography and Environmental Engineering, United States Military Academy, West Point, New York 10996

Brandon K. Herl, Department of Geography and Environmental Engineering, United States Military Academy, West Point, New York 10996

Preface

Military geography is the application of geographic information, tools, and techniques to military problems. The subfield links geography and military science, and in one respect is of an applied nature, employing the knowledge, methods, techniques, and concepts of the discipline to military affairs, places, and regions. In another sense, military geography can be approached from an historical perspective (Davies, 1946; Meigs, 1961; Winters, 1998), with emphasis on the impact of physical or human geographic conditions on the outcomes of decisive battles, campaigns, or wars. Traditionally, both the applied and historical approaches have focused on the military's wartime role. This rigid interpretation of the scope of military geography has long neglected the plethora of opportunities that occur during "military operations other than war" (MOOTW) and/or peacetime endeavors. In general, militaries spend far less time at war than they do in peace. As I have advocated elsewhere (Palka, 1995; 2000), it is time to overcome inertia and broaden the scope of the subfield to address the wide range of problems that militaries experience during MOOTW and peacetime scenarios. The latter two contexts have long been fertile ground for military geographic inquiry, but have become most apparent since the end of the Cold War.

Expansion into the MOOTW or peacetime arenas does not necessitate abandoning traditional themes in military geography. What is called for is a broadening of the military geographic perspective to keep pace with contemporary military concerns. To *not* expand the scope of military geography accordingly, would severely constrain the continued growth of the subfield and ignore the unmistakable trend that has occurred since the end of the Cold War.

This book introduces and showcases a military geography that is broader in scope, essentially devoting the information, tools, and techniques of the discipline to the three distinct contexts within which the U.S. Military (and the militaries of virtually all other countries of the "Western World") has been required to operate. The organization of the book implements the model that was previously introduced (Palka, 2000) to conceptualize the scope of military geography across the spectrum from peacetime to war (see Figure 1).

Figure 1 reflects a conceptually broader military geography, capable of addressing military problems across a full spectrum of employment scenarios, or *contexts*, from peacetime to war. The scheme enables one to identify the magnitude, or *scale* of the operation, and the geographic *approach* involved (approach). By also indicating whether the perspective is *applied* or *historical*, this model provides a comprehensive, yet useful method for organizing and classifying research in military geography.

Previous attempts have been made to develop generic classification schemes to distinguish or categorize work in the subfield. One scheme differentiated between topical, systematic or regional endeavors, based on the nature of geography (Peltier and Pearcy, 1966). Another common practice classified work as either tactical or strategic military geography, based on the level of warfare. The former scheme has limited utility, and the latter has long been obsolete.

The current model, which considers context, scale, approach, and perspective, will facilitate organizing research themes and help to clarify the boundaries of the subfield. The latter will serve to reduce traditional ambiguities and enhance coherence. Perhaps most important, military geography will now be able to address an assortment of military problems that have previously occurred outside the range of the traditional, yet narrowly construed military geography.

Based on the structure depicted in Figure 1, this book is organized into three sections. Part I is devoted to military geography during war. We begin with the wartime context because this is the most traditional forum for military geography. As in the case of the other two contexts (MOOTW and Peacetime), the focus can range from the tactical to the strategic level of analysis, using either a systematic, topical, or regional approach. And, the work can be presented from either an applied or historical perspective. The chapters of Part I are not all-inclusive of the possibilities that exist within this context, but merely representative of the types of Military geography that address the military's problems during wartime. We have included battles and campaigns from the

American Civil War, World War II, and the Korean War to address different timeframes, as well as varied geographical realms.

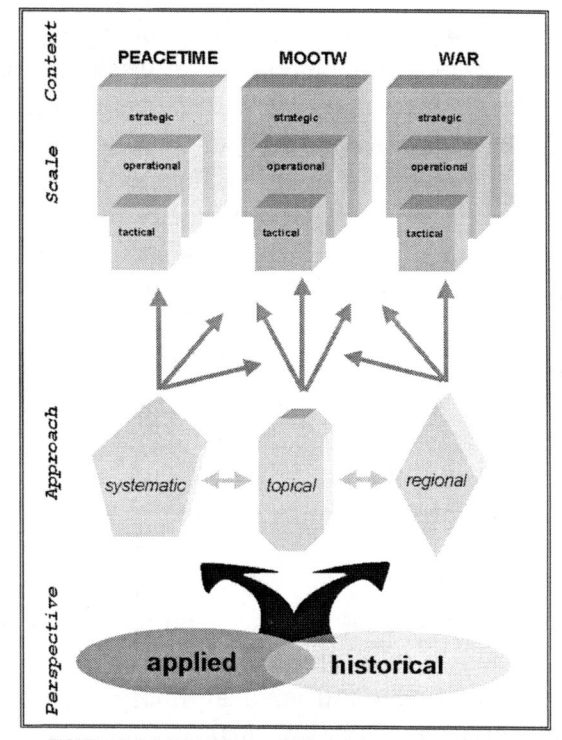

Figure 1. The Scope of Military Geography (Palka, 2000).

Part II is devoted to military operations other than war (MOOTW). This context has been at the center of the stage for the U.S. Military (as well as for other "Western Militaries") since the end of the Gulf War. Indeed over the past decade, the U.S. Military has conducted more missions within the MOOTW realm than it had throughout the entire Cold War Era. MOOTW span the globe. Some are nearly spontaneous in response to natural disasters, while others can be planned in great detail for several months before any formal undertaking. Operations can range from a couple of days (in the case of refugee evacuations for example) to several years (such as operations in Bosnia or Kosovo). Rather than focus on specific operations within the MOOTW context as we have done during the wartime section, we have included several chapters that present a broader perspective and address the general applications of military geography to the MOOTW context.

Part III addresses military problems during peacetime. Virtually all militaries spend far more time involved in peacetime training and operations than they do in the other two contexts combined. Surprisingly, however, this is the least developed of the three contexts. Again, we provide only a sample of problem areas with which the Military must contend within the peacetime arena. The chapter on recruiting, a current dilemma for the U.S. Military represents an age-old problem with which all volunteer militaries have had to contend. Other chapters on land-use planning and training area management constitute areas that have become increasingly important over the past two decades, after years of neglect.

A quick glance at the table of contents might prompt one to conclude that this book is an eclectic assortment of readings loosely linked under the umbrella of military geography. Yet a closer examination of Figure 1 will reveal that every chapter fits well within the scope of its section, and the sections collectively provide organization and coherence to an expanded military geography.

The broadened scope for the subfield, envisioned in Figure 1 and presented in this book, serves as an "enabler," rather than a constraint. Military geography requires the additional latitude to pursue complex problems at various scales across multiple disciplinary and subdisciplinary boundaries, from either an applied or historical perspective (Palka, 2000). Recent works by Corson and Minghi (1996a, 1996b, 1997) and Shaw et al (2000) clearly illustrate the positive benefits of this mindset.

Wartime endeavors will continue to necessitate the use of geographic information, tools and techniques, as long as battles are waged on the surface of the earth or in its atmosphere. But military geography must continue to keep pace with technological developments, global instability, and the changing role of the Military. Geographic information, principles, and tools that have great application to wartime problems have equal, if not more relevance to MOOTW and peacetime scenarios.

Bibiliography

Corson, Mark W. and Minghi, Julian V. 1996a. *Powerscene*: Application of New Geographic Technology to Revolutionize Boundary Making. IBRU Boundary and Security Bulletin. Summer 1996.

_____. 1996b. The Political Geography of the Dayton Accords. *Geopolitics and International Boundaries*, 1/1: 77-97.

_____. 1997. Prediction and Reality in Bosnia: Two Years after Dayton. *Geopolitics and International Boundaries,* 2/3: 14-27.

Davies, A. 1946. Geographical Factors in the Invasion and Battle of Normandy. *Geographical Review,* 36/4: 613-631.

Meigs, Peveril. 1961. Some Geographical Factors in the Peloponnesian War. *Geographical Review,* 51/3: 370-380.

Palka, Eugene J. 1995. The US Army in Operations other than War: A Time to Revive Military Geography. *GeoJournal*, 37/2: 201-208.

_____. 2000. "Military Geography: Its Revival and Prospectus." In *Geography in America at the Dawn of the Twenty-First Century.* Edited by Gary Gaile and Cort J. Willmott. Oxford: Oxford University Press.

Peltier, Louis C., and Pearcy, G. Etzel. 1966. *Military Geography*. Princeton, NJ: D. Van Nostrand Company, Inc.

Shaw, Robert B., Doe, William W. III, Palka, Colonel Eugene J., and Macia, Thomas E. 2000. Where Does the U.S. Army Train to Fight? Sustaining Army Lands for Readiness in the 21st Century. *Military Review*, LXXX/5, (September-October 2000.

Winters, Harold A. 1998. *Battling the Elements: Weather and Terrain in the Conduct of War*. Baltimore, MD: The John Hopkins University Press.

The Wartime Context

Introduction

The use of geographic information to facilitate military decision-making in battle pre-dates written history. One of the earliest examples can be traced to Megiddo (near present day Haifa, Israel) in 1479 BC (Thompson, 1962b). Later historical writings are replete with instances where military leaders recognized the constraints imposed by geographic factors and planned their actions accordingly. Thucydides' history of the Peloponnesian War from 431-404 BC (Meigs, 1961), Xenophon's account of the march of 10,000 Greek mercenaries across Asia Minor to the shores of the Black Sea in 400 BC (Rouse, 1959), and Ceasar's geographic insights during the Gallic Wars from 60-55 BC (Edwards 1939), constitute a small sample of the hundreds of early writings that include a military geographic perspective (Palka, 2000).

Military geography evolved into a formal field of study in Europe during the nineteenth and early twentieth centuries. The first publication exclusively devoted to military geography was Theophile Lavallee's (1836) *Geographie Physique, Historique et Militaire*. A year later, the field gained additional credibility when Albrech von Roon (1837), a captain on the Prussian General Staff, detailed the physiographic descriptions of military regions of Europe.

During the late nineteenth and early twentieth centuries, military geography was linked to the strategies necessary to accomplish national objectives (Peltier and Pearcy, 1966). Within the United States, Alfred T. Mahan (1890) provided the first widely recognized American contribution related to the field and laid the foundation for what was later to become strategic geography. American publications, however, were sparse at the turn of the twentieth century, but the British continued to develop a significant body of literature. Landmark works by Maguire (1899), Mackinder (1902), May (1909), and MacDonnell (1911) were characteristic of the maturity of the subfield in Great Britain at that time, and clearly reflective of the country's global vision.

Throughout the twentieth century, professional and academic geographers made enormous contributions to the U.S. Military's understanding of distant places and cultures. The vast collection of 'Area Handbooks' found in most university libraries, serves as testament to the significant effort by geographers during wartime. Although some of the work remains hidden by security classification, a casual glance at Munn's (1980) summary of the roles of geographers within the Department of Defense (DoD) enables one to appreciate the discipline's far-reaching impact on military affairs.

Among academic and professional geographers, the first formal demand for military geography in the United States surfaced during World War I. American geographers answered the call by providing written descriptions of the physical landscapes surrounding major training camps throughout the country. Descriptions were subsequently printed on the backsides of training maps and were used to teach basic terrain analysis skills to leaders and soldiers. D.W. Johnson (1917) published *Topography and Strategy in the War*, but his later work, *Battlefields of the World War, Western and Southern Fronts: A Study in Military Geography* (Johnson, 1921), became the most widely recognized military geographic publication stemming from the war years.

During World War II, the Joint Army and Navy Intelligence Studies (JANIS), provided regional geographies of selected theaters. Regional studies were helpful not only to commanders and military planners, but also to the Army

Quartermaster and the research and development teams who were responsible for designing uniforms, vehicles, equipment, weapons, and materials. A plethora of articles and studies were published on various aspects of military geography from both applied and historical perspectives (Palka and Lake, 1988). Arthur Davies' (1946) 'Geographical Factors in the Invasion and Battle of Normandy,' evolved as one of the best known American publications during the period and continues to be regarded as a classic analysis.

After the war military geography continued to thrive and the Cold War era and the Korean War served as a catalysts just as World War II had a few years prior. Russell's (1954) treatment of military geography in *American Geography: Inventory and Prospect* confirmed the healthy status of the subfield among academic and professional geographers, and reinforced the perception that the scope was restricted to wartime concerns.

On the eve of the Vietnam era, several noteworthy publications emerged. 'The Potential of Military Geography' (Peltier, 1961) was an emphatic statement of the legitimacy and utility of the subfield. Meigs (1961) examined geographical factors that influenced the outcome of the Peloponnesian War, and drew parallels with World War II operations in North Africa and Sicily. Jackman's (1962) article entitled, 'The Nature of Military Geography,' provided a clear, concise, theoretical perspective and became the most frequently cited paper on military geography. Although unpublished, Thompson (1962b), under the mentorship of Preston James, provided exceptional coverage of the history of military geographic thought and superbly developed the theoretical underpinnings. *Military Geography* (Peltier and Pearcy, 1966) was published a few years later, and until recently, was the most comprehensive and focused publication on the topic, essentially serving as the 'guide-post' for the subfield between the end of World War II and the mid-1990s. The legitimacy of the subfield was reinforced in the AAG's (1966) special bulletin, *Geography as a Professional Field*, which featured a section entitled 'Careers in Military Geography.'

The demise of military geography among universities and academics coincided with the widespread social and political unrest that occurred in America during the mid-1960s and early 1970s. During that era, anti-war sentiments and a general mistrust of the federal government prompted geographers to become increasingly concerned with being socially, morally, and ecologically responsible in their research efforts and professional affiliations with government agencies. Contributing to the war effort in Vietnam came to be regarded as irresponsible by many members of the AAG. The controversy surrounding the Vietnam War cast a persistent shadow on military geography as an academic discipline throughout the 1970s. In retrospect, the adverse effect of the unpopular war was understandable, if not predictable. The applied wartime focus of military geography, perceived for many years as its *raison d'etre*, came under intense scrutiny, and without alternative applications, the subfield lost its appeal within academia. Understandably, however, the field continued to evolve within military circles.

Despite the success of American military operations in Grenada and Panama during the early and mid-1980s, the justification for those incursions was controversial among academics so there was little renewed interest in military geography. Nevertheless, several publications still managed to surface. Some employed military geographic studies as a means to analyze regional conflicts (Soffer 1982), or to assess the security implications of land-use patterns (O'Sullivan, 1980; Soffer and Minghi, 1986). Others such as Munn (1980) reinforced the Military's continued interest in the subfield and outlined the role of geographers within DoD. Among military publications, Palka (1987, 1988) examined the routine requirement for geographic perspectives and tools in military planning, and Garver (1981, 1984) and Galloway (1984, 1990) compiled an excellent melange of contemporary readings for use in their military geography course at West Point. More recently, Winters (1991) discussed the influence of geomorphologic and climatologic factors in his historical account of a Civil War battle, and O'Sullivan (1991) provided extensive treatment of the relationships between terrain and small unit engagements.

Recent works by Collins (1998) and Winters (1998) confirm the renewed interest among academic geographers in applying geography to military operations in wartime. In *Military Geography for Professionals and the Public*, Collins (1998), considers a wide range of environmental conditions, employs historical vignettes, and addresses warfare at the tactical, operational and strategic levels. In *Battling the Elements*, Harold Winters (1998) highlights the impacts of weather, climate, terrain, soils, and vegetation on the outcomes of important military operations.

It would be virtually impossible to overstate the importance of military geography during wartime. The reliance on geographical information has long been fundamental to training, equipping, and deploying forces overseas throughout the twentieth century. Pertinent regional and systematic geographies routinely provide fundamental informational considerations to support decision-making processes at all echelons. As geographic tools and techniques have evolved, so have their military applications. Ironically, most of the geographer's current inventory of tools has a military origin. Regardless, LANDSAT imagery, aerial photographs, computer assisted cartography, the global positioning system, geographic information systems, and a wide range of map types and scales are routinely used to develop plans and conduct operations at all echelons. Suffice it to say that within military circles, military geography is an imperative within the wartime context as long as battles are waged on the earth's surface or in its atmosphere.

In Chapter one, World War II in the Aleutian Islands, Eugene Palka employs a historical perspective to examine the impacts of weather, climate, terrain and remote location on the battle for Attu, one of the most costly battle of the entire Pacific Campaign. Francis Galgano explains the significance of understanding local relief and drainage patterns to one's advantage in Chapter two, "Streams and the Civil War Landscape." In Chapter three, "Military Geography of the American Civil War," Joe Henderson describes how two of the physiographic regions of the United States shaped military strategy, campaign planning and operations.

Francis Galgano reveals how detailed geomorphical analysis in "Operation Chromite: Amphibious Turning Movement at Inchon." Continuing with this amphibious theme, Jon Malinowski reveals the dire consequences of conducting an attack on a remote Pacific island without a complete geographic understanding of the physical environment in Chapter four, "Peleliu: Geographic Complications of Operation Stalemate." Malinowski details the human suffering that resulted from inaccurate and incomplete intelligence estimates. Finally, in Chapter six, "Medical Geography and the Buna Campaign," Palka and Galgano describe the necessity of understanding the health risks that certain regions pose to non-indigenous soldiers. The Buna Campaign of W.W. II underscore the age-old lesson that the physical environment often has the potential to inflict more casualties and suffering than hostile fire.

Collectively, these chapters demonstrate the understandable utility, if not necessity, of military geography within a wartime context. As long as war is waged on the earth's surface or in its atmosphere, geographic variables must be considered as an integral part of all military plans and operations.

Bibliography

Association of American Geographers. 1966. *Geography as a Professional Field*. Bulletin 1966, number 10. Edited by Preston E. James and Lorrin Kennamer. Washington, DC: US Department of Health, Education and Welfare.

Collins, John M. 1998. *Military Geography for Professionals and the Public*. Washington, DC: National Defense University Press.

Davies, A. 1946. Geographical Factors in the Invasion and Battle of Normandy. *Geographical Review,* 36/4: 613-631.

Edwards, H.J. 1939. Translated from Ceasar, *Gallic Wars*, Loeb Classical Library. London: William Heinemann.

Galloway, Colonel Gerald E., Jr., compiler. 1990. *Readings in Military Geography*. West Point, NY: Department of Geography, US Military Academy.

Garver, Colonel John B., Jr., compiler. 1981. *Readings in Military Geography*. West Point, NY: Department of Geography & Computer Science, US Military Academy.

Garver, John B., Jr., and Galloway, Gerald E., Jr., compilers. 1984. *Readings in Military Geography*. West Point, NY: Department of Geography & Computer Science, US Military Academy.

Jackman, Albert. 1962. The Nature of Military Geography. *The Professional Geographer,* 14: 7-12.

Johnson, D.W. 1917. *Topography and Strategy in the War*. New York: Henry Holt.

_____. 1921. *Battlefields of the World War, Western and Southern Fronts: A Study in Military Geography*. New York: American Geographical Society, Research Series No. 3.

Lavallee, T.S. 1836. *Geographie Physique, Historique et Militaire*. Paris.

MacDonnell, Colonel A.C. 1911. *The Outlines of Military Geography*. London: Hugh Rees, LTD.

Mackinder, Halford J. 1902. *Britain and the British Seas*. New York: D. Appleton and Co.

Maguire, T. Miller. 1899. *Outlines of Military Geography*. Cambridge: The University Press.

Mahan, Alfred Thayer. 1890. *The Influence of Sea Power upon History, 1660-1783*. Boston: Little, Brown and Co.

May, Colonel Edward Sinclair. 1909. *An Introduction to Military Geography*. London: ?

Meigs, Peveril. 1961. Some Geographical Factors in the Peloponnesian War. *Geographical Review,* 51/3: 370-380.

Munn, Alvin A. 1980. The Role of Geographers in the Department of Defense. *The Professional Geographer,* 32/3: 361-364.

O'Sullivan, Patrick. 1980. Warfare in Suburbia. *Professional Geographer,* 32/3: 355-360.

_____. 1991. *Terrain and Tactics*. New York: Greenwood Press.

Palka, Eugene J. 1987. Aerial Photography. *Infantry Journal* (May-June): 12-14.

_____. 1988. Geographic Information in Military Planning. *Military Review,* 68/3: 52-61.

Palka, Eugene J. 2000. "Military Geography: Its Revival and Prospectus." In *Geography in America at the Dawn of the Twenty-First Century*. Edited by Gary Gaile and Cort J. Willmott. Oxford: Oxford University Press.

Palka, Eugene J., and Lake, Dawn. 1988. *A Bibliography of Military Geography*, four volumes. West Point, NY: US Military Academy Press.

Peltier, Louis C. 1961. The Potential of Military Geography. *Professional Geographer,* 13/6: 3-4.

Peltier, Louis C., and Pearcy, G. Etzel. 1966. *Military Geography*. Princeton, NJ: D. Van Nostrand Company, Inc.

Roon, Capt. Albrecht Theodor Emil Graf von. 1837. Militarische Landerbeschreiburg von Europa, Vol I., 'Mittel-und Sud-Europa.' Berlin.

Rouse, W.H.D. 1959. *The March Up Country*. A translation of Xenophon's *Anabasis*. Ann Arbor, MI: The University of Michigan Press.

Russell, J.A. 1954. 'Military Geography.' In *American Geography: Inventory and Prospect*, by James, P.E., and Jones, C.F., eds. Syracuse, NY: Syracuse University Press.

Soffer, Arnon. 1982. Topography Conquered: The Wars of Israel in Sinai. *Military Review,* 62/4: 60-72.

Soffer, Arnon, and Minghi, Julian V. 1986. Israel's Security Landscapes: The Impact of Military Considerations on Land Uses. *Professional Geographer,* 38/1: 28-41.

Thompson, Edmund R. 1962b. The Nature of Military Geography: A Preliminary Survey. Ph.D. Dissertation. Syracuse, NY: Syracuse University.

Winters, Harold A. 1991. The Battle That Was Never Fought: Weather and the Union Mud March of January 1863. *Southeastern Geographer,* XXXI/1: 31-38.

_____. 1998. *Battling the Elements: Weather and Terrain in the Conduct of War*. Baltimore, MD: The John Hopkins University Press.

World War II in the Aleutian Islands:
Physical Geographic Challenges in the Battle for Attu

By

Eugene J. Palka

Introduction

The war in the Aleutian Islands has been described as "the forgotten war" (Chandonnet, 1993), "our hidden front," (Gilman, 1944), and "the thousand-mile war," (Garfield, 1969). Each of the characterizations alludes to the uncelebrated nature of the Aleutian Campaign and its relatively insignificant impact during World War II. Indeed these references stem from the paltry number of publications that shed light on the bloody, yet little known World War II struggle between the United States and Japan in the North Pacific.

It is a little known fact that one of the most extraordinary island battles of World War II occurred on American soil. As a percentage of the total forces involved, the battle to retake Attu was second only to Iwo Jima as the most costly battle of the Pacific Campaign (Garfield, 1969). This information is further complicated by the difficulty in locating the tiny Island of Attu on most maps. In some respects, the entire campaign in the Aleutians was nothing more than a footnote within the context of global war, or even war in the Pacific. Nevertheless, the battle to retake Attu came at a cost of 549 American lives, 1,148 wounded, and 2,100 other casualties who suffered from sickness, exposure or non-battle injuries (Hutchison, 1994). Not to be ignored were the billions of dollars of infrastructure and the multitudes of personnel, weapons, and equipment necessary to retake the Island. Although Japan initially captured the island unopposed, 2,350 Japanese soldiers were later killed and the remaining 29 were taken prisoner (Hutchison, 1994). Throughout the Aleutian campaign, physical geographic conditions exacted a heavy toll on Japanese and American ground, sea, and air forces. Indeed, the remote location and extreme conditions prompted routine actions to require excessive time and resources. In the end, the harsh nature of the physical environment and the relative location limited the utility of Attu as key terrain, and questioned the logic of the Japanese incursion in the first place. Moreover, the battle for Attu clearly demonstrated the difficulty of conducting military operations in such forbidding environments in remote locations and prompted immediate changes in Army doctrine, weapons, uniforms, and equipment.

This chapter is an example of historical military geography explained within an operational and strategic context (Palka, 2000). My intent is to describe the impact of the unique physical environment and remote location on military operations and to elaborate on some of the lessons that were learned and subsequently applied elsewhere.

The Physical Setting

Location

The Aleutian Islands extend westward from the Alaskan Peninsula for more than 1,200 miles, separating the North Pacific and the Bering Sea (see Map 1). The arcuate chain of volcanic islands extends from east to west from about 165° west longitude to about 170° east longitude, and on the average sits astride the 53° north latitude parallel. Even a cursory glance at a map enables one to draw a number of inferences about the climate, vegetation, and geomorphologic characteristics of the Aleutian Islands.

While the absolute location of the Aleutians may lead one to inferences regarding the physical nature of the place, the relative location of the chain, and specifically of Attu, is far more important to explaining why this site became a contested place during W.W. II. Within the chain, Attu is located at the western

Physical Geographic Challenges in the Battle for Attu

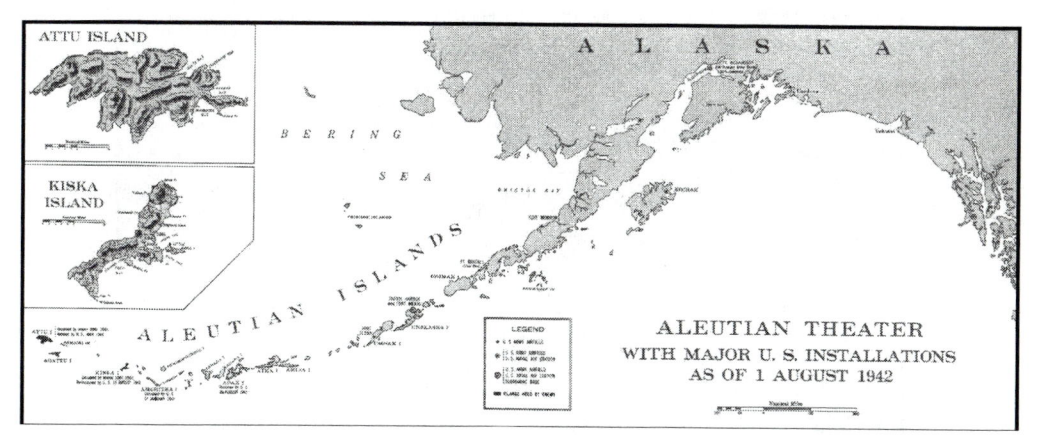

Map 1. The Aleutian Islands.
Source: Naval Historical Center, 1993.

extreme, and is technically the eastern-most point of the U.S., crossing the 180° line of longitude (although the international dateline has been adjusted to curve around Attu). To the casual observer, the Aleutians, and especially Attu, are remotely located from both the continental United States and Japan. The straight-line distance, however, is significantly distorted by many of the widely used map projections. Using a polar projection provides an illuminating perspective, and reveals the importance of the relative location of Attu to both the United States and Japan (see Map 2). Following a great circle route depicted on the previous map, one might be surprised to note that the island of Attu is only approximately 650 miles from the Japanese Kuriles (see Map 3). Most maps suggest that the logical and most direct route from the West Coast of the United States to Japan would pass through the Hawaiian Islands. In reality, if traveling from San Francisco to Tokyo, it is significantly shorter to travel via a route that passes along the Aleutian Islands than to travel via Hawaii. The former route is actually about 1,000 miles shorter than the latter (see Map 4). Thus, the significance of Attu's relative location was far more important than its absolute location during the War in the Pacific.

Attu was uniquely located about mid-way between the more populated areas of the Japanese and American homelands, and was perceived to have utility as a forward base for limited incursions into the remote territories of either country. Moreover, the island could potentially provide a base to support operations to control the North Pacific.

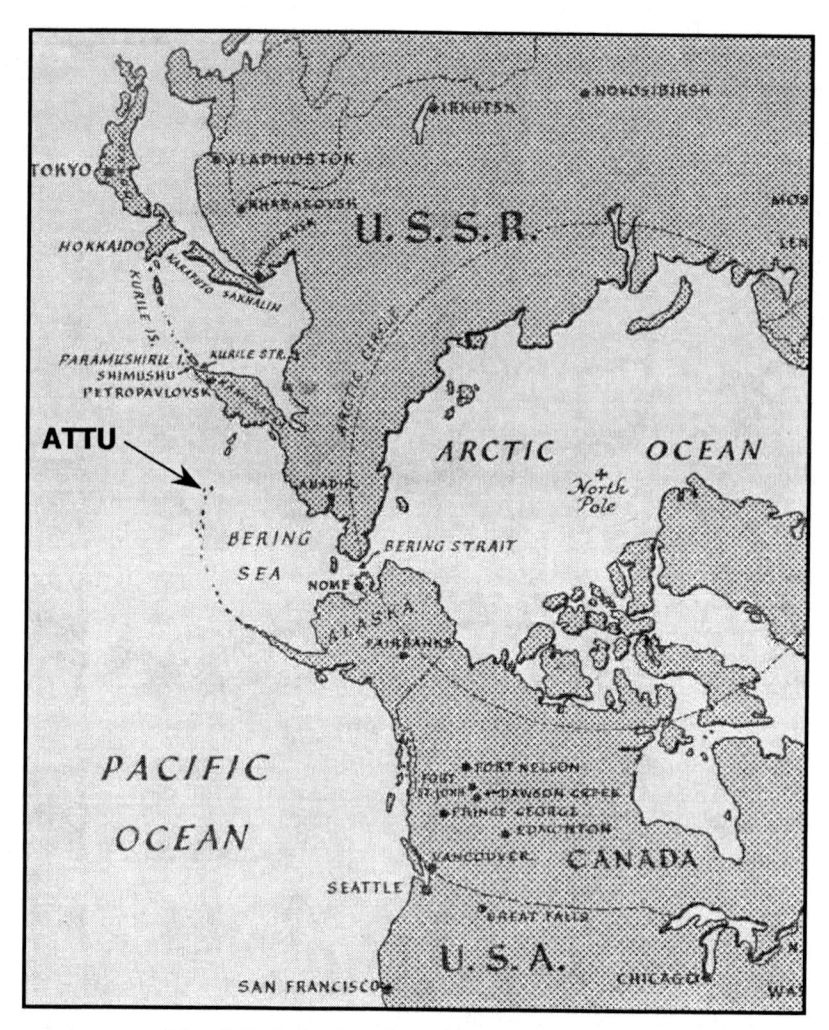

Map 2. Relative location of the Aleutian Islands.
Source: Modified from Gilman, 1944.

Physical Geography

The physical environment found on Attu and throughout the Aleutians is one that presents a plethora of problems to military operations. The nature of the terrain, climate, weather, vegetation, soil, and sea conditions combined to produce one of the most forbidding operating environments ever experienced by U.S. military forces.

Physical Geographic Challenges in the Battle for Attu

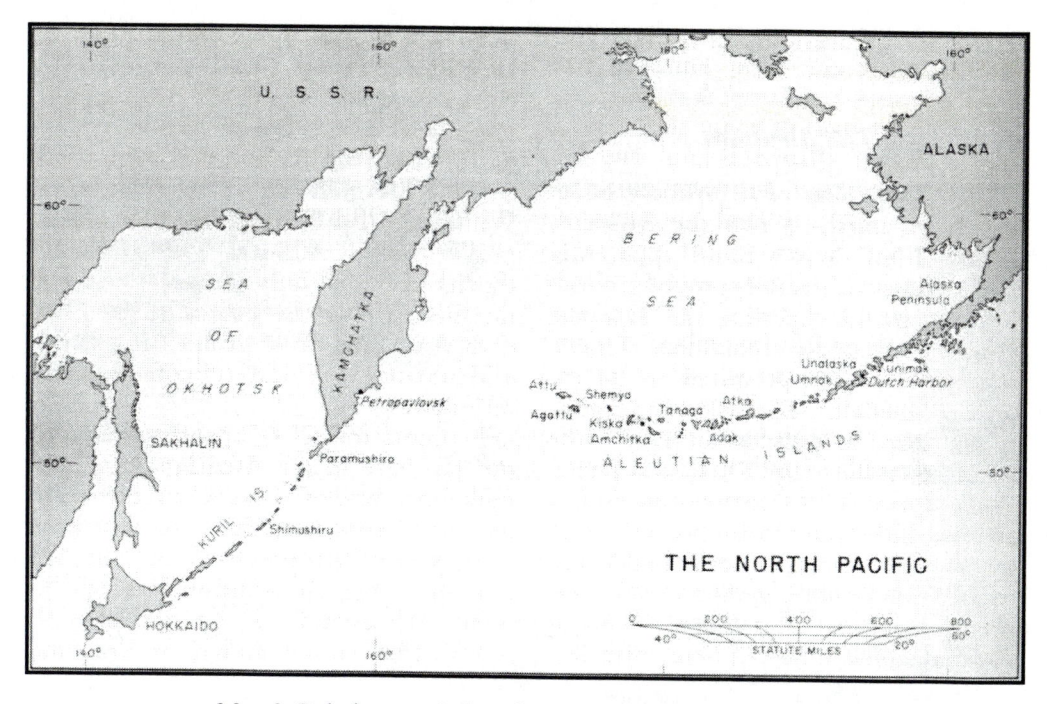

Map 3. Relative proximity of Attu to Japan's Kurile Islands.
Source: Morton, 1989.

Terrain

The Aleutian Islands are the world's longest small-island archipelago, extending nearly 1,200 miles. The Aleutians are comprised of approximately 120 islands, all of which are uniformly rocky and barren, and have volcanic origins (Department of the Navy, 1993). The main islands are the exposed peaks of a submerged mountain range that separates the Bering Sea from the Pacific Ocean, and includes 57 volcanoes, 27 of which are still active and exceed heights of 5,000 feet (Faust and Bailey, 1995). The inland mountains are conical in shape and are covered with volcanic ash, while the shorelines are comprised of jagged and submerged rock formations (Department of the Navy, 1993).

Climate

The principal controls that combine to shape the Aleutian climate include latitude, ocean currents, air masses, and mountainous terrain. Average summer temperatures are 45°F, while winter temperatures average around 30°F (Faust and Bailey, 1995). Although precipitation averages only about 50 inches annually, the chain averages 90 percent cloud cover and more than 200 days of measurable precipitation (Faust and Bailey, 1995). A common characteristic is the ever-present high wind, which has a significant "wind chill effect" on temperatures and contributes to turbulent seas.

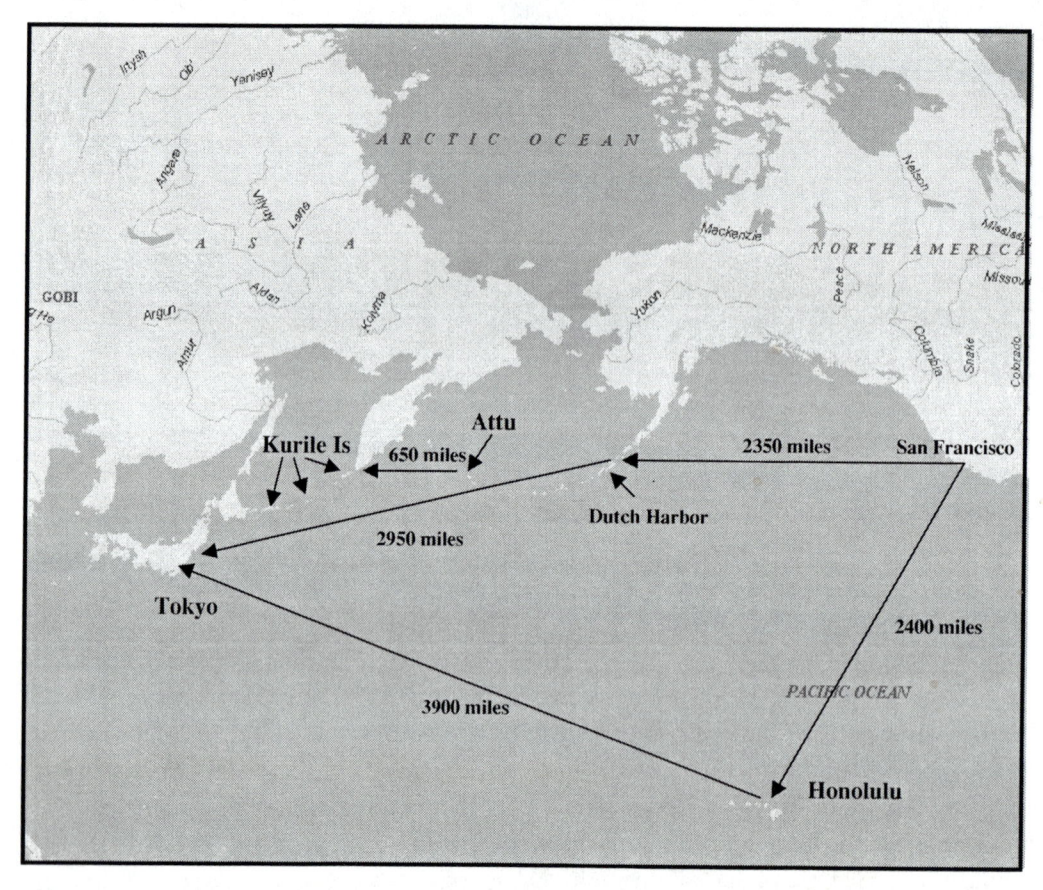

Map 4. Actual distances between key points in the North Pacific.
Source: Alaska Geographic Society, 1995.

While the general climate of the Aleutians does not exhibit extremes of temperature or precipitation, the weather (regarded as the atmospheric conditions at a specific place and time) is highly localized and often unpredictable. The high latitude and rugged, mountainous terrain are subjected to the convergence of the tropical Japan Current and the Siberian air masses. This unique combination results in violent storms and extreme weather conditions, including blizzards, dense fog, turbulent seas, bone-chilling temperatures, and winds that can reach gusts of more than 140 miles per hour (Garfield, 1969). "Williwaws" or sudden squalls that can reach gale force within half an hour are unique local phenomena that pose considerable problems for air and sea navigation.

Vegetation

Because of its proximity to Asia, Attu hosts some flora that is common to both Asia and the rest of the Aleutians (Faust and Bailey, 1995). Like the majority of the Aleutian Islands, however, Attu is a treeless expanse of tundra void of any significant range of vegetation. Temperatures, high winds, and thin soils impose significant constraints and limit vegetational growth.

Most of the tundra is of the high-alpine type (as opposed to low-tundra found throughout much of the lowland areas of interior Alaska), void of any significant brush or shrubbery. Lowland areas on Attu and the other islands are covered with muskeg, a spongy carpet-like layer of vegetation that is underlain with discontinuous permafrost. During the warmer months, the muskeg presents a boggy, shifting surface that is difficult for even foot soldiers to traverse. The tundra, however, does display an assortment of wildflowers, fireweed, lichens, mosses, and wild grasses, that combine to produce a colorful tapestry during the late summer and early fall.

Utility of Attu and The Aleutians?

Within the context of strategic geography, William Jacobs (1993) referred to the North Pacific as a "strategic backwater." He noted that the Aleutians were situated along a great circle route from the West Coast of the United States to Japan, and that their occupation "appeared" to offer strategic advantages (Jacobs, 1993). Many strategists, however, recognized the enormous challenges to military operations imposed by the natural environment of the North Pacific and questioned any rationale for securing the Aleutians. Even after the Japanese occupied Attu and Kiska, many argued that the wisest policy would have been to bypass the islands, just as we did later with many other Japanese-held islands in the South Pacific (Allard, 1993).

Of all the islands in the Aleutians, Attu appeared to hold the most potential for both the United States and Japan, simply because it was the westernmost island in the chain and only 650 miles from the major Japanese military base at Paramushiro on the northern tip of the Kurile Islands. Yet the value of Attu was still controversial among military and political leaders of both countries. The island is located at great distance from any sizable concentration of people in either the United States or Japan; it has no significant natural resources; nor has the island ever been populated by more than a few dozen Aleuts, who continue to practice a traditional, subsistence lifestyle. Finally, access to Attu is possible only via prolonged and hazardous travel across treacherous seas or through turbulent air.

Japanese Strategy

There are a number of plausible, if not debatable, explanations for Japan's invasion and occupation of Attu and Kiska. Some sources conclude that the Japanese had long coveted Alaska and the Aleutians and that many of the military leaders considered the poorly defended outposts in the chain as easy targets and a logical route for invading North America (Department of the Navy, 1993). Various intelligence sources anticipated Japan's seizure of the Aleutians to prevent the United States from invading the Japanese homeland from the North (Hutchison, 1944). Other intelligence estimates pointed to Japan's goal of thwarting U.S.-Soviet communications in the North Pacific (Hutchison, 1944). Still others dismissed both offensive and defensive strategic explanations and looked to psychological reasons. Garfield (1969), for example, claimed that the Japanese tried to psychologically offset their loss at Midway by seizing the undefended American territory in the Aleutians and draw attention to the latter. There is significant support for the explanation that by seizing Attu and Kiska after the disastrous loss at Midway, Japan's political and military leaders were able to "save face," maintain high morale within the military, and retain popular support on the home front (Russell, 1993; Nishijima, 1993; Garfield, 1969).

American Strategy

In American political and military circles, the strategic importance of Alaska and the Aleutians dates back to the early 1920s (Braisted, 1971). By the 1930s, military planners considered Alaska to be part of a strategic triangle linking Hawaii and Panama, and forming the basic framework for the U.S. military strategy in the Pacific (Allard, 1993). Despite the historical interest, however, there were no significant plans to fortify the Western Aleutians at the outbreak of hostilities with Japan.

The strategic situation began to change when the Japanese bombed Dutch Harbor on June 3rd and 4th, 1942, then subsequently invaded Kiska and Attu on June 6th and 7th. Even after the uncontested occupations of Attu and Kiska, however, the Joint Chiefs of Staff considered the western extremes of the Aleutians to be insignificant within the context of global war in Europe and the Pacific (Morton, 1989). The Joint Chiefs and U.S. military strategists did not foresee the need to undertake substantial military operations in the North Pacific to retake either Attu or Kiska. Resources were deemed more critical elsewhere, the conditions for military operations were considered hazardous, and mainland Alaska did not appear threatened since the islands were more than 1,0000 miles from the state's heartland, and even more distant from the Continental U.S. (Morton, 1989).

After persistent prodding from political officials and military leadership in Alaska, however, the Joint Chiefs began to entertain notions of retaking Attu and Kiska. Public pressure mounted with the perceived threat to Alaskan security. Another concern was the potential need to maintain sea and air lines of communication with the Soviet allies across the North Pacific, especially to ensure the continuous flow of lend-lease supplies (Jones, 1969). Perhaps most important, however, was the American pride that was manifested in the pressing call to reclaim America's territory (Allard, 1993). Hence, just as the Japanese incursion was motivated by psychological factors, so was the impetus for the American effort to retake the Attu and Kiska.

Historical Summary of The Battle for Attu

Prior to the U.S. mission to retake Attu and Kiska, Japanese troops occupied the two islands for approximately eleven months, becoming well entrenched in the process. Kiska was regarded as the primary objective because it provided potentially better air facilities, possessed a more satisfactory harbor, and included terrain that was more suitable for a base (Department of the Navy, 1993). Additionally, U.S. intelligence estimates considered Kiska to be more heavily fortified, so it posed the more imminent threat to Alaska. Nevertheless, the U.S. plan called for Attu to be taken first because of a lack of logistics, sealift, and manpower necessary to assault the more heavily fortified Kiska (Coox, 1993; Department of the Navy, 1993).

The Joint Chiefs approved the U.S. plan on 3 March 1943. A concerted aerial bombing campaign was initiated almost immediately to "soften-up" the objectives. Emphasis was directed towards Kiska rather than Attu in an effort to conceal the U.S. intentions, achieve the element of surprise, and further degrade the defenses at Kiska prior to invading. Aerial bombing shifted to Attu on 1 May in support of a scheduled D-day of 7 May 1943.

The U.S. plan called for extensive air and naval gunfire support prior to and during the amphibious landings at five different beaches along the western half of Attu (Department of the Navy, 1993; see Map 5). Continuous reconnaissance flights were also planned. The U.S. invasion force was scheduled to depart from Cold Bay in the Aleutians on 3 May, but poor weather and turbulent seas delayed sailing until 4 May (Department of the Navy, 1993). Accordingly, D-day was also postponed until the 8th of May. With the assault force already at sea, D-day was again postponed because of dense fog, high winds, and hazardous surf conditions until 11 May.

Although they encountered difficulties with the dense fog and surf conditions, the U.S. Seventh Division landed on the north shore around Holtz Bay and on the south shore at Massacre Bay with minimal resistance. The Japanese commander, Colonel Yasuyo Yamasaki, had concentrated his defense inland and had placed only coastal guns and a dozen antiaircraft guns to oppose the anticipated landings (Coox, 1993). Moreover, despite initial problems, the dense fog proved to be a mixed blessing to U.S. forces. Although it virtually eliminated any air, naval gunfire, or artillery support, it concealed the landings from Japanese observation. By the end of the first day, about 3,500 American soldiers had landed at Attu (Campbell, 1995). Approximately 400 troops landed at Beach Scarlet, 1,100 hundred at Beach Red, and 2,000 at Beaches Blue and Yellow (Chandonnet, 1993).

With the Japanese defense located on the high ground and well inland between Holtz and Massacre Bays, it quickly became apparent that the U.S. operation would take much longer than the three days that were initially anticipated. Moreover, the rugged nature of the terrain, boggy tundra and mud, and extreme weather ensured an unexpectedly slow rate of advance for U.S. troops. By 17 May, more than 12,500 American troops had landed on Attu, suffering more than 1,100 casualties in the process (Hutchison, 1994). By the 18th of May, however, the U.S. northern and southern forces had linked up in the vicinity of Jarmin Pass and surrounded the Japanese defenders in northeastern Attu (Hutchison, 1994; see Map 6).

Physical Geographic Challenges in the Battle for Attu

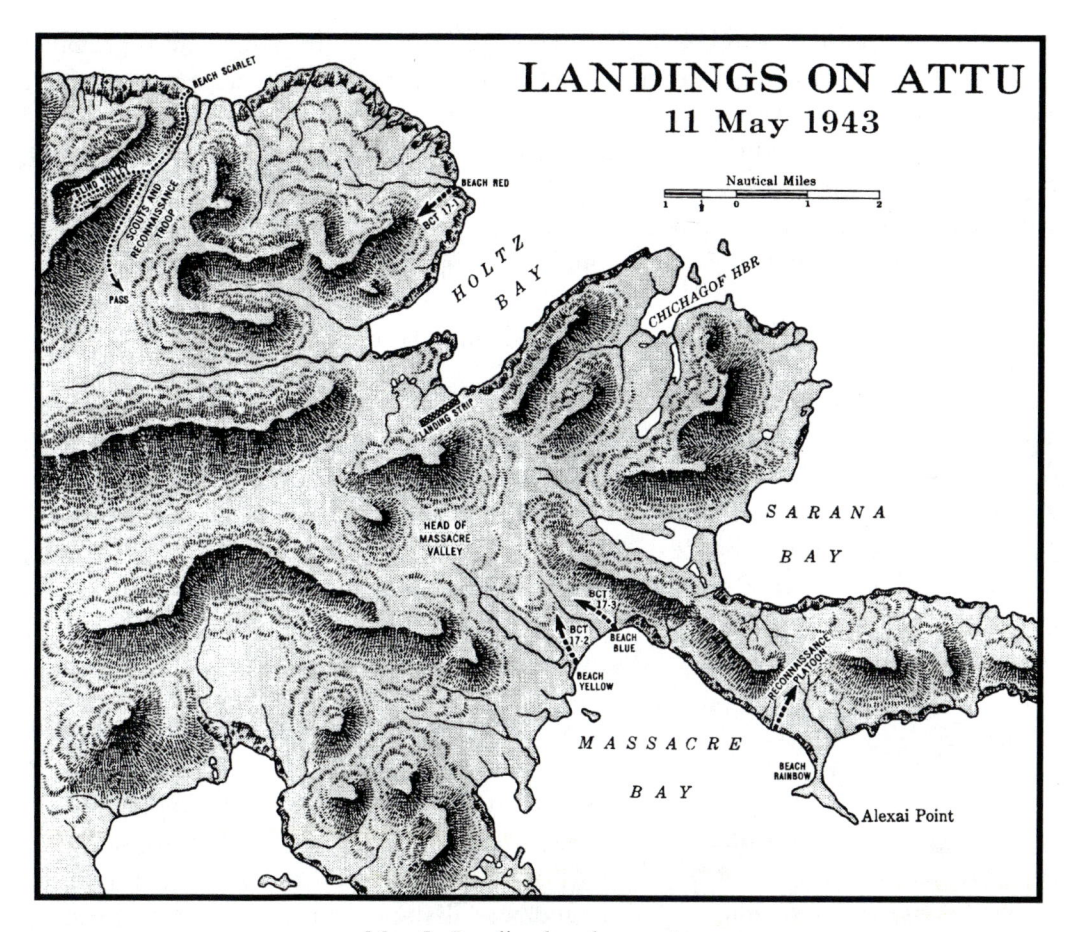

Map 5. Landing beaches on Attu
Source: Department of the Navy, 1993.

Given the reinforced nature of the Japanese defensive positions, severe weather that limited visibility and air and artillery support, and the rugged nature of the terrain, the U.S. attack was literally "reduced to a crawl" from 18 to 28 May. Nevertheless, persistence prevailed, and by the 29th of May, Japanese forces on Attu were completely encircled. In a last ditch effort; the remaining Japanese troops (estimated to be from 800-1,000) conducted a suicidal counterattack against American positions. The initial counterattack was repelled, yet the Japanese conducted successive assaults, believing it was better to die with honor than to surrender (Campbell, 1995). In the hand-to-hand combat that ensued, virtually every Japanese soldier was killed. The remaining survivors allegedly committed suicide by blowing themselves up with hand grenades (Handleman, 1943) or died from injections of morphine that were ordered by the Japanese commander (Campbell, 1995).

On 29 May, Attu Island was secured. The cost of recapturing the tiny, rugged and remote island came at a cost of 549 Americans killed, 1,148 wounded, and 2,100 suffering from exposure or disease (Hutchison, 1994). Japanese losses totaled 2,350 dead and 29 prisoners of war (Hutchison, 1994).

Geographic Factors Influencing the Battle

The physical geography of the Aleutian Islands, and of Attu in particular had a profound impact on the conduct of military operations. A plethora of factors, especially those relating to location, coastlines, weather, surf conditions, vegetation, and relief, posed continual challenges for the attackers and defenders alike. Throughout the battle to retake Attu, Americans and Japanese troops battled the elements and terrain, suffering immeasurably in some respects, and in other instances, modifying their plans and tactics accordingly to overcome the adversity imposed by the physical environment.

Remote Location

Attu's remote location was equally troublesome for both the Japanese and Americans. From the outset, operations were hindered by the lack of reliable maps of the island and charts of its adjacent waters; shortcomings attributed to the remoteness of the place. The island was distant from any military base or population center in either country. Once the Japanese attacked and seized the island, they encountered extreme difficulty as they tried to sustain operations and support the occupation force. Similarly, the U.S. military encountered a logistics nightmare as it attempted to muster, transport, and sustain the necessary forces, equipment and supplies to retake the island.

Coastline and Sea Conditions

Attu's coastline is extremely rugged due to the volcanic origin of the island. Sea cliffs, rugged escarpments, and rocky outcrops along the coast presented major obstacles to amphibious landings of any type, especially during rough seas (see Photo 1). To make matters worse, the remote coastline was poorly mapped and much of the adjacent waters were uncharted. Moreover, the occasional sand beaches were often saturated with extensive piles of driftwood-eight to ten feet high and situated 200 to 300 yards inshore of the normal beach as a result of recurring "superstorms" (Campbell, 1995). Even its most ideal harbors were unsuitable for large naval vessels (Department of the Navy, 1993).

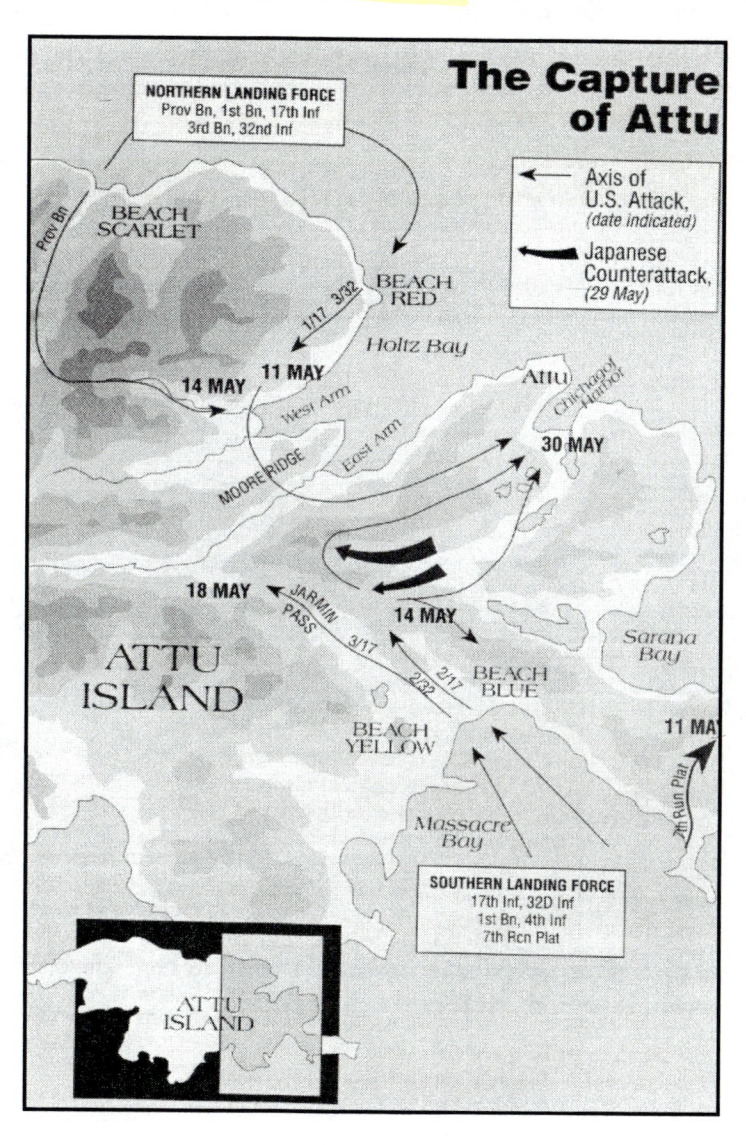

Map 6. The tactical plan on Attu.
Source: Chandonnet, 1993.

Sea conditions throughout the Aleutians had been considered unfavorable for naval operations long before the operation to resecure Attu. Turbulent seas were made even more treacherous by the occasional incidence of *rogue waves* (Campbell, 1995). The latter are great solitary waves or "super-crests" that tower over all others and occasionally include extra-deep troughs into which ships can drop and become subsequently submerged by the next crest (Bascom, 1980). These conditions, coupled with extreme temperatures, limited visibility, high winds, and unreliable navigational aids, presented the ultimate environmental challenge to even routine naval operations.

Photo 1. Attu's rugged coastline
Source: Command Historian's Office, Alaskan Command.

Weather

Aspects of the island's weather that proved to be problematic included cold temperatures, extensive cloud cover, high winds, and dense fog. Inclement weather was decisive from the outset of the Aleutian Campaign, adversely impacting on sea, air, and ground operations. D-Day was postponed on several occasions as a result of severe weather conditions. Cloe (1990) noted that of the 225 allied aircraft destroyed, 184 were attributed to adverse weather.

Dense fog proved to be a mixed blessing. While the fog effectively concealed the U.S. amphibious landings from Japanese observation on D-Day, it also contributed to the initial U.S. casualties. Four soldiers drowned when a transport prematurely dropped its ramp before hitting the beach, filled with water and sank (Campbell, 1995). Other injuries were incurred when the *Sicard* and *Macdonough* collided during the initial landing (Department of the Navy, 1993). After the landing, the persistent fog served to conceal ground maneuvers for U.S. troops, but it inhibited the use of air or artillery support. On occasion, the fog dissipated at inopportune times, exposing U.S. troops that were out in the open to Japanese fire. At one point during a period of dense fog, a tragic battle was fought among U.S. troops between their own sentries and litter bearers (Gilman, 1944). Perhaps most important, dense fog limited the use of air or naval gunfire in support of the landing and subsequent offensive operations.

Persistent cloud cover had its greatest effect on air operations. Clouded skies reduced visibility and prevented the use of aerial reconnaissance during both the planning and execution stages of the operation. Throughout the attack, it also dictated when aerial fire support and/or resupply could be employed. Needless to say, breaks in the skies rarely coincided with those instances when support was needed most.

High winds proved to be hazardous to both air and naval operations, contributing to turbulent seas and untenable flying conditions. Ground troops felt the impact in the form of personal discomfort from the increased wind chill factor, and a disruption in the flow of logistics via sea or air. During stormy conditions, Aleutian gales reached speeds of up to 140 miles an hour (Garfield, 1969). Under such adverse conditions, it was virtually impossible for ground troops to effectively operate weapons and equipment, or even erect and maintain suitable shelters. *Williwaws* (incredibly violent and impromptu gusts of cold wind) wreaked havoc and inflicted considerable damage to equipment caches, maintenance tents, first aid shelters, communication and weather stations, aircraft, and ships (see Photo 2).

Cold temperatures proved to me a menace to both Japanese and American troops. Poorly equipped U.S. troops suffered from more incidents of trench foot, immersion foot, and frostbite than they did from enemy fire (Handleman, 1943). Vehicles, aircraft, weapons, and equipment functioned poorly or became inoperative in the extreme cold. To exacerbate matters, the treeless tundra and partially frozen ground afforded little opportunity for ground troops to find shelter from the cold (see Photo 3).

Terrain & Relief

The rugged terrain posed a considerable challenge to U.S. troops during the attack. Steep slopes, rugged cliffs, extensive gullies, and elevations in excess of 4,000 feet were made more difficult by the nature of the ground surface. Mud and boggy, shifting tundra ensured poor traction for vehicles and soldiers alike (see Photo 4). Ground movement was slow and exhausting for foot soldiers, and the displacement of vehicles, weapons systems, and supplies was

Photo 2. A *williwaw* rolling up an 8 ton section of an airfield landing strip.
Source: Command Historian's Office, Alaskan Command.

especially difficult. The evacuation of casualties was particularly troublesome. Because extensive use of mechanized or wheeled vehicles was simply impractical, or severely limited to isolated areas, the burden of transporting wounded and supplies literally fell on the shoulders of the foot soldier. Rugged relief, however, did not prove to be equally burdensome to the defenders who were situated in static positions. The Japanese had cached the supplies, weapons, and equipment necessary to defend the island. Moreover, they selected key inaccessible locations that afforded protection, as well as long range observation and fields of fire.

Conclusion

Attu represented "uncharted waters" and *terra incognitae* to American military forces. Prior to the Aleutian Campaign, the U.S. Military simply did not have the necessary training, weapons systems, vehicles, supplies, and equipment specifically designed for regions characterized by cold-wet climates, high winds, limited visibility, and rugged terrain. Consequently, the lessons learned were costly in terms of resources and casualties. In proportion to the number of troops employed, Attu ranked as the second most costly battle in the Pacific Theater, second only to Iwo Jima (Garfield, 1969). Of the 3,829 American casualties, a substantial number were classified as cold-weather injuries, and were the first combat cold injuries suffered by U.S. troops in WW II (Garfield, 1969).

The recapture of Attu and the subsequent seizure of Kiska did give the Western Aleutians back to America. Control of the Aleutian Chain offered a "jumping-off" point for an invasion of Japan. Moreover, occupation of the archipelago facilitated control of the North Pacific. It is debatable, however, whether the battles for Attu, and later Kiska, were strategically important, or whether their significance was merely psychological. Regardless, the lessons learned were later applied to operations in the Pacific and in Europe and contributed immensely to success in both theaters.

The battle for Attu clearly revealed the profound, if not decisive impact that physical geographic conditions can have on military operations at high latitudes. Fortunately, the experience prompted additional research and developments in a wide range of areas. Feedback from Army doctors studying Attu veterans resulted in various changes to Army footgear, clothes, tents, bedrolls, and food (Garfield, 1969). Recovery of an intact Japanese Zero

Photo 3. Attu's cold, partially frozen, high alpine tundra.
Source: U.S. Army Center of Military History, 1988.

fighter plane facilitated training American fliers in combat strategy (Campbell, 1995). In an effort to deal with the chronic problem of poor visibility, the Aleutians became an experimental proving ground for airborne search radar (Russell, 1993). Perhaps most important, amphibious landing tactics were revised and subsequently applied elsewhere within the Pacific and European theaters with great success. Throughout the Aleutian Campaign and afterwards a wide range of technical modifications were made to aircraft and vehicles to enhance operability, durability and survivability during weather and climate extremes. Finally, a number of different innovations were born out of American ingenuity at the lowest echelons during attempts to overcome challenges imposed by the physical environment. Examples included wicker mats to prevent howitzers from sinking into the mud when firing; "Marston mats" for use in building temporary runways in areas of boggy tundra or mud; and an A-frame cargo hoist (developed in Massacre Bay) used to unload cargo from shallow draft boats in austere bays. These innovations and others were subsequently refined, adopted and employed elsewhere under similar circumstances. Thus, Attu proved to be a catalyst within the military's research and development community.

The battle for Attu has never been a celebrated part of American military history. Nor was the seizure or retaking of Attu of great military importance during World War II. It was, however, the only World War II battle fought on American soil, and the cost of seizing and retaking the tiny, volcanic island was unquestionably high to both Japanese and Americans. In hindsight, the military utility of such a place remains questionable given the extensive resources and preparation necessary for any military force to seize, defend, and operate in such an austere and physically challenging environment.

Photo 4. Typical Aleutian mud.
Source: U.S. Army Center of Military History, 1988.

Bibliography

Allard, Dean C. 1993. "The North Pacific Campaign in Perspective." In *Alaska at War, 1941-1945: The Forgotten War Remembered*, edited by Fern Chandonnet. Anchorage: Papers from the Alaska at War Symposium, November 11-13, 1993, 3-11.

Alaska Geographic Society. 1995. *World War II in Alaska*. Special edition of *Alaska Geographic* 22 (4): 1995.

Alaska Geographic Society. 1995. *The Aleutian Islands*. Special edition of *Alaska Geographic* 22 (2): 1995.

Bascom, Willard. 1980. *Waves and Beaches*. New York: Anchor Press.

Braisted, William R. 1971. *The United States Navy in the Pacific, 1909-1922*. Austin, TX: University of Texas Press.

Campbell, L. J. 1995. "Arsenal of Democracy." *Alaska Geographic* 22 (4): 4-45.

Chandonnet, Fern, ed. 1993. *Alaska at War, 1941-1945: The Forgotten War Remembered*. Anchorage: Papers from the Alaska at War Symposium, November 11-13, 1993.

Cloe, Jolin Haile. 1990. *The Aleutian Warriors*. Anchorage: Anchorage Chapter of the Air Force Association.

Coox, Alvin D. 1993. "Reflecting on the Alaska Theater in Pacific War Operations, 1942-1945." In *Alaska at War, 1941-1945: The Forgotten War Remembered*, edited by Fern Chandonnet. Anchorage: Papers from the Alaska at War Symposium, November 11-13, 1993.

Department of the Navy. 1993. *The Aleutians Campaign, June 1942-August 1943*. Washington, D.C.: Naval Historical Center.

Faust, Nina and Bailey, Edgar. 1995. "The Aleutians: Tiny Islands in Turbulent Seas." *Alaska Geographic* 22 (2): 4-33.

Garfield, Brian. 1969. *The Thousand-Mile War: World War II in Alaska and the Aleutians*. Garden City, NY: Doubleday.

Gilman, William. 1944. *Our Hidden Front*. New York: Reynal & Hitchcock, Inc.

Handleman, Howard. 1943. *Bridge to Victory: The Story of the Reconquest of the Aleutians*. New York: Random House.

Hutchison, Devin Don. 1994. *World War II in the North Pacific: Chronology and Fact Book*. Westport Conn.: Greenwood Press.

Jacobs, William A. 1993. "American National Strategy in East Asian and Pacific War: The North Pacific." In *Alaska at War, 1941-1945: The Forgotten War Remembered*, edited by Fern Chandonnet. Anchorage: Papers from the Alaska at War Symposium, November 11-13, 1993, 13-17.

Jones, Robert H. 1969. *The Road to Russia: United States Lend-Lease to the Soviet Union*. Norman, OK: University of Oklahoma Press.

MacGarrigle, George L. 1992. *Aleutian Islands*. Washington, D.C.: US Army Center of Military History.

Mills, Stephen E. 1971. *Arctic War Birds: Alaska Aviation of WWII*. Seattle: Superior Publishing Company.

Morton, Louis. 1989. *United States Army in World War II: The War in the Pacific. Strategy and Command: The First Two Years*. Washington, D.C.: United States Army Center of Military History.

Nishijima, Teruo. 1993. "Recalling the Battle of Attu." In *Alaska at War, 1941-1945: The Forgotten War Remembered*, edited by Fern Chandonnet. Anchorage: Papers from the Alaska at War Symposium, November 11-13, 1993, 109-112.

Palka, Eugene J. 2000. "Military Geography: Its Revival and Prospectus." In *Geography in America at the Dawn of the Twenty-First Century*. Edited by Gary Gaile and Cort J. Willmott. Oxford: Oxford University Press.

United States Army Center of Military History. 1988. *United States Army in World War II: Pictorial Record. The War Against Japan.* Washington, D.C.: United States Army Center of Military History.

2

Streams and the Civil War Landscape:
THE BATTLE FOR ISLAND NO. 10, MISSISSIPPI RIVER, 1862

By

Francis A. Galgano Jr.

Introduction

Rivers have always played a decisive role in warfare. A river, with its wide looping meanders and attendant floodplain represents a unique set of tactical problems to an Army. For example, Allied planners began preparing for the Rhine River crossings many months before the actual operation in March 1945 (MacDonald, 1984); and the stunning success of the German Blitzkrieg into the Soviet Union in June 1941 was supported in large measure by carefully planned and executed river crossing operations that employed specially adapted tanks (Ziemke and Bauer, 1987). The scope of military operations on the fluvial landscape is very broad. Military operations on rivers and inland waterways are typically brought together under the designation 'riverine operations' (Department of the Army, 1998), and for that reason embrace a wide range of environments and sometimes confounding geographic problems. Riverine operations have played a crucial role in military operations throughout history and in the context of Military Operations Other Than War, this role may be elevated (Department of the Army, 1993). In this chapter, we shall come to understand how the fluvial landscape can influence military operations on the strategic and tactical scale and how a combined Federal naval and land force overcame physical and tactical problems along the Mississippi River during the spring of 1862.

Wide, swift rivers presented a nearly impassable obstacle to ancient armies. With improvements in technology, the ability to conduct river crossings improved as well; although these innovations did not guarantee success. The disastrous river crossing operations along the Rapido River in Italy during January 1944 (D'Este, 1992), and failed Allied airborne operation to cross the Rhine in September 1944 (Ryan, 1974) are testaments to this truth. More recently the American 1st Armored Division's difficulty crossing the Sava River into Bosnia in January 1996 (see Photograph 1) illustrate the complications associated with river crossing operations even when uncontested. Although rivers are ordinarily thought of as obstacles, they have been used as important highways by invading armies. Nowhere was this more evident than in the Western Theater of the American Civil War.

Photograph 1. An M1A1 crosses the Sava River into Bosnia-Herzegovina in January 1996.
Source: U.S. Army Photograph.

The Role of Rivers in Military Operations

The typical perception of a riverine operation is that of a deliberate river crossing such as the crossing of the Rhine River during World War II, or more recently during the Egyptian assault crossing of the Suez Canal during the 1973 Yom Kippur War (Herzog, 1984). A deliberate crossing of a defended river is an operation that most soldiers would prefer not to risk. Nonetheless, these are the most common of large-scale military operations. River crossings are associated with trepidation because they offer a capacity for defense that is equaled only by a beach landing or perhaps high ground (Patton, 1947).

A heavily defended river represents a nightmare for those who are tasked to fight their way across to the other side. Notwithstanding the relative strength or disposition of an enemy force, operations on fluvial landscapes can easily be foiled by nature alone. A skillful enemy defending a crossing point can magnify the slightest terrain or hydrologic impediment such as a swift current or flood conditions. The attacker typically must contend with heavy enemy fire and numerous natural and manmade obstacles at the crossing point. Marshy approaches, steeply sloped banks, turbulent currents, unstable riverbeds, dense vegetation along the banks and steep bluffs that delimit the floodplain can easily confound any operation along a river. Furthermore, the impact of any one of these factors can be exaggerated in a matter of hours by a passing rainstorm or early season thaw. All of these factors, taken together with difficult traffic control problems and confusing command and communication relationships must figure into the detailed planning for this sort of operation (Spiller, 1992).

Riverine operations are usually part of a larger land campaign and integrated into overall strategy. There are however, numerous examples of campaigns that have been designed around riverine operations as the focal point of the war such as the campaign along the Ebro River mounted by the Spanish Republican Army in 1938 (Macdonald, 1985). Although operations on the fluvial landscape are typically thought of in the context of a crossing operation, they have other equally important roles in warfare. A river may be a part of a strategically vital region, and as such may represent key terrain such as the Hudson River during the American Revolution. The strategic geometry of the war dictated that in order to be victorious, the winning side must retain control of the waterway (Palmer, 1969). Similarly, Harpers Ferry, a vital river gap through the mountains along the Potomac River, played a pivotal role in the early campaigns of the American Civil War (Foote, 1958).

The more "typical" role of a river in military operations is that of an obstacle that must be crossed by deliberate assault or hasty crossing. In this circumstance the river serves as a barrier to cross-country movement and can be exploited by a defending army to reinforce the defense of a front or flank. In this scenario the river channelizes and compartmentalizes movement and usually presents a clear set of advantages to the defender. History is replete with examples of rivers fulfilling the role of an obstacle to movement. By way of example, the Rhine River has played a pivotal role throughout European history from the time of the Romans to World War II (Howard, 1976). The grueling World War II Italian Campaign included of a series river obstacles which the Germans very adroitly exploited to bog down Allied operations for the better part of two years (Doughty et al., 1996). In Korea the final line of defense along the Pusan Perimeter depended heavily on defending the line of the Naktong River (Center for Military History, 1989). The fortuitous seizure of the Remagen Bridge in March 1945 (MacDonald, 1984) and the Israeli crossing of the Suez Canal during their counterattack in the Sinai in 1973 are excellent examples of hasty river crossings (Herzog, 1984).

A river may also serve as a line of operation as part of a major campaign or war. A line of operation can be defined as *"a directional orientation that connects the force with its base of operations and its objective"* (Department of the Army, 1993). In this situation the river is the axis of the military operation and enhances mobility, logistics and communications. A classic military operation of this nature occurred during the 1898 British campaign against the Khalifa in Egypt and The Sudan. The British used the Nile River to link their principal base at Alexandria to their main objective at Khartoum (Macdonald, 1985). In this same vein, operations on the Cumberland, Tennessee and Mississippi Rivers during the American Civil War demonstrate the numerous advantages associated with rivers as lines of operations against a defending force. Union forces exploited the inherent mobility and tactical advantage of north-south river lines to penetrate deeply into the Confederate hinterland (Coombe, 1996).

Armies have also used the fluvial landscape to augment power projection operations as well. In this sort of operation the river serves as the medium upon which a nation projects its power. In this case too, the geographic variety of stream channels can enhance or degrade operations. During the period of the early Egyptian Empire, the Pharaoh's used the Nile River as a strategic mobility asset to project power and maintain dominion over conquered peoples in adjacent lands. The American Navy used the Yangtze River patrols after the Boxer Rebellion to project power and enforce American political will in China until the late 1930s (Doughty et al., 1996).

The American Experience

American land and naval forces have conducted operations on fluvial landscapes since the Revolutionary War. The American experience since that time has involved almost every conceivable type of riverine operation. The nature and frequency of these operations have largely depended on geography — influenced by the climate and topography of the area of operations — as well as the operational/strategic setting. For example, during the American Civil War, Union and Confederate forces in the west were inexorably tied to rivers because of the nature of the Mississippi basin and its drainage pattern, but less so in the Eastern Theater. Although operations on fluvial landscapes have not always been operationally decisive, they typically have made a significant contribution to the successful prosecution of the campaign or war in question (Fulton, 1985).

Perhaps the most noteworthy military operations of the Revolutionary War was the fortification and battles for control of the Hudson River, which eventually led to the creation of Fortress West Point at the site of the modern United States Military Academy. These fortifications became necessary because the Hudson River was the strategic center of gravity and key terrain of the American Revolutionary War, and the colonists' retention of this important feature was a decisive factor in the attainment of victory (Palmer, 1969).

Not long after the opening battles of the war, both British and American forces sought to gain control of the strategic Hudson River-Lake Champlain-St. Lawrence River waterway system. This system was a vital factor in the strategic geometry of the American Revolution principally for two reasons. First, the Hudson River was the natural dividing line between New England and the mid-Atlantic colonies. By controlling the Hudson, the British could drive a wedge between the manufacturing and agricultural centers of the colonies, thus fracturing the already fragile economy and war effort. Second, this waterway physically connected the British military centers in New York City and Canada at Montreal (see Map 1). Possession of the waterway by the colonist was necessary to prevent concentration of British military effort and fragment their ability to act in unison in New York and New England (Palmer, 1969).

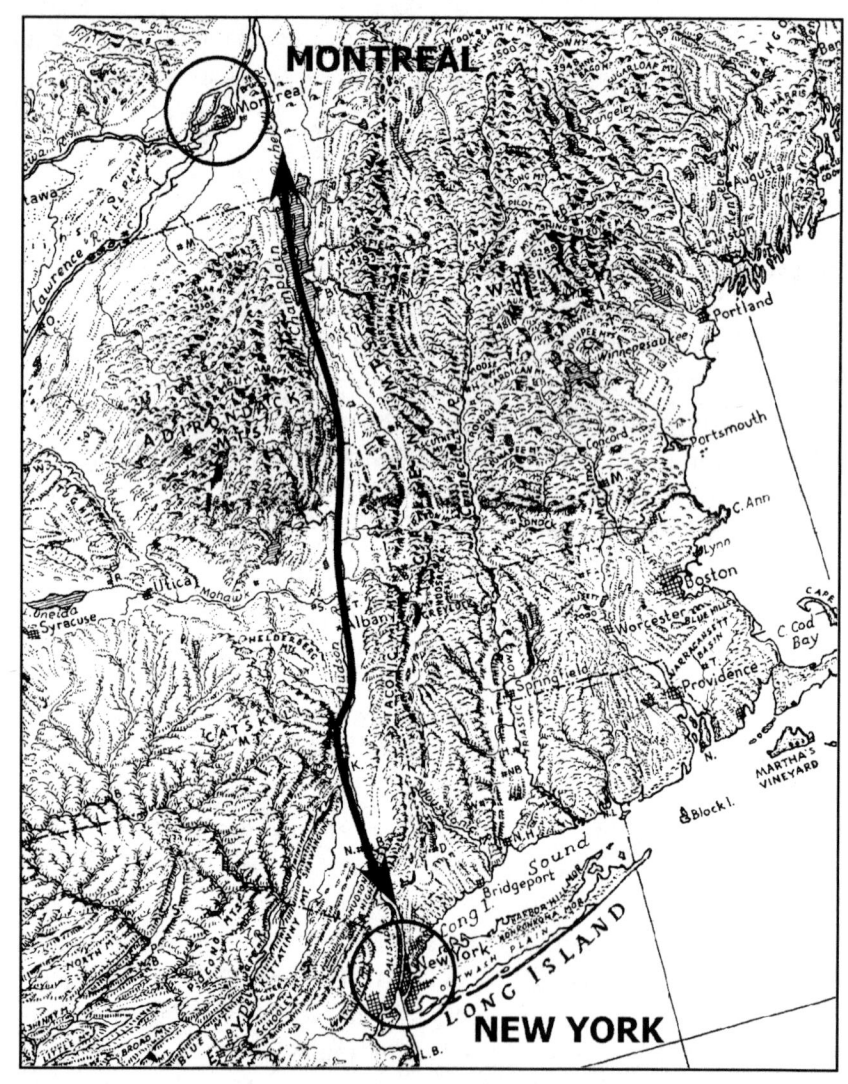

Map 1. This map illustrates the strategic position of the Hudson River. The river was the geographic link between the principle British bases in New York City and Canada, and separated the New England Colonies from those in the Mid-Atlantic and South.
Source: Raisz, 1957.

During the War of 1812 American and British forces were once again engaged on North American waterways, and much like operations during the War of Independence, inland waterways served as avenues to prosecute the war. Warfare on the fluvial landscape witnessed decisive engagements that

The Battle for Island No. 10, Mississippi River, 1862

substantially aided American efforts in three strategically important regions: Lake Erie and the St. Lawrence River, then later on the Chesapeake Bay and finally near New Orleans. In 1813, Commodore Perry assembled a fleet on Lake Erie to counter a British threat from Canada. He led this fleet to a series of important victories over British naval forces and gained control of the Great Lakes and St. Lawrence River, permitting a subsequent American advance into Canada (Fulton, 1985). The loci of the war shifted to the Chesapeake Bay and estuaries of the Potomac and Patapsco Rivers estuary around Washington and Baltimore respectively. In August 1814, the British were stymied by American naval forces in this area and were compelled to settle for a landing near Benedict, Maryland and burning the Nation's Capital in Washington, D.C.. This impasse, combined with the defeat of a British effort in Lake Champlain, caused a shift in major operations to the region around New Orleans (Fulton, 1985).

The British moved against New Orleans in December 1814 using the tributaries of the Mississippi delta. As British naval forces assembled in the Gulf, Americans under Commodore Patterson pulled together a small riverine force to oppose their advance. Although Patterson made excellent use of terrain and the fluvial landscape, he was ultimately defeated by a larger, better-equipped British force (Fulton, 1985). The small American river flotilla retreated and joined forces with General Andrew Jackson to defend the city of New Orleans. During the land battles around the city, Commodore Patterson used his two largest river gunboats, the *Carolina* and *Louisiana*, to harass British naval forces and aided in their defeat at the gates of this strategically important city (Doughty, et al., 1996).

The American Army and Navy fought a protracted and frustrating form of low intensity combat during the Seminole Wars (1835-1842) against the Creek and Seminole Indians in Florida. During this campaign, land and naval forces operated jointly using the inlets, creeks, rivers and swamps of the Florida Everglades to prosecute a largely indecisive and wearisome war. Once again, the fluvial landscape would serve as a means of transportation and line of operation. The Seminole War also witnessed the building of highly specialized river craft, constructed to fight exclusively in the unique riverine environment -- a unique characteristic of river warfare that would reoccur in the American Civil War and later in Vietnam. Rivers require shallow draft, stable watercraft capable of supporting heavy armament and armor (Coombe, 1996). Accordingly, American forces under Colonel W.J. Worth and Navy Lieutenant J.T. McLaughlin assembled a specialized fleet of flat-bottomed boats, floating batteries, canoes and specialized troop transports to operate in the low-lying swamps and rivers of the region. Although this "war" can not be called an unqualified victory, the mobility and striking power of the special river fleet combined with the characteristics of the fluvial landscape permitted the Americans to realize their major objective of reducing the perceived Indian threat in the region (Fulton, 1985).

The War with Mexico (1846-1848) witnessed a more modest role in terms of operations on fluvial landscapes. The United States' Navy conducted two river raids that were not decisive as stand-alone operations, but contributed to the ultimate American victory. Perhaps the most important outcome of these relatively minor operations was the participation of a number of junior naval officers who would plan and execute the grand river offensives of the American Civil War as senior naval officers a decade later. These young officers observed the operational possibilities of this form of warfare, and the lessons learned during these minor actions would serve them well during the War of Rebellion. The two naval raids were conducted approximately 75-miles up the Tabasco River against a Mexican crossing point and base at San Juan Bautista. Once again a river flotilla consisting of specialized naval craft was assembled to fight on the Tabasco River. The naval craft supported infantry landings and overall victory was assured because of the speed and maneuverability of this force (Fulton, 1985).

The American Civil War witnessed riverine operations of all types. It has been argued that the Civil War accounted for the most extensive riverine operations of any war past or present (Coombe, 1996). In the Eastern Theater, rivers tended to serve as obstacles to maneuver, requiring careful planning and extensive crossing operations (e.g., Fredericksburg). In the Western Theater the Mississippi River and its major tributaries commonly flowed parallel to Federal lines of advance and in due course served as highways for invasion. However, riverine operations were not exclusive to the Western Theater. Federal forces conducted important joint operations on the inland waterways of the Carolina sounds (e.g., Roanoke Island) and the Chesapeake Bay (e.g., The Peninsula Campaign). The Navy operated in estuaries and coastal rivers while enforcing the blockade of Southern ports (Foote, 1958). The American Civil War would also witness the evolution of very specialized river craft, sometimes designed to deal with the unique physical characteristics of a specific river (Coombe, 1996). Many of the most important Union victories (e.g., Forts Henry and Donelson, and

Vicksburg) were river-based campaigns and perhaps more than any other battles, precipitated the collapse of the Confederacy (Miles, 1994).

The Civil War represented the zenith of American operations on fluvial landscapes until World War II and the Vietnam War. The most notable river operations during the intervening period were the Navy's Yangtze River patrols in China that followed the Boxer Rebellion (Fulton, 1985). The immobile Western Front of World War I occasioned only minor, small-scale operations that had no considerable impact on the outcome of the war. However, operations during World War II would necessitate a number of important river crossing operations, sometimes involving entire Corps and Armies. The Second World War manifested river operations on a grand scale, and some of the greatest victories and bloodiest defeats of the war are associated with riverine operations. In Vietnam, operations on the Mekong River delta necessitated the creation of a joint Army-Navy Riverine Force. This force was composed of unique watercraft designed to content with the operational requirements of the fluvial landscape (Fulton, 1985). Not surprisingly many of the Vietnam-era rivercraft looked remarkably similar to their Civil War-era cousins, and many retained the same types of names (e.g., monitor and mortar barge).

Streams and the Civil War Landscape

At the start of the American Civil War, Federal strategy dictated a concerted drive down the Mississippi River to split the Confederacy and open it to Union commerce (the so-called "Anaconda Plan, conceived by General Winfield Scott). In fact, each of the Western Theater's great rivers—Mississippi, Cumberland and Tennessee—represented highways for invasion, and Federal armies would use them to pierce the left wing of the Confederate line (see Map 2). No one was quicker to realize the importance of these rivers than General U.S. Grant and President Lincoln (Catton, 1989). Lincoln saw the enormous geographic advantage this river system afforded the Northern cause all too clearly. Ownership of one or more of these rivers would bring about a major penetration of the western flank of the Rebel line and split the Confederacy (see Map 2). Further, he saw possession of the Mississippi River as perhaps the singular objective of the war. Lincoln instructed MG Henry W. Halleck (General of the Armies) to impart this vision to the western commanders in a general memorandum in early February 1862 which stated, *"The President regards the opening of the Mississippi River as the first and most important of all our military and naval operations*" (The War of Rebellion, 1883).

Grant understood that the Cumberland and Tennessee Rivers pointed directly into the heart of strategically important Tennessee (Grant, 1885). A rapid penetration of this state and control of these rivers would unhinge General Albert Sidney Johnston's western flank and open Nashville and Chattanooga to easy Federal attack (see Map 2). Grant acted on this plan and with the help of a powerful Union river flotilla, seized Forts Henry and Donelson in early February 1862. This was a remarkable victory because Grant forced Johnston to withdraw the Confederate line south to Corinth, Mississippi – a retreat of 200 miles giving the Union Army Nashville and all of western Tennessee (Catton, 1989). Grant quickly followed up this success with a rapid advance down the Tennessee River to Pittsburg Landing, where the bloody Battle of Shiloh would be fought on 6 and 7 April 1862. While Grant's army fought for its life along the river bluffs at Shiloh, another equally important battle was reaching its climax at Island No. 10 near the town of New Madrid, Missouri along the Mississippi River (see Map 2).

Physical Setting: Eastern Versus Western Theater

Riverine operations in the Eastern and Western Theaters of the Civil War were very distinct in character because of differences in their physiography. The Eastern Theater, which was largely confined to the region between Washington D.C. and Richmond is composed of an Upland-Piedmont-Coastal Plain configuration (Fenneman, 1938). This landscape encompasses a trellis drainage pattern with rivers that flow perpendicular to the coast, directly across the major line of north-south operations (see Map 3). Consequently, in the Eastern Theater rivers were obstacles and the majority of riverine operations were confined to deliberate river crossing operations or securing fords to ease crossings by large numbers of infantry.

The geographic implications of the fluvial landscape were very different in the Western Theater. The Mississippi River and its great tributaries flowed across a large inland basin physiographic province (Fenneman, 1938). This region includes a large area of horizontally bedded strata dipped slightly to the Gulf of Mexico with a characteristic dendritic drainage pattern. In this geographic scenario, large rivers tend to flow south, and toward the Mississippi River, perpendicular to Confederate lines. Therefore, the rivers could be employed to easily breach and unhinge the Confederate defensive strategy

(Catton, 1989). The magnitude of the geographic dilemma is clearly illustrated on Map 2. The Cumberland River offers a direct line of advance to Nashville, which served as a major transportation hub and manufacturing center during the Civil War. Once the Confederate line was breached along this river at Fort Donelson, Nashville was doomed and there was little the Rebel generals could do to prevent Union ironclads from wrecking railroad bridges and river ports as they freely rampaged down the Cumberland and ultimately captured the city.

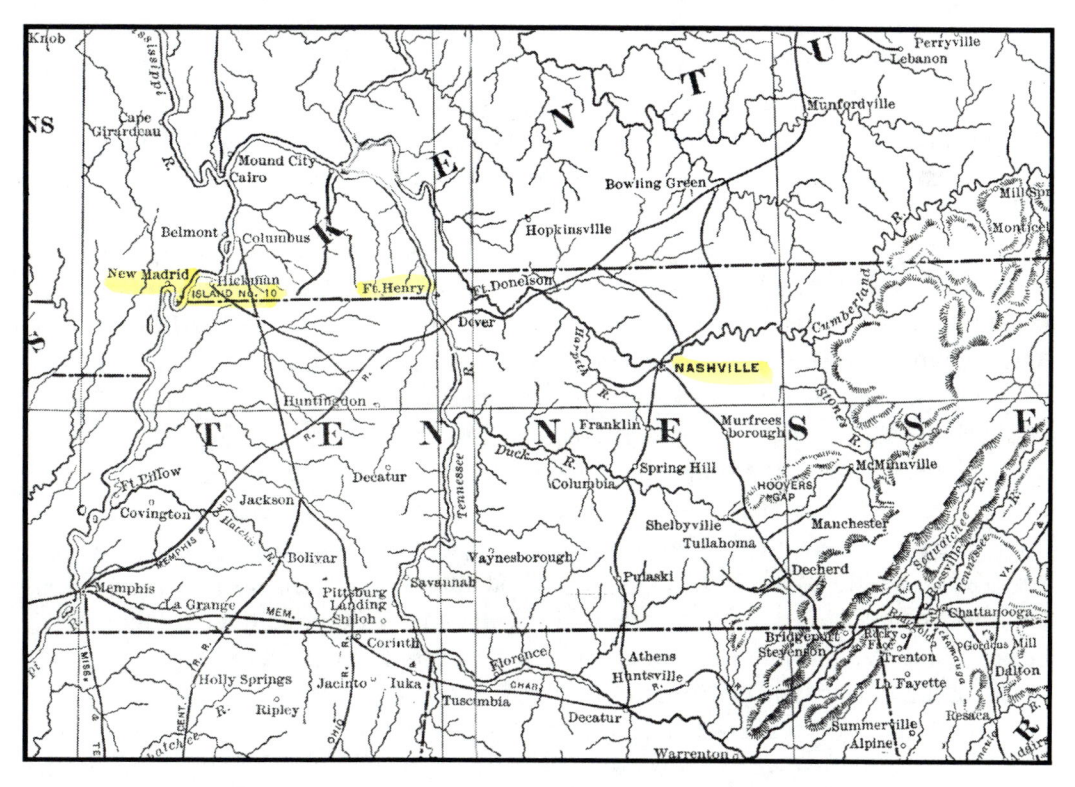

Map 2. The rivers of the Western Theater. The map illustrates the dendritic drainage pattern of the rivers in the Mississippi River basin and their suitability as lines of operation for Union forces. Source: Adapted from Dodge, 1887.

The loss of Fort Henry was even more problematic (see Map 2). The Tennessee River penetrates central Tennessee and northern Alabama like a gigantic fishhook (Foote, 1958). Once Grant's forces seized Fort Henry in February 1862, the entire center of the Confederate line in the west was open to a Union river-borne advance. Furthermore, Union gunboats traveled as far south as Muscle Shoals, Alabama wreaking havoc as they went, destroying railroad bridges, river craft and factories (Coombe, 1996). Grant exploited this geographic advantage and rapidly advanced along the river to Shiloh in March and April 1862 (Catton, 1989).

Simultaneously, as part of the Union Army's general southward advance along the western rivers, another Federal army mounted an assault to seize the Confederate strongpoint blocking a large meander—or "S" bend—in the Mississippi at Island No. 10 near New Madrid, Missouri (see Map 2). The Mississippi here makes one of its gigantic meanders (see Map 4). The objective of the Federal operation was Island No. 10, so-called because it was the 10th island south of the Ohio River's junction with the Mississippi (Nevin, 1994). Island No. 10 was selected as a fortified zone on the Mississippi River because the Rebels were forced to abandon their major fort at Columbus, Kentucky (see Map 2), which was outflanked after the loss of Forts Henry and Donelson in February 1862.

Island No. 10 was the only defensible terrain on the Mississippi, north of Memphis, and General P.T.G. Beauregard considered the retention of this fort essential to Confederate fortunes in the west (The War of Rebellion, 1883). Without question he saw Island No. 10 as the crucial strategic point along the Mississippi because it stood squarely in the way of the anticipated Federal advance south along the axis of the river. The location of Island No. 10 at the base of the meander combined with the bluffs behind it and flooded banks along the Tennessee shore sealed off river operations from New Madrid to Memphis (see Maps 2 and 4). This compelled Beauregard to order the construction of a chain of forts along the bluffs above and on the island itself so that heavy guns could cover every possible approach along the river and the adjacent banks. The largest guns available to the Rebels in the Western Theater were emplaced in the fortifications (The War of Rebellion, 1883).

Island No. 10 was located (it no longer exists) at the base of a southerly meander, and the town of New Madrid at the top of the northerly loop of the Mississippi in 1862 (see Map 4). The straight-line distance from Island No. 8 to New Madrid was 6 miles, while by river it was 15 miles. Likewise, the overland distance between Island No. 10 and Tiptonville was 5 miles, but 26 by water (The War of Rebellion, 1883). To the rear of Island No. 10, the mainland was inundated and swampy, severely limiting access to and exit from the fort. Reelfoot Lake (Map 4) is an artifact from the 1811 New Madrid earthquake and was an uncrossable barrier. The lake was choked by cypress trees that prevented boat traffic (Coombe, 1996). Therefore, an attack to capture the fort was possible only by crossing the river, through New Madrid, or from the south by the road that passed through Tiptonville (see Map 4) (Dodge, 1898). The fort itself—deemed impregnable by many—consisted of a series of large earthen works with 49 heavy guns and 9,000 veteran troops, supported by an additional five batteries on the Tennessee bank of the river (The War of Rebellion, 1883). A floating battery and a small flotilla of gunboats (see Map 4) augmented the land defenses. Therefore, Federal land and naval forces would have to fight a difficult twin battle. One against this formidable array of Confederate power and the other, a function of geography, against inundated land, treacherous currents and sandbars within the meander loop adjacent to the fort (Nevin, 1994).

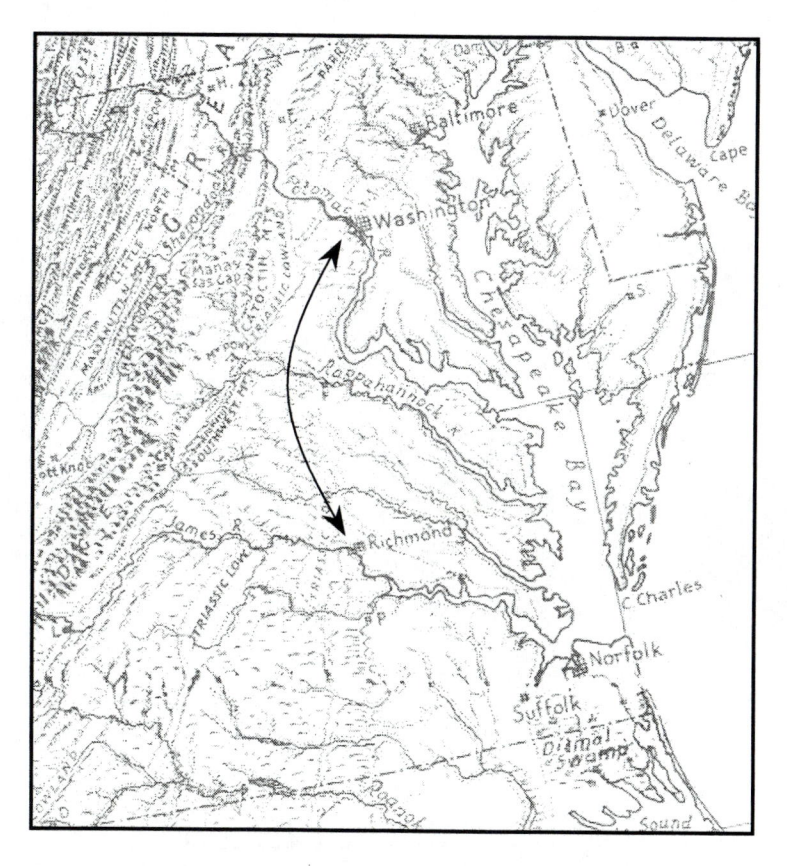

Map 3. Trellis drainage pattern of the Eastern Theater. The region's major streams flow perpendicular to the coast and across the principal lines of movement.
Source: Adapted from Raisz, 1957.

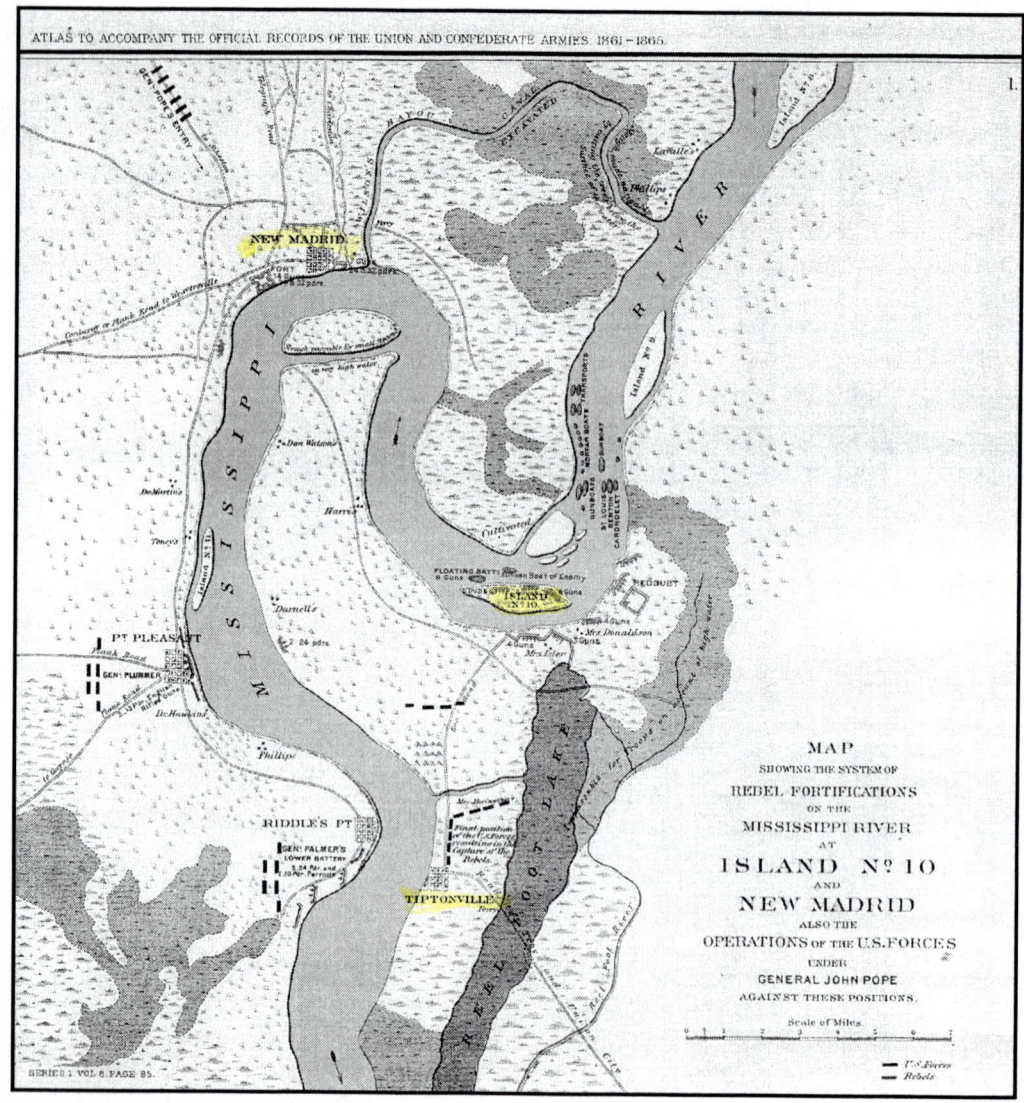

Map 4. Confederate positions around New Madrid and Island No. 10 in March 1862. Source: Cowles, 1895 (Plate No. 10).

Synopsis of the Battle for Island No. 10

General Pope commanded a Federal army of 20,000 men, supported by the River Squadron under Flag Officer Foote (Nevin, 1994). Foote was no stranger to this type of operation. He was the naval commander who successfully captured Forts Henry and Donelson with Grant in January 1862 (Catton, 1989). Although the Federals greatly outnumbered the Rebels, the combination of the fortifications and geography—the swift Mississippi, its meanders and swampy, flooded valley—presented daunting physical obstacles to the Union commanders. The only practicable land approach to the fort was along the western bank of the Mississippi via New Madrid. In order to get at the fort; however, Pope's army needed transport craft and gunboats to silence defending shore batteries; but the powerful fortifications on Island No. 10 stood between the Federal army and these naval craft (see Map 4).

The battle began well enough for Pope's army. They quickly seized New Madrid on 14 March, and Pope fittingly wanted to send his soldiers across the river south of Island No. 10 to encircle and capture its garrison using the River Squadron's transports and gunboats for covering fire (Dodge, 1898). Notwithstanding his seizure of New Madrid, Island No. 10 remained safe unless he crossed the river because two avenues of supply and escape remained open to the Rebels. Flag Officer Foote on the other hand, decided it was too dangerous to push his ironclads—and certainly his transports—within range of the heavy guns in the Confederate fort (Official Records, 1884). His squadron was badly shot up and himself seriously wounded at Fort Donelson in February. This ordeal may have clouded his thinking and certainly diminished his aggressiveness (Foote, 1958). Foote's plan was to begin a long-range bombardment of the fortifications on Island No. 10, but it quickly proved to be ineffective (Official Records, 1884).

In the interim, Pope's infantry attempted several methods to get at Island No. 10, but with little effect on the tactical situation. General Pope grew tired of Foote's ineffective bombardment and decided to attempt a novel plan to bypass the formidable island. His formula was to cut a canal across the base of the flooded meander north of the "S" bend (see Map 4). The proposed canal would directly connect the river from the area between Islands No. 8 and No. 9 and New Madrid. This shortcut would use the geography of the meander to bypass the island with the light-skinned transports he needed to cross the river (The War of Rebellion, 1883).

Engineers and a regiment of Wisconsin timbermen toiled at digging the canal through flooded fields and tangled bottomlands, and devised ingenious tools to saw down trees below the level of the floodwaters held behind the natural levee (The War of Rebellion, 1883). Some 600 soldiers eventually hacked a 50 foot-wide, eight-mile long canal in three weeks (Map 4). Finally, on

4 April, the levee above Island No. 9 was pierced and the transports rode the surge of floodwater driven by the Mississippi's swift current safely to New Madrid. Simultaneously, two Union ironclads completed successful nighttime runs past the guns of Island No. 10. Now Pope had transports to ferry his army across the river and the guns of the ironclads *Carondelet* and *Pittsburgh* to defeat shore batteries and cover his crossing (Dodge, 1898).

Figure 1. Union mortar boats bombard Island No. 10 during Flag Officer Foote's abortive attempt to pound the island's defenders into submission.
Source: Harper's Weekly.

The drive to take Island No. 10 from the rear began in earnest. On 6 April the ironclads cleared Rebel batteries on the Tennessee shore and those opposite Point Pleasant (see Map 4). The *Carondelet* and *Pittsburgh* attacked Watson's Landing on the morning of 7 April, then covered the crossing of Pope's Army to the eastern bank of the river. Pope's troops quickly sealed off the escape routes by seizing Tipptonville and the narrow isthmus between the Mississippi and Reelfoot Lake. By noon, Island No. 10's garrison was cut off, and by 2100 hours many of them surrendered. Early on 8 April (just as Grant's Army was gathering itself after the Battle of Shiloh) the last Confederate stragglers surrendered to General Pope's soldiers (Nevin, 1994). This was a stunning victory, in which the Union Army took advantage of the geography of the area to compel the Confederates to surrender. More than 5,000 Confederates, including three generals were taken prisoner along with in excess of 100 captured guns and a mountain of ammunition (Foote, 1958). The anchor of the Confederate left flank in the west was smashed – all with the loss of a handful of Federal soldiers and sailors.

Summary of Geographic Factors

Unquestionably, geography had a profound effect on the battle for Island No. 10. The inundated terrain and meanders of the Mississippi River seemingly afforded the Confederates with an ideal, almost impregnable fortress secure from land attack. The river channel here was wide and at the highest levels in years, and could not be bridged without considerable effort. In order to cross the Mississippi, Pope's infantry required transports, which were blocked on the opposite side of the meander. To assist Pope, Union naval commanders were forced to contend with strong currents and sandbars as well as fighting the formidable fortifications. Hence, the "s" shaped meander played a diabolical geographic trick on the Union plan to seize the fort because the thin-skinned transports remained bottled up north of the island, separated from New Madrid by only 6 airline miles where the infantry they were intended to carry remained idle. Union infantry commanders could only nibble at the periphery of the fort, blocked by swampy bottomland and the river. Island No. 10 was approachable only by way of a single road through Tiptonville, and herein lies one irony of this battle. Once general Pope's soldiers gained a foothold on the Missouri banks of the river, this geographic advantage was reversed and the single access road through Tiptonville became an Achilles heel of the defenders.

Figure 2. Panorama of the Union bombardment of Island No. 10.
Source: The Navy and Marine Corps Museum.

Clearly the most important geographic influence on the battle was the Mississippi River. The position of the twin meanders presented a unique tactical puzzle to the Union Army. Moreover, the position of Island No. 10 at the base of the meander was fortuitous for the Rebels because the main river current would force Foote's ponderous gunboats directly against the river batteries (Map 4). This problem was compounded because the river was at its highest floodstage in recent memory (The War of Rebellion, 1883). The high water generated a current so great that Union ironclads and gunboats had to anchor themselves to the shore because their engines were not powerful enough to maintain position in the river. Foote's biggest problem however was the direction of flow. At Forts Henry and Donelson the Tennessee and Cumberland Rivers flowed to the north, hence a ship that took a disabling shot or experienced a mechanical breakdown would drift harmlessly back to Union lines. At Island No. 10 the situation was reversed. Here Foote had to contend with a swift southerly current which would sweep a disabled ship south, into Confederate hands. This factor more so than any other caused him to rely on the seemingly timid long-range bombardment of the island (The War of Rebellion, 1883).

Flooded banks and swamps were a key security factor early on for the Confederate forces assembled at Island No. 10. A land assault from the Tennessee side of the river was out of the question because of impassable swamps and bogs (see Map 4). This inundated land was reinforced by Reelfoot Lake, which connected with the Mississippi River some 14 miles to the south. Reelfoot Lake was not easily crossed because it was thoroughly obstructed by dense swamp vegetation and cypress trees. Further, Pope's efforts to occupy the meander opposite Island No. 10 were similarly foiled by flooded land and bogs (The War of Rebellion, 1883).

However, the security offered by the flooded land adjacent to the river and single road approach quickly turned into a major disadvantage for the Confederates. After capturing New Madrid on 14 March, Pope placed artillery at Riddles Point and south along the natural levee (see Map 4). This was significant for two reasons, and ultimately reversed the relative advantage presented by the inundated land. First, the Union batteries at Riddles Point commanded the river below Island No. 10, all but preventing resupply by river craft and forced the Rebels to withdraw their small fleet of gunboats from the bend in the river. The withdrawal of the Confederate gunboats from the river bend would later facilitate the passage of Union ironclads past the fort on the evenings of 5 and 6 April (Coombe, 1996). Second, the Riddles Point batteries controlled the land approaches to Tiptonville from the south. Since the land south of Tiptonville was flooded, movement had to take place on the single elevated road or along the natural levee -- both in easy range of Union artillery.

Conclusions

The battle for Island No. 10 was a key Union victory that opened the Mississippi River to Memphis, Tennessee 150 miles to the south, and crushed Confederate plans to defend forward along the river. The battle is an excellent illustration of the influence of fluvial landscapes on military operations and portrays the geographic obstacles to be overcome in conjunction with riverine operations. Union forces were able to overcome a heavily defended fort protected by bluffs and flooded land along a very wide, swift river. This victory resulted in the capture of the entire Confederate garrison, several general officers and a huge bag of heavy guns and ammunition -- all with the loss of handful of Union soldiers. The battle witnessed remarkable ingenuity in the construction of the canal to bypass the heavy Confederate guns on Island No. 10. In terms of joint operations, it is nearly a textbook operation illustrating cooperation between land and naval forces during a time when there were no formal regulations mandating joint action of any sort. Ironically, this important battle remains virtually unknown. Perhaps the battle was simply overshadowed by the bloody fighting at Shiloh, which was very controversial and captured the lions-share of the headlines in the Northern press. Furthermore, General Pope, the battle's principal architect, slipped quickly into oblivion after his defeat in the Battle of the Second Bull Run during the fall of 1862. Notwithstanding its relative obscurity, the battle for Island No. 10 was the first major engagement on the Mississippi River and was the model for a number of battles to come. In fact, when faced with a similar geographic problem at the meander opposite Vicksburg, Mississippi, Union forces quickly dusted off the solution used at Island No. 10 and attempted to bypass its river batteries in much the same way as they did the year before. Unfortunately they met with failure and were forced into a protracted battle because the conditions of the Mississippi River could not support the scouring and maintenance of the canal long enough to float the transports across the meander.

Bibliography

Catton, B.C. 1989. *Grant Moves South*. New York, New York: Ballentine Books.

Center for Military History. 1989. *Korea, 1950*. Department of the Army, Center for Military History Publication 21-1, Washington, D.C.: United States Government Printing Office.

Coombe, Jack, D. 1996. *Thunder Along the Mississippi*. N.Y., New York: Bantam Books.

Cowles, Calvin, D. 1895. *Atlas to Accompany The Official Records of the Union and Confederate Armies*. Series 1, vol. 8, p. 85. Washington, D.C.: U.S. Government Printing Office.

D'Este, Carlo. 1992. *Fatal Decision: Anzio and the Battle for Rome*. New York, New York: Harper Collins Publishers, Inc.

Department of the Army. 1980. *FM 30-10: Military Geographic Intelligence (Terrain)*. Washington, D.C.: United States Government Printing Office.

Department of the Army. 1990. *FM 5-33: Terrain Analysis*. Washington, D.C.: United States Government Printing Office.

Department of the Army, 1993. *FM 100-5: Operations*. Washington, D.C.: United States Government Printing Office, 276 pp.

Department of the Army. 1993. *FM 90-13: River Crossing Operations*. Washington, D.C.: United States Government Printing Office.

Dodge, T.A. 1897. *A Bird's-Eye View of Our Civil War*. New York, New York: Da Capo Press.

Doughty, R.A., Flint, R.A., Lynn, J.A., Grimily, M., Howard, D.D., and Murray, W. 1996. *Warfare in the Western World*. Lexington, Massachusetts: D.C. Heath and Company.

Fenneman, N.M. 1938. *Physiography of the Eastern United States*. New York, New York: McGraw-Hill Book Company, Inc.

Foote, Shelby. 1958. *The Civil War, A Narrative: Fort Sumter to Perryville*. New York, New York: Random House.

Fulton, William, B. 1985. "Riverine Operations." 1965-1969. *Department of the Army, Vietnam Studies*, Washington, D.C.: United States Government Printing Office.

Grant, Ulysses, S. 1885. *Personal Memoirs of U.S. Grant*. New York, New York: Webster.

Herzog, Chaim. 1984. *The Arab-Israeli Wars*. New York, New York: Random House.

Howard, Michael. 1976. *War in European History*. New York, New York: Oxford University Press.

MacDonald, Charles, B., 1984. "The Last Offensive." *U.S. Army Center for Military History, The United States Army in World War II, The European Theater of Operations*, Washington, D.C.: Government Printing Office.

Macdonald, John. 1985. *Great Battlefields of the World*. New York, New York: Macmillian Publishing Company.

Miles, J., 1994. *A River Unvexed*. Nashville, Tennessee: Rutledge Hill Press.

Nevin, D. 1994. *The Road to Shiloh, Early Battles in the West*. Alexandria, Virginia: Time Life Books.

Official Records of the Union and Confederate Navies in the War of Rebellion, 1884. *Volume 22: Operations on Western Waters*. Washington, D.C.: U.S. Government Printing Office.

Palmer, Dave. 1969. *The River and the Rock*. New York, New York: Hippocrene Books.

Patton, George, S. 1947. *War as I Knew It*. Boston, Massachusetts: Houghton Mifflin Co.

Raisz, Erwin. 1957. *Landforms of the United States*.

Ryan, Cornelius. 1974. *A Bridge Too Far*. New York, New York: Simon & Schuster, Inc.

Spiller, R.J. 1992. "Crossing the Rapido." In *Combined Arms in Battle Since 1939*. Fort Leavenworth, Kansas: Command and Staff College Press.

The War of Rebellion: A Compilation of the Official Records of the Union and Confederate Armies. 1883, *Volume VIII: Operations in Missouri, Arkansas, Kansas and the Indian Territories*. Washington, D.C.: United States Government Printing Office.

Winters, Harold. 1998. *Battling the Elements*. Baltimore, Maryland: The Johns Hopkins University Press.

Ziemke, E.F., and Bauer, M.E. 1987. *Moscow to Stalingrad: Decision in the East*. Washington, D.C.: Center for Military History, U.S. Army.

3

Military Geography of the Civil War:
THE BLUE RIDGE AND VALLEY & RIDGE PROVINCE

By

Joseph P. Henderson

Introduction

In the American Civil War, the folded terrain of the Ridge and Valley province of the Appalachian Highlands played a key role in shaping the strategies of both Union and Confederate (C.S.A.) forces. The long, linear ridges of the Ridge and Valley Province compartmentalized troop movements (see Map 1), making the widely-spaced mountain passes key avenues for movement between these compartments. While valleys provided high-speed corridors for maneuver, ridges were barriers that shielded forces from observation by the enemy. General Braxton Bragg of the C.S.A. succinctly summed up the complications of fighting in this terrain when he stated,

> "It is said to be easy to defend a mountainous country, but mountains hide your foe from you, while they are full of gaps through which he can pounce upon you at any time." (Cozzens, 1992, p. 54).

To highlight the influence of this type of terrain on military operations, we will examine General Stonewall Jackson's 1862 Valley Campaign and the movements leading to Battle of Chickamauga. Both examples offer interesting insights into terrain compartmentalization and the relative advantages and disadvantages associated with operating in this terrain at the tactical level of warfare. In Jackson's case, he masterfully used the terrain to achieve his objective. In the Chattanooga example, both sides were able to make advantageous use of the terrain, but the latter occasionally caused near disasters for each side.

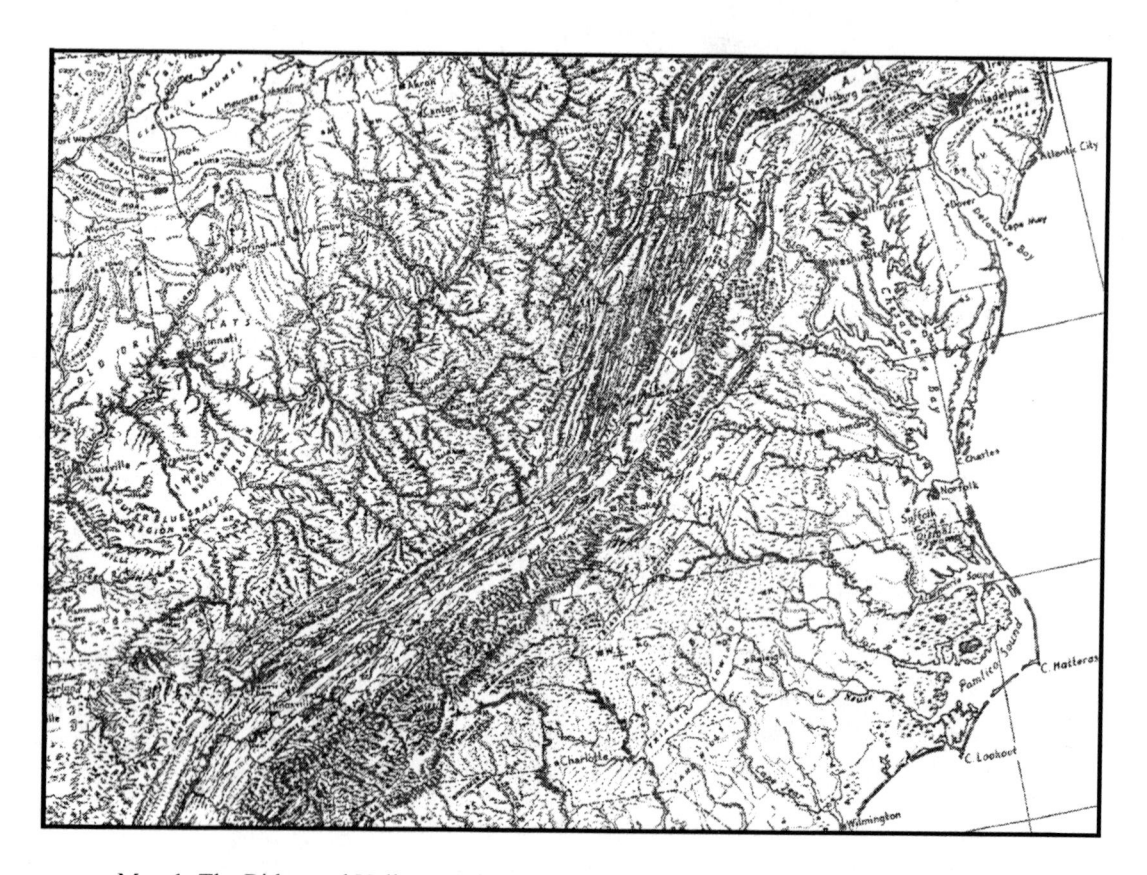

Map 1: The Ridge and Valley terrain of the Eastern Theater of the American Civil War. The northeast-southwest trending terrain dictated the flow of major campaigns.
Source: Adapted from Raisz, 1957.

Valley and Ridge Geography

The Valley and Ridge province lies within the major geomorphic division known as the Appalachian Highlands and extends from the East Gulf Coastal Plain in Alabama to the St. Lawrence Lowland of Canada at its northeastward limit. To the east lies the crystalline igneous and metamorphic rocks of the Blue Ridge Mountains, and to the west lies the slightly folded, rugged terrain of the Appalachian Plateau (see Map 2). The region is characterized by numerous elongated ridges and intervening valleys, all trending in a northeast/southwest direction with much folding and fracturing present (Miller, 1974). The trend of the ridges is caused by folding of the underlying bedrock as a result of continental convergence of Africa and North America about 350 million years ago. The drainage pattern is predominantly trellis, indicative of the underlying structure.

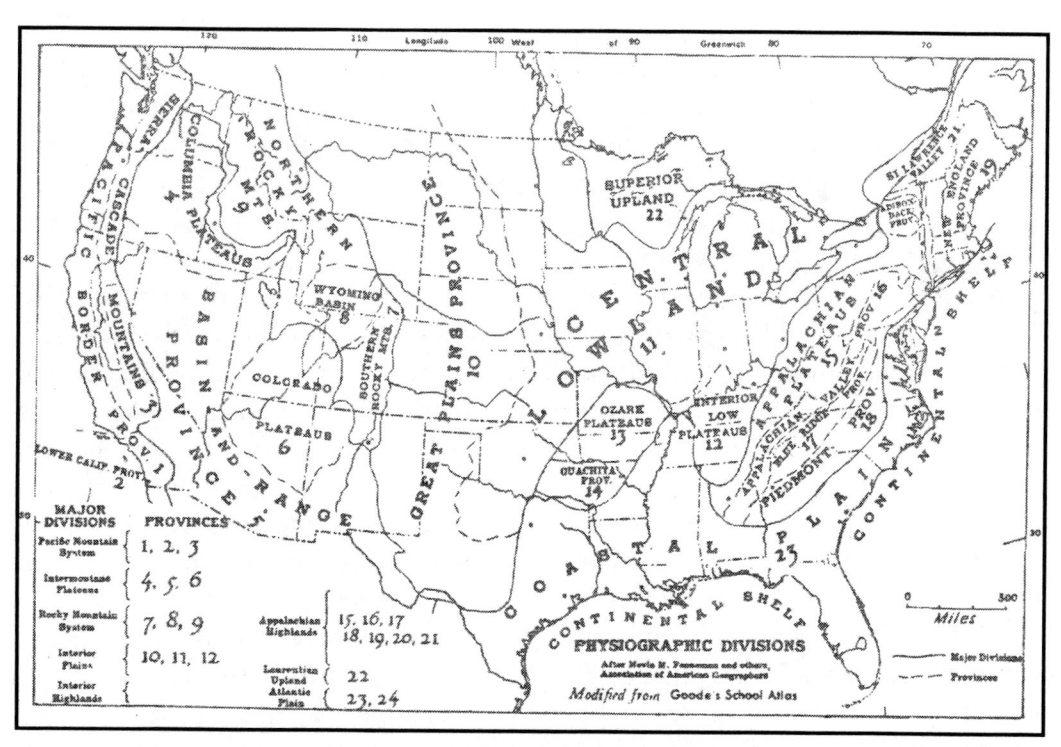

Map 1. Valley and Ridge and adjacent Physiographic Provinces.
Source: Raisz, 1957.

The lithology and structure of the Ridge and Valley consist of an extensive belt of folded and faulted clastic and carbonate sedimentary rocks. The ridges are composed of predominantly sandstone and quartzites that are quite resistant to erosion. The sandstones were formed from sediments eroded off ancient mountain ranges lying in the present Piedmont region east of the fold belt. Hundreds of wind and water gaps cut across these sturdy ridges. Less resistant carbonate rocks underlie the valleys, particularly in the Great Valley Subprovince. These carbonates consist of limestones and dolomites formed on a vast continental shelf on the margin of North America about 500 million years ago (Fenneman, 1938).

General Affects on Military Operations

The unique topography of the Ridge and Valley influenced military operations in the Civil War in a variety of ways. First, rapid troop movements were only possible along the axis of the main valleys, where most of the trails and all of the improved surface roads were located. Hence, commanders mounted most of their major offensive actions through these corridors, which generally ran southwest to northeast throughout the region (See Map 3).

Secondly, while the high relief, steep-sided ridges were largely barriers to east-west maneuver, these mountains served to conceal troop movements from enemy observation. Troop formations located on the opposite side of a ridgeline from the enemy could be repositioned up or down-valley in relative secrecy. Then, using a small gap in the ridgeline, the unit could cross over the barrier and surprise their foe, often attacking an unprotected flank. These gap crossing were often extremely hazardous and physically demanding due to the unimproved, steep mountain roads; many gaps were gaps in name only and required extensive engineer support to allow heavy equipment such as artillery pieces to traverse the steep inclines. Nevertheless, the mountain passes were key terrain, and control and observation of these passes was crucial for both sides.

The final significant impact of the ridges involves cross-compartment offensive operations by large-sized forces. An army on the offense, seeking to move rapidly east-west over a series of ridgelines, could become widely dispersed, as the available mountain passes are widely separated (See Map 3). In those situations, the army operated on a broad front with subordinate units utilizing independent avenues over the narrow gaps. Consequently, units were no longer in supporting distance of each other, communication was extremely problematic, and the isolated forces were dangerously vulnerable to piecemeal attack. The ridges concealed the enemy to their front, and as a maneuvering unit emerged from a gap, a well-positioned enemy could surprise and rapidly overwhelm a unit with superior combat power.

LOS comms

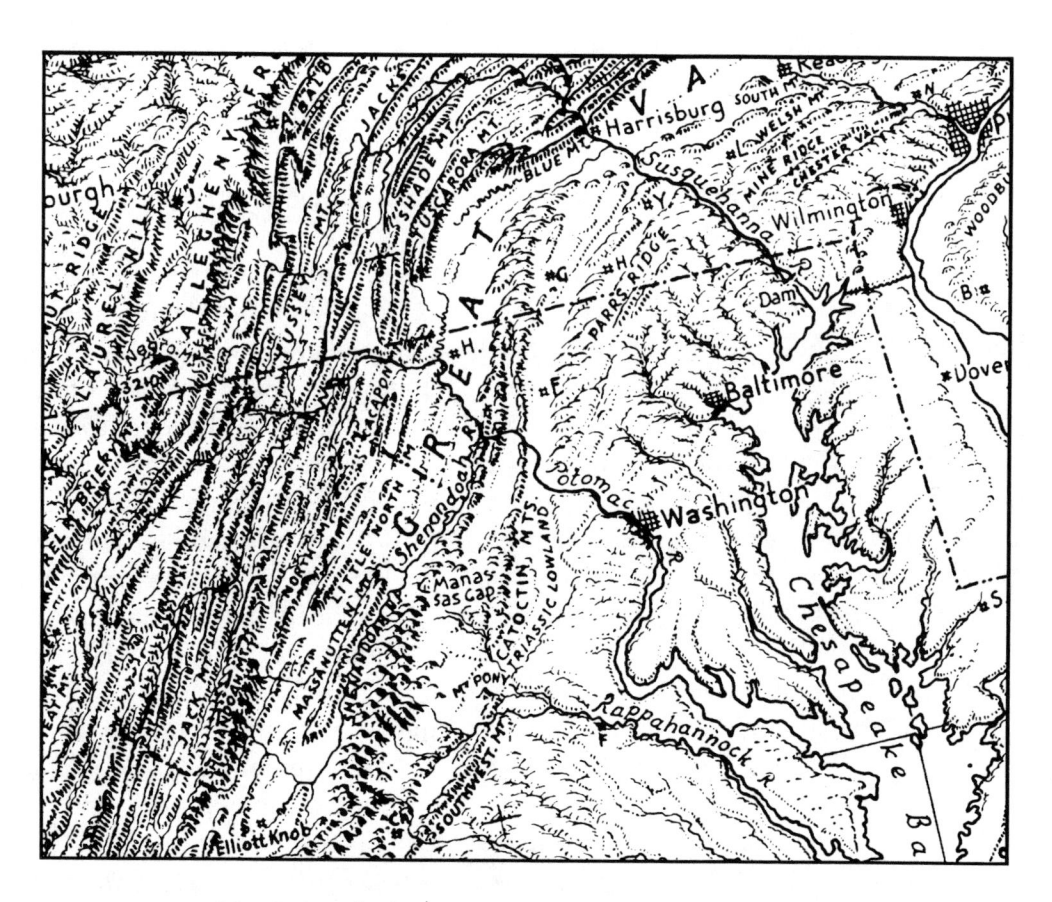

Map 3. Detail of the Ridge and Valley Province in Virginia.
The linear nature of the terrain and compartmentalization are clearly evident.
Source: Adapted from Raisz, 1957.

The Valley Campaign of 1862

Geography of the Shenandoah Valley

In the middle section of the Ridge and Valley is found the Shenandoah Valley, the focal point of Stonewall Jackson's Valley Campaign of 1862. The wide, fertile Shenandoah Valley, extending from its headwaters near Lexington, Virginia to the Potomac River at Harper's Ferry, makes up a large portion of an expansive lowland region known as the Great Valley. The major drainage feature in the valley is the north-flowing Shenandoah River, whose tributaries are the North and South Forks. A single prominent ridge, Massanutten Mountain, dominates the center of the valley and splits the North and South Forks (See Map 4). The North Fork rises in the Appalachian Plateau to the west, flows eastward to New Market, then northward along the western flank of Massanutten Mountain. The South Fork, originating near Port Republic, flows northward along the eastern flank of the Massanutten.

Massanutten Mountain is a 45-mile long synclinal ridge rising 3000 feet above the valley from Harrisonburg to Front Royal (See Map 4). The topography reflects the local geologic structure, with resistant sandstones and shales of the Massanutten synclinorium standing above the limestone and dolostone of the adjacent valleys (Lesure, 1985). Karst sinkholes, caverns, swallow holes, and springs are prevalent in the lowlands. High relief areas flank the Shenandoah Valley, with the Allegheny Mountains to the west and Blue Ridge to the east.

During the Civil War, the Shenandoah Valley was primarily agricultural as the rich alluvium of the lowlands produced valuable crops such as wheat, corn and fruits. This "Breadbasket of the Confederacy" provided valuable sustenance to the army of General Robert E. Lee. Not only was the region prime farmland, but woolen, flour, and lumber industries also operated throughout the valley.

While the strategic importance of food supplies and industry in the valley cannot be overemphasized, equally crucial to the Confederacy were the transportation routes within the Shenandoah. The valley was a vital corridor into the northern states between two rugged mountain ranges, the Blue Ridge and Alleghenies. The Blue Ridge helped to screen and protect the right flank of Confederate forces from the Army of the Potomac as they moved northward. General Lee used the valley as an invasion route into Maryland in 1862 and Pennsylvania in 1863.

The most significant transportation route through the valley was the Valley Turnpike, a macadamized surface that allowed rapid all-weather travel. This veritable superhighway for the horse and wagon was bedded with cement and buttressed on either side with limestone shoulders (Tanner, 1976). The Turnpike ran from Winchester down to Lexington, and coincides with the present day Interstate 81 (See Map 4). The town of Winchester was a key transportation hub, as the turnpike and eight other roads intersected here, and the town was connected to the east by a leg of the Baltimore and Ohio railroad.

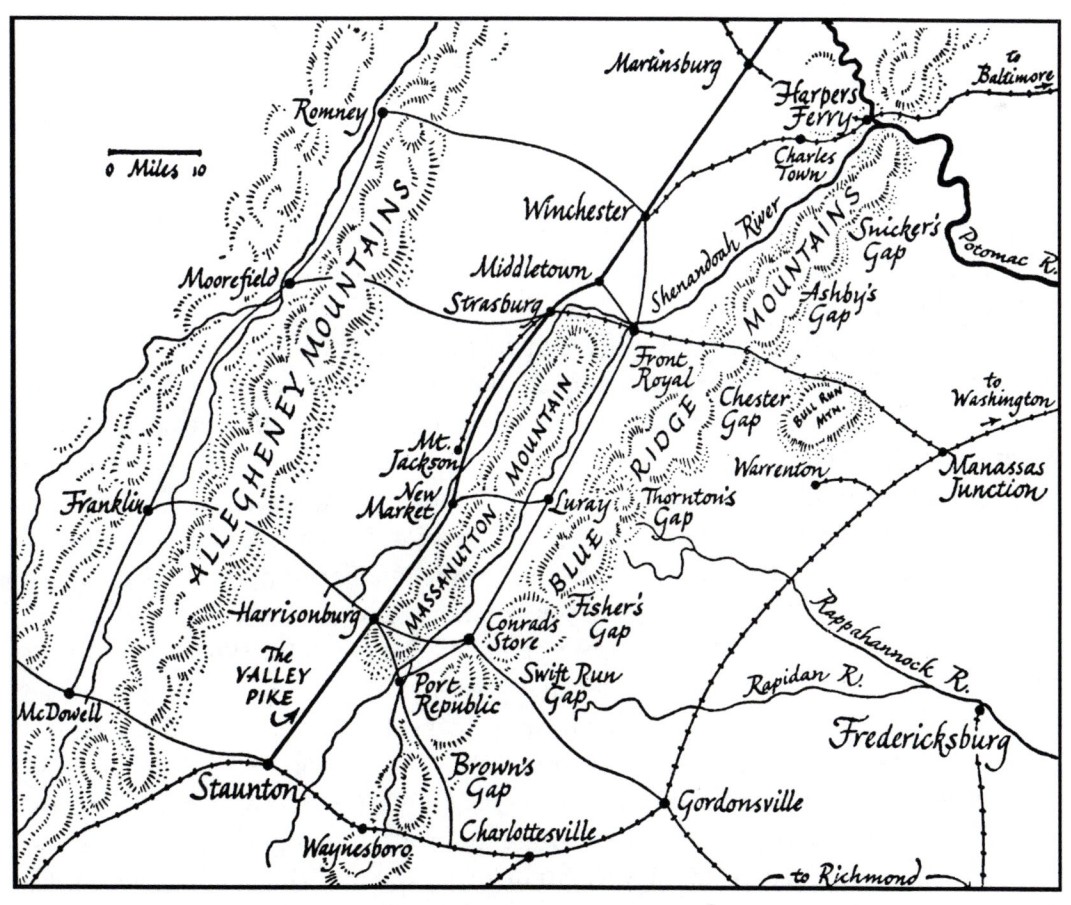

Map 4. The Shenandoah Valley during Jackson's 1862 Valley Campaign.
Source: Adapted from Foote, 1963.

Due to its strategic location, Stonewall Jackson made his headquarters in Winchester. In addition to the Valley Turnpike, a number of unimproved roads ran along the main axis of the valley. The few east-west routes crossed through gaps in either Massanutten Mountain or passes in the Allegheney Mountains and Blue Ridge to the west and east, respectively. The gaps of primary concern for this case study are Luray Gap, which split the center of the Massanutten, and the Manassas Gap at Front Royal. The Manassas Gap was of critical importance because the Manassas Gap Railroad ran through it, connecting Strasburg to Alexandria to the east.

Strategic Military Situation in 1862

Throughout the Valley Campaign, the primary focus of both Union and Confederate forces was not the Shenandoah Valley, but Richmond, the Confederate capital. General George B. McClellan's Army of the Potomac was conducting the Peninsular Campaign to take Richmond against General Robert E. Lee's Army of Northern Virginia. As a result, the bulk of both armies were located in eastern Virginia. In the Shenandoah Valley, Stonewall Jackson commanded a small force of roughly 10,000 Confederate soldiers, while General Nathaniel P. Banks led approximately 23,000 Union troops. Between Jackson and Lee, the Confederate Army totaled slightly less than 100,000 men in this theater, while the forces of McClellan and the other Union commanders in the region totaled more than 200,000. Given the considerable advantage in numbers enjoyed by the Union, Lee hoped to occupy as many Union troops as possible by conducting diversionary operations within the Shenandoah Valley. This daunting mission was given to General Jackson, and by any measure of success, he executed the mission to a level of professional mastery (Foote, 1963).

Stonewall Jackson was a commander well-suited for the challenges of the Valley Campaign. No stranger to the terrain, Jackson served as a professor at the Virginia Military Institute in Lexington at the southern end of the valley. Moreover, having grown up in the mountains of West Virginia where there are but a few wagon roads, he had a keen awareness for the topographical features of the valley which aided in his appreciation of their military significance. Like Napoleon, Jackson was very particular about securing the most accurate topographical information possible. Prior to the start of the Valley Campaign, Jackson required his skilled topographical engineer, Major Hotchkiss, to produce a complete map of the valley from Harper's Ferry to Lexington (Henderson, 1949). Jackson intently studied the map to devise his campaign strategy. In addition to possessing a thorough understanding of the terrain and its importance, General Jackson also manifested another important trait: audacity. His ability to skillfully maneuver his small but formidable army across this landscape, while taking great risks in the process was vital to Confederate success. An army inferior in numbers must exploit secrecy and surprise, and Jackson's small band of Rebels used rapid movement and cloaked maneuver through the ridges and valleys to achieve victory (Tanner, 1976).

An excellent example of Jackson's agility and audacity in the valley occurred in May 1862, when he chased General Bank's forces out of the valley and caused a sizable force under McClellan to withdraw northward from the gates to Richmond in defense of Washington. Jackson's bold maneuvers, which prompted the Union retreat, began at New Market on 20 May. With General Bank's force occupying Strasburg, Jackson employed his cavalry south of the town to demonstrate and occupy Bank's attention. Simultaneously, he marched the bulk of his force rapidly eastward through Luray Gap in the Massanutten ridge (See Map 4). Once on the eastern side of the mountain, Jackson joined forces at Luray with a subordinate command under General Ewell on 21 May. Using the ridgeline to screen his movements northward, Jackson's troops marched along the South Fork of the Shenandoah, intending to attack Bank's flank or rear from the east, once beyond the northern limit of the Massanutten.

Overwhelming a small Federal outpost at Front Royal on 23 May, the Confederates cut all telegraph communications through Manassas Gap to the east, effectively isolating Banks' force at Strasburg. When Banks was finally made aware of Jackson's position, he feared that his unit would be outflanked and retreated to Winchester on 24 May. Jackson's "Foot Cavalry" followed in pursuit and eventually drove the Union forces northward across the Potomac at Harper's Ferry. Despite the fact that Jackson did not destroy Banks' army, his adroit maneuvering caused much concern in the Federal camp and affected McClellan's operations in the Peninsula. Specifically, the threat Jackson's army now posed to Washington caused Union leaders to recall General McDowell to the north side of the Rappahannock River after he had already crossed it enroute to Richmond (Deaderick, 1956). In this series of maneuvers, Jackson took advantage of the ridges, gaps, and valleys in an offensive action. In the sequence of battles that followed, the Jackson demonstrated equal aptitude in the defense, as he avoided the pincers of an attempted Union ambush south of Massanutten.

In late May, Jackson was still positioned in the northern part of the Valley near Harper's Ferry. Union commanders, who had tired of Jackson's recent success and desired to regain control of the valley, were determined to annihilate his force by converging in his rear from two sides. The focal point of the Union trap was Strasburg (see Map 4), with Major General John C. Fremont's unit approaching from the Alleghenies to the east, and Brigadier General James Shield's division advancing from Front Royal to the west (see Map 4).

Once Jackson learned of the planned Union trap, he used the high-speed avenue of the Valley Turnpike to withdraw quickly to Strasburg before the Union forces arrived, effectively slicing through a vise of 50,000 enemy troops. The safe passage of Jackson's force was aided by token elements of Confederate cavalry and infantry that slowed the movements of Fremont and Shields. Once clear of Strasburg, Jackson's next concern was to maintain a clear axis of retreat down the Valley Turnpike to the south. Recalling how his own forces employed Massanutten Mountain as a covering shield, Jackson was troubled by the possibility of Shields advancing down the east side of the ridge, crossing over the Luray Gap, and bottling up his troops between New Market and Fremont's force advancing from the north (Tanner, 1976).

While Jackson fought a delaying action against Fremont, he dispatched his cavalry to burn the bridge across the South Fork that enabled passage westward to Luray Gap (see Map 4). He likewise ordered his cavalry to destroy the bridge at Conrad's Store that would have provided Shield's division another route westward on the southern end of the Massanutten. Jackson's anticipation thwarted Shield's plans, as he had indeed intended on using the gap to close with Jackson's forces. Shields' column was forced southward, buying valuable time for the Confederate defense to consolidate.

On 6 June, Jackson finally massed his army just south of the Massanutten ridgeline above Cross Keys. Here, he prepared to defeat the two Union columns in detail, as Fremont and Shields were not in communication and their advance was not synchronized. Jackson's plan was to hold Fremont with a brief demonstration at Cross Keys, defeat Shields near Port Republic, then turn and finish off Fremont. The overly ambitious plan failed, largely because Port Republic was located at the convergence of the South Fork of the Shenandoah and the South River, and Shields' division was on the opposite bank from Port

Republic (see Map 4). Both streams were at high flow due to recent heavy rains, and in this instance, Jackson failed to foresee the impact of these major obstacles in the valley. His master plan required crossing both rivers to attack Shields, then reversing the army's line of operation back over the rivers to engage Fremont. The chaos and confusion resulting from the first series of river crossings disrupted the synchronization of his attack on Shields. Although his Valley Army defeated Shields in a desperate battle, he was unable to regroup and attack Fremont's force, which had established artillery batteries on the west bank of the South Fork (Tanner, 1976).

Fortunately for Jackson, he planned an escape route eastward into the Blue Ridge through Brown's Gap, through which his weary army retired (see Map 4). Once again, Jackson skillfully used the ridges, rivers and mountain passes to his advantage, but in this instance a wind gap offered an avenue for retreat. Despite the overall failure of his planned counterattack, Jackson avoided entrapment by a superior Union force and saved his Valley Army to fight another day.

The battle at Port Republic ended the 1862 Valley Campaign, which is perhaps one of the most remarkable in military history. Through bold maneuver and an unparalleled understanding of the terrain, Jackson succeeded in occupying and besting a much larger Union army. He brought hope to the Confederacy, and most importantly, he bought time for the South by stalling McClellan's offensive against Richmond. The Shenandoah Valley again became a focal point of the war in 1864, when the Federals launched a campaign to destroy Lee's valuable food supply in the valley. Under General Sheridan, the Union decimated the valley's agricultural base and permanently crippled the Confederate war effort in the region.

The geography of the Ridge and Valley would likewise profoundly influence the conduct of Civil War campaigns in other arenas. In the summer and fall of 1863, the fighting in the Ridge and Valley shifted southwestward to Georgia and Tennessee at the lower limit of the province. While Stonewall Jackson's operations in the Valley demonstrated the advantages offered by the ridge and valley terrain for an aggressive offense, the events leading up to the Battle of Chickamauga provide a case study of the drawbacks of a poorly coordinated attack in compartmentalized terrain.

The Battle of Chickamauga, 1863

Geography of Northwestern Georgia and Southeast Tennessee

Chickamauga Battlefield is located in the northwestern corner of Georgia. In this region, the Ridge and Valley Province, with its folded and faulted mountains tends to blend together with the more flat-lying, mildly folded Appalachian Plateau to the west (see Map 5). Indeed, the anticlines and synclines in the eastern half of the Cumberland Plateau in southeast Tennessee and northwest Georgia represent the propagation of the compressive stress by which the Ridge and Valley province was formed (Fenneman, 1938). Notwithstanding the clear lines separating the two provinces in certain publications, for the purposes of this analysis the area in which the Chickamauga battle was fought will be considered to be in the Ridge and Valley due to the pervasive synclinal ridges and anticlinal valleys (see Map 5).

The major city in the region was Chattanooga, which was the focal point of the events immediately leading up to, and following the battle. While small, scattered farms characterized the cultural landscape surrounding the city, Chattanooga was a center of industry, and the surrounding mountains of the plateau produced valuable coal for the southern war effort. More importantly, Chattanooga was a key railroad hub, with rail lines radiating outward toward Atlanta to the southeast, Nashville to the west, and Virginia to the northeast. Indeed, Chattanooga was literally and figuratively the door to the Deep South and a gateway leading to Atlanta. Atlanta was critical because it was the site of vitally important quartermaster, commissary, and ordnance depots for the Confederate Army. Hence, Chattanooga was strategic terrain and a primary goal of the Union army -- it quickly became the center of attention in the Western Theater in the fall of 1863.

The mountainous terrain around Chattanooga favors an army in the defense. As in the Shenandoah Valley, northeast-southwest trending ridges and valleys are the major terrain features that affected maneuver (see Map 5). The Tennessee River flows through the city of Chattanooga from the north. From Chattanooga, the river flows southwestward, passing through a narrow defile and incising the Cumberland Plateau along its course. Just south of the river, three elongate ridges parallel one another and extend southwestward (see Map 5). The most prominent ridgeline is Lookout Mountain, in the center, which runs a distance of some one hundred miles into northwestern Georgia. Lookout Mountain rises over 2,000 feet in elevation and has a local relief of over one thousand feet, a rather imposing precipice overlooking the Tennessee River below. The synclinal ridge contains layers of limestone, shale, and sandstone, and the former layers explain the presence of karst caverns and subterranean falls within the mountain. The ridge is extremely steep-sided on its upper flanks,

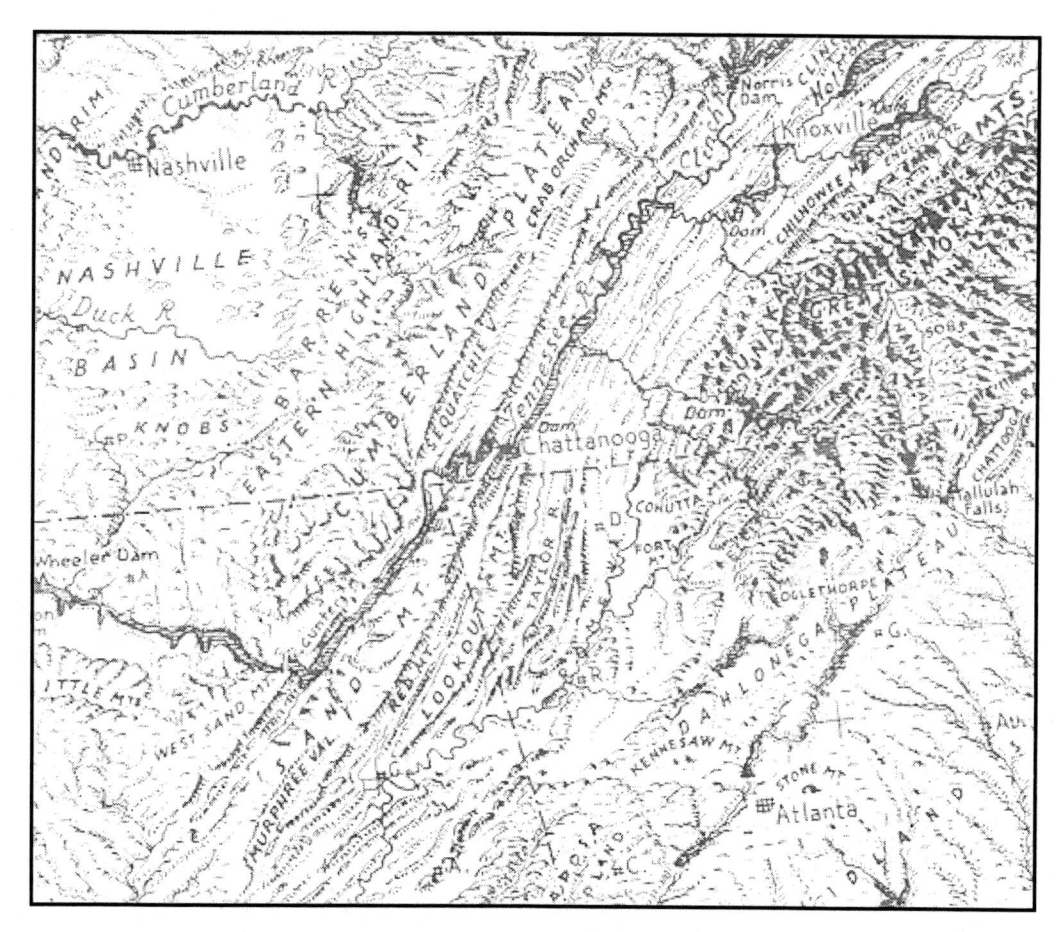

Map 5. Detail of the Valley Ridge terrain near Chattanooga and scene of the maneuvering by Civil War armies in the summer and fall of 1863.
Source: Adapted from Raisz, 1957.

while the summit is nearly flat and rather wide, ranging from one to seven miles in breadth. The two lesser ridges are Sand Mountain and Missionary Ridge to the west and east, respectively. While Sand Mountain was an imposing barrier, Missionary Ridge (see Map 6) was no more than a rugged, heavily forested hill five or six hundred feet in relief that could be crossed anywhere along its crest with relative ease (Cozzens, 1992).

In the fall of 1863 navigation through this rugged terrain in an east-west direction was a difficult endeavor, to say the least (see Map 6). Few roads traversed these ridges, and the few routes which did exist, were little more than wagon trails that were "rough, stony, and ascended in steep zig zags" (Cozzens, 1992). The gaps through which these routes negotiated were widely separated. Along Lookout Mountain, the passes were not true notches in the mountain, but in reality only small breaks in the rugged escarpment along the rim. The two most important passes in this battle were Steven's Gap to the north (26 miles from Chattanooga) and Winston's Gap in the south (42 miles from Chattanooga), both astride Lookout Mountain (Maps 6 and 7).

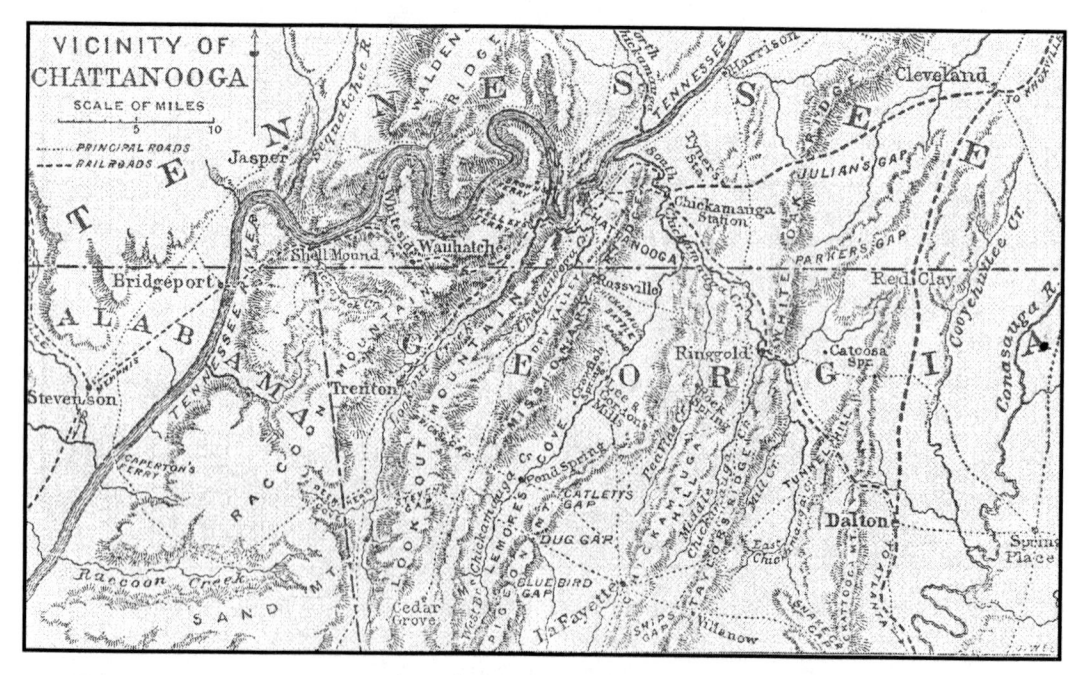

Map 6. Map of the region around Chattanooga illustrating the ridge and valley sequence.
Source: Battles and Leaders, 1887.

Strategic/Military Situation in August 1863

In August of 1863, Major General William S. Rosecrans led his Army of the Cumberland on a successful Union campaign through Middle Tennessee (see Map 5), achieving a number of impressive victories against Major General Braxton Bragg of the C.S.A. Bragg's Army of the Tennessee was in a defensive posture at Chattanooga, and his force of about 40,000 confronted Rosecrans' 66,000 troops approaching from the northwest. Rosecrans strategy was to take the city by a large turning movement, encircling the Confederate rear from the southwest and cutting off Bragg's line of communication with Atlanta (see Map 7). The plan called not only for an assault crossing of the formidable Tennessee River, but a rapid movement across the imposing ridges south of the city on three separate axes (Tucker, 1961).

Notwithstanding its risks, the flanking maneuver was a tremendous success, largely due to Rosecrans' deception plan and weak security and reconnaissance efforts by the Confederates. The offensive began in mid-August with a noisy and highly successful feint north of Chattanooga by Major General Thomas L. Crittenden's corps. Believing that the Union crossing of the Tennessee would come from the north, Bragg was unprepared for the river crossings near Bridgeport (see Map 7) by two corps under Major Generals Alexander McCook and George H. Thomas. In fact, Bragg did not learn of Rosecrans' crossing downstream from his own scouts, but from a citizen of Caperton's Ferry, who brought him the intelligence on 31 August (Tucker, 1961). That the river crossing was a great success was a tribute to Rosecrans' skill and planning and Bragg's ineptitude. In reality, the Union forces had no reasonable expectation that they would achieve such an astonishing success since the mountains on the south side of the river provided easy observation of any crossing. Only Bragg's poor disposition of his army permitted an uncontested crossing (Foote, 1963).

Once on the south side of the river, the ridges to the south and east of Chattanooga provided an excellent screen behind which the two Union corps moved undetected. Bragg made no provisions to guard the passes over Sand Mountain, nor did he outpost Lookout Mountain until 2 September. Bragg's failure to see or anticipate Rosecrans' movements and fortify these gaps was a monumental blunder (Tucker, 1961).

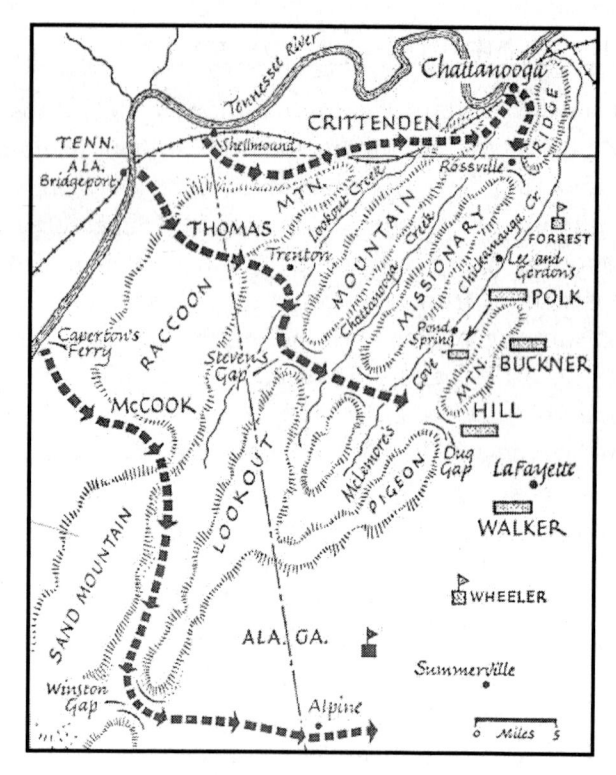

Map 7. Union movements and Confederate dispositions leading to the Battle of Chickamauga. Source: Adapted from Foote, 1963.

In their movement southeastward, Thomas and McCook faced the daunting task of crossing the steep slopes of Sand and Lookout Mountains. Toiling up and down the hillsides in ninety degree heat, Union troops had to expend considerable effort widening and improving the existing mountain trails. Hauling heavy equipment such as field artillery pieces up the ridges was a tremendous undertaking. Many wagons slipped down the rocky slopes, killing mules and scattering supplies. Eventually, Thomas' corps reached Steven's Gap, and McCook positioned his corps in Winston's Gap on Lookout Mountain. Meanwhile, Crittenden crossed the Tennessee at Shellmound at headed for Chattanooga along the southern bank of the river (see Map 7). Although it appeared that Roscrans' grand envelopment was working to perfection, the Union army was now dangerously spread out in the mountains south of Chattanooga, in terrain where they could not easily come to each other's assistance (Foote, 1963).

Fearing that his line of communication to Atlanta was threatened by the Union envelopment, Bragg evacuated Chattanooga on 9 September and moved toward Lafayette, Georgia (see Map 7). Up to this point, Rosecrans' strategy worked to perfection. He managed to cross a major river and negotiate a series of major ridges using a primitive road network and compelled Bragg to surrender a strategically vital city with virtually no pitched battles. However, Bragg would soon turn the tables as he realized the weakness of the Union dispositions in the mountainous terrain.

In his attempt to rapidly maneuver from west to east and cut off Bragg's retreat, Rosecrans had to disperse his three corps over a 40-mile wide front, largely due to the limited number of roads and the passes over Sand and Lookout Mountain. Thomas argued for concentrating in Chattanooga and improving communications rather than pushing three corps over isolated mountain passes. However, Rosecrans ignored this sage advice and pressed onward. He was convinced that a deep envelopment was necessary to ensure that Bragg could be defeated. However, by dangerously separating his units, Rosecrans violated the principles of mass, unity of effort and security—no unit was in supporting distance of the other. McCook's corps Winston's Gap was approximately 17 miles from Thomas' at Steven's Gap, much too far apart for a rapid concentration. Moreover, communications were extremely slow from Rosecrans' headquarters to each corps. For example, on 9 September, it took nine hours for orders to reach McCook from Army Headquarters near Trenton (see Map 7). Lateral communications between the center and right of the Army of the Cumberland were tentative at best; between the left and right, they were virtually nonexistent (Cozzens, 1992).

Once Rosecrans was informed that Bragg was heading southward toward Rome, he began to aggressively pursue the Confederates to cut off their retreat. Crittenden was sent from Chattanooga southward toward Rossville with the mission to harass the Confederate rear. Thomas was directed to attack into Lafayette, and McCook was ordered to take Summerville. When McCook reached Alpine, he separated from Thomas by over 30 miles of uncertain mountain trails (see Map 7).

Recognizing that Rosecrans' isolated corps were vulnerable to counterattack, Bragg revised his plans and decided to go on the offensive. He halted his forces and concentrated near the town of Lafayette. There he could lie in wait and pounce on whichever unfortunate corps came out of a gap to his front. Missionary Ridge, Pigeon Ridge, and Lookout Mountain shielded his forces from early detection by the Union, so Rosecrans was unaware of Bragg's sudden halt in his trek toward Rome. Bragg also quickly mounted a deception plan in which volunteers were sent out to "dessert" into Federal lines and talk of the imminent collapse of the Confederate Army. Also unbeknownst to Rosecrans, a large Confederate force of approximately 40,000 soldiers would soon arrive from Lee's Army of Northern Virginia to bolster the Southern ranks, putting Rosecrans at a sizable disadvantage.

    The first Union corps to fall into Bragg's trap belonged to Thomas as he emerged from Steven's Gap (see Map 7). On 10 September, Major General James S. Negley's division under Thomas advanced through the gap and into an area just east of Missionary Ridge called McClemore's Cove. Negley was completely unaware of the disposition of Confederate forces to his front, which was four times larger than his division. The nearest support to Negley was a day's march away on the west side of Lookout Mountain. Bragg attacked, but indecisively, hence that the bulk of Negley's division was able to escape westward into Steven's Gap. Not only did Brag fail to spring his trap decisively and cripple Rosecrans' army deep in southern territory, but he now alerted his antagonist to his mortal danger. Rosecrans' acted resolutely—it was a matter of life and death to the Army of the Cumberland to concentrate their forces, as Bragg now turned to attack Crittenden to the north.

    Because of the distances separating the corps across Lookout Mountain, the plea for help was slow to arrive. Orders from Thomas to McCook to move to McClemore's Cove as rapidly as possible took 18 hours in transit. For the three days prior to this dispatch, McCook had been operating virtually on his own, completely out of touch with events to the north (Cozzens, 1992). McCook's union with Thomas on the battlefield was slow. He marched 46 miles back through Winston's Gap, to Valley Head, up Lookout Valley, and through Steven's Gap (see Map 7). The grueling forced march took McCook over four days to complete. Had McCook reconnoitered the summit of Lookout Mountain he would have discovered a trail along the rather flat summit from Alpine to McClemore's Cove via Dougherty's Gap, a distance of only 15 miles. In his defense, his own scouts assured him that no such route existed.

    Without adequate intelligence and accurate maps, McCook took an extra two days to join the fight. This delay likely would have been fatal to the Army of the Cumberland had Bragg's forces aggressively attacked when given the opportunity. Despite the slow concentration of Union forces, Bragg's efforts against the Union center were weak and indecisive, so he could not take advantage of his local superiority in numbers while the Army of the Cumberland was divided.

By the evening of 17 September, all of the Union corps were finally in supporting distance of each other, and the terrible Battle of Chickamauga commenced on 19 September. The battle was a Confederate victory; however, Bragg permitted the Union army to withdraw into the safety of Chattanooga. Unquestionably the ridge and valley terrain aided in Rosecrans' retreat from the field (Tucker, 1961). Therefore, in a sense, Rosecrans made the greatest gains in this campaign as he succeeded in obtaining his major objective of wresting Chattanooga from the Army of Tennessee notwithstanding his defeat at Chickamauga.

Conclusion

As these two case studies indicate, Civil War Battles in the Blue Ridge and Valley & Ridge Provinces presented the opposing commanders with unique challenges. Commanders who appreciated the importance of the topography and knew the terrain in detail had a pronounced advantage over their opponents. General Jackson's successes in the Valley Campaign were in no small measure attributable to his superior appreciation for key terrain features and how to best use them to his advantage in the attack and defense. In the Chickamauga Campaign, neither commander seemed to have a firm grasp of the terrain nor how to effectively employ forces on this landscape. As a result, the mountains were like a fickle temptress, promising conquest and threatening destruction in equal measure (Cozzens, 1992). While the ridges afforded secrecy to Rosecrans in his sweeping envelopment, they likewise shielded the retreat of Bragg from Union observation. Reckless pursuit over the ridgelines south of Chattanooga was nearly fatal for the Army of the Cumberland. The widely-spaced gaps stretched Rosecrans forces, and the treacherously steep ridgelines made road marching tiresome and dangerous for the troops of the Union Army. Only Bragg's inept generalship enabled the Union army to escape destruction.

Bibliography

Battles and Leaders of the Civil War. 1887. *Volume III*. Washington, D.C.: U.S. Government Printing Office.

Berger, Carl. 1986. *The Korea Knot*. Westport, Connecticut: Greenwood Press.

Cozzens, Peter. 1992. *This Terrible Sound*. Urbana, Illinois: University of Illinois Press.

Davies, W.E. 1968. "Physiography." in *U.S. Geological Survey and U.S. Bureau of Mines, Mineral Resources of the Appalachian Region*: U.S. Geological Survey Professional Paper 580, p. 37-48.

Deaderick, Barron. 1951. *Strategy in the Civil War*. Harrisburg, Pennsylvania: The Military Service Publishing Company.

Emiliani, Cesare. 1987. *Dictionary of the Physical Sciences*. New York, New York: Oxford University Press.

Fenneman, Nevin M. 1938. *Physiography of Eastern United States*. New York, New York: McGraw-Hill Book Company, Inc.

Fitch, John. 1864. *Chickamauga, The Price of Chattanooga*. Philadelphia, Pennsylvania: J.B. Lippincott and Company.

Flint, Roy K., Kozumplik, Peter W., and Waraksa, Thomas J. 1981. *Selected Readings in Warfare Since 1945*. Department of History, United States Military Academy, West Point, New York.

Foote, Shelby. 1963. *The Civil War, A Narrative: Fredericksburg to Meridian*. New York, New York: Vintage Books.

Henderson, George. 1949. *Stonewall Jackson and the American Civil War*. New York, New York: Longmans, Green and Company.

Lesure, Frank G. 1985. *Geologic Map of the Southern Massanutten Roadless Area, Page and Rockingham Counties, Virginia*. Reston, Virginia: Department of Interior, U.S. Geological Survey.

Lewis, Thomas A., et. al. 1987. *The Shenandoah in Flames: The Valley Campaign of 1864.* Alexandria, Virginia: Time-Life Books.

Miller, Robert A. 1974. *The Geologic History of Tennessee.* Nashville, Tennessee: State of Tennessee Department of Conservation Division of Geology, Bulletin 74.

Orndorff, Randall C. and Goggin, Keith E. 1994. *Sinkholes and Karst-Related Features of the Shenandoah Valley in the Winchester.* 30 x 60 minute Quadrangle, Virginia and West Virginia. U.S. Department of Interior, U.S. Geological Survey.

Raisz, Erwin. 1957. *Landforms of the United States.*

Tanner, Robert G. 1976. *Stonewall in the Valley.* Garden City, New York: Doubleday and Company.

Thornbury, William D. 1965. *Regional Geomorphology of the United States.* New York, New York: John Wiley and Sons.

Tucker, Glenn. 1961. *Chickamauga: Bloody Battle in the West.* Dayton, Ohio: Bobbs-Merrill Company.

PANORAMA FROM LOOKOUT MOUTAIN, TENNESSEE - 1864
Chattanooga is at the base of the river's bend.

Source: Library of Congress Historical Collection.

THE SUMMIT OF LOOKOUT MOUNTAIN, 1863

Source: Library of Congress Historical Collection.

4

Amphibious Warfare:
OPERATION CHROMITE: TURNING MOVEMENT AT INCHON

By

Francis A. Galgano Jr.

"The history of war proves that nine out of ten times an army has been destroyed because its supply lines have been cut off... We shall land at In'chon, and I shall crush them."
General Douglas MacArthur (Appleman, 1992, p. 489)

Introduction

On 15 September 1950 the U.S. X Corps landed at Inchon, Korea. In a single decisive stroke, the tide of the Korean War was reversed. Operation Chromite, planned under the direction of General Douglas MacArthur, was perhaps the most decisive amphibious turning movement in military history, and dramatically illustrates the enormous advantages associated with this form of warfare. Notwithstanding its astonishing success, the amphibious assault at Inchon was fraught with peril. Amphibious assaults are perhaps the most difficult of military operations to plan and execute; and Operation Chromite was certainly no exception. The Americans had to overcome complex tidal conditions, strong currents, difficult bathymetry and poor beaches. For precisely those reasons the North Koreans all but discounted a landing there. Therefore, if an assault there could succeed, the payoff would be incalculable. It was predicable that MacArthur would think in these terms, and he was perhaps the only military leader of his day with the experience and credentials to pull off this operation in the face of its geographic difficulties.

The landing was confronted with hydrographic difficulties that would have turned away all but the most intrepid military leaders. Clearly, MacArthur understood the inherent advantages associated with amphibious operations; and the devastating effect of a deep turning movement against an unsuspecting enemy. Under his direction the operation was planned to last degree and was predicated on the first-rate Army, Navy and Marine cooperation that he came to trust during his island hopping campaigns against the Japanese (Utz, 1994). The amphibious planners learned their deadly business during countless operations during World War II—it was unquestionably the graduate team. Between their expertise and his unshakable nerve, the operation seemed destined to succeed if Chromite's timing could exploit the tides. The North Koreans expected and guessed at landings elsewhere, but to them Inchon appeared to be an impossibility (Weintraub, 2000).

This review is an example of historical military geography explained within a strategical and tactical framework. In this chapter, we will examine the geographic considerations that have an effect on the planning and conduct of amphibious strategy and operations. This assessment will first consider these factors topically at the tactical level of war. The historical geographic analysis of Operation Chromite—the Inchon landing—will illustrate how the geography of amphibious landings drives the strategic decision making and tactical execution of amphibious assaults. It is my goal to demonstrate the influence of the unique set of problems associated with operations in the coastal zone, and their impact on military operations.

Amphibious Operations

Three geographic spheres come together at the shoreline—the land; the atmosphere and the ocean—making the zone of land-sea contact one of the most dynamic environments on earth. Because of this, an amphibious assault against a defended beach is perhaps the most difficult of military operations (Brown, 1992). The opening scene of the movie *Saving Private Ryan* clearly endorses this statement. As was evident in that film, during an amphibious assault the attacker crosses the line of contact at the shoreline after contending with rough sea conditions, waves, currents and a host of other hydrographic problems; then must immediately face the enemy across an open beach.

The geographic reality of amphibious warfare is that there are a finite number of suitable beaches on which a force can land. Hence, the defender typically has all advantages of time and space. The defender knows the terrain and surf conditions, while the attacker's knowledge is often imperfect. The defender more often than not has the luxury of time, and therefore can develop prepared defenses, usually in depth with heavy weapons, ample reserves, elaborate fortifications and extensive minefields and obstacles (Brown, 1992). The army defending the beach also has the benefit of being able to develop zones of fire to cover every approach and pre-register targets. Meanwhile the attacker is limited to small windows of time when the tides can support a landing and restricted to a relatively small number of lightly armed soldiers in the assault force. All of this is further complicated by fact that attacker is at mercy of weather, surf and beach conditions.

Although physical obstacles make an amphibious landing inherently risky, these operations do present an attacker with three intrinsic advantages that sometimes offset those of the defender. First, an amphibious force has unprecedented mobility. This mobility compels the defender to fortify and garrison long stretches of coastline and numerous possible landing sites. For example, the Germans expected a landing somewhere in northern France in 1944, but were forced to defend the coast from Brittany to Calais (Harrison, 1993). Second, the attacker holds the initiative and can employ the element of surprise. Clearly, the inherent advantage of being able to appear unexpectedly from over the horizon and quickly land on a stretch of beach cannot be overstated. Finally, the heavy guns and carrier-borne aircraft of the naval covering force can neutralize the heaviest fortifications. Repeatedly, the battleships and heavy cruisers of the naval gunfire group proved to be the decisive factor in a number of landings in the Mediterranean and Pacific during the Second World War. By way of example, naval gunfire was all that stood between German armor and the beach during the most critical moments of the Salerno and Anzio landings (D'Este, 1991).

American amphibious units would maximize these advantages repeatedly as they perfected amphibious doctrine during the Second World War. This was punctuated at Inchon, which was conceivably the biggest payoff. MacArthur appeared to truly grasp this when he insisted that Inchon remain as Operation Chromite's operational objective when virtually everyone tried to warn him off such a risky landing site (Utz, 1994). Operations in the littoral and forced entry operations such as Inchon are expected to be a central component in future combat scenarios for the American military establishment (Department of the Army, 1993). Therefore, analyses such as this one are relevant and important for a force projection Army.

Amphibious Warfare

Amphibious warfare is an inherently joint endeavor. Any type of landing requires the integration of naval, land and air units together with special operations forces; and today, space-based platforms. Amphibious operations are defined as attacks launched from the sea by naval and landing forces against a hostile shore (Department of the Army, 1993). They are conducted for three purposes: 1) prosecute further land-based operations against an enemy (e.g., Normandy, Anzio and Inchon); 2) obtain bases and anchorages for future operations against an enemy (e.g., Tarawa, Saipan and Guam); and 3) to deny a base or region to an enemy (e.g., Guadalcanal, Attu and Kiska).

Amphibious operations can be broken-down into five discrete types of operations: assaults, landings, withdrawals, demonstrations and raids. The most difficult and costly is the *amphibious assault*. An *amphibious assault* is a tactical landing on a defended beach. In this type of operation, the enemy defends the beach, and line of contact is the shoreline. Historic examples of *amphibious assaults* are Tarawa (Russ, 1975) and Peleliu (Costello, 1991). In contrast, an *amphibious landing* is typically uncontested and accomplished against an undefended shoreline. In this case, the line of contact is some distance inland, such as the Marines experienced at Guadalcanal (Miller, 1989). Dunkirk is perhaps the most celebrated example of an *amphibious withdrawal* (Polmar and Mersky, 1988). In this type of operation, the amphibious action is designed to extract a force. *Amphibious demonstrations* are conducted to deceive an enemy and hold defensive forces in place so that they cannot be used against a land or amphibious operation elsewhere. The demonstration by a Marine division off the Kuwaiti coastline during the Gulf War pinned down sizable Iraqi reserves and is a classic example of this type of operation (Department of Defense, 1992). Finally, *amphibious raids* are conducted against land targets to gather intelligence, destroy vital targets and deceive the enemy. The raid against Dieppe in 1942 by a Canadian division is perhaps one of the more infamous examples of this type of operation (Polmar and Mersky, 1988). The fundamental constraint in an *amphibious raid* is that the landing force is not intended to remain ashore once its task in completed.

Characteristics of Amphibious Operations

Because of their complex nature, amphibious operations have distinctive characteristics not shared with other types of military operations. Not surprisingly, geography plays a central role in amphibious warfare because of distance, hydrography, weather and terrain. The invasion force is by necessity self-contained, and must be capable of independent action. For this reason, *distance* is a decisive characteristic. Frequently, an amphibious force is projected hundreds, and sometimes thousands of miles from its staging point and primary logistics base. By way of example, during Operation Torch in November 1942, American amphibious units were transported from Norfolk, Virginia across the Atlantic for landings in French North Africa (Morrison, 1960).

The landing force is usually transported in relatively slow, lightly armored ships (see Photograph 1). Consequently, air and naval *superiority* are essential characteristics of amphibious operations. For instance, the entire focus of the Battle of Britain in the summer of 1940 was the destruction of the British R.A.F., so that the German Army could successfully cross the English Channel. The German high command did not think that a landing could succeed unless the R.A.F was smashed (Addington, 1984). Similarly, most amphibious operations are conducted only under the umbrella of friendly air cover. Anzio was selected as the landing beach for Operation Shingle in January 1944, even though Civitavecchia was a better landing beach, in part because Anzio could be covered by Allied aircraft (Galgano, 1993).

Finally, a significant attribute of amphibious operations is "terrain analysis." In this case, "terrain" includes the ocean surface, sea floor, beach and inland areas. A successful amphibious operation must account for waves, tides, bathymetry, sea floor material, beach slope and sediment type, nearshore currents, weather and terrain beyond the beach. History is replete with examples in which the amphibious force struggled against the natural elements. For example, Naval planners misjudged the tides at Tarawa and the 2nd Marine Division was forced to wade ashore through hundreds of yards of deadly enemy fire (Russ, 1975).

Photograph 1. A troop ship departs for Casablanca in November 1942.
Source: U.S. Army Photograph.

Amphibious Doctrine: A Geographic Perspective

The amphibious doctrine employed at Inchon, had its roots in the period just after the end of the First World War (Morrison, 1963). Amphibious warfare as we know it was developed, refined and perfected by the United States Marines during the inter-war period. Although we take the contemporary mission of the U.S. Marines as our principal amphibious force for granted, this was not always the case. The Marines' role in amphibious warfare was codified by a Joint Army-Navy Board decision in 1927. This document directed that the Marines would be the proponent for developing amphibious doctrine and conduct " . . . land operations in support of the fleet for initial security and defense of advance *bases and for such auxiliary landing operations as are essential for the prosecution of the naval campaign."* (Parker, 1970, p. 47).

The amphibious doctrine refined by the Marines during the inter-war period was the most highly evolved of its kind at the start of World War II. The Marines perfected a doctrine that was based on the careful study of the geography of beaches, islands and oceans in the Pacific. During the decade prior to World War II they researched and conducted practice landings in a variety of geographic settings to include atolls, mainland beaches and barrier islands (Parker, 1970). The end result was a comprehensive doctrine, coupled with detailed plans for specialized equipment, boats, weapons and tactics to support the capture of heavily defended beaches by forces landed from the sea (Costello, 1981). This doctrine (Fleet Marine Force, 1934 and 1944) was adopted by the Army and our Allies and it was refined to a high military art form between 1942 and 1945. The decisive assault landings at Inchon was perhaps the *denouement* of the evolution of this doctrine (Knox, 1985).

Background and Evolution

"The landing of vast bodies of men and horses, with the artillery and stores, even under the most favorable circumstances and with the most perfect organization, presents great difficulties. If, then, we complicate these difficulties by subjecting the troops to the fire of a determined enemy ... what but pre-eminent disaster can be the result?"
Captain Asa Walker, USN, 1900 (Weigley, 1973, p. 256)

Today, the concept of an amphibious attack against a defended shoreline is firmly established and the term *amphibious assault* is an accepted component of our military lexicon. This was not always the case, however, especially in the post-World War I era. As Captain Walker's comments demonstrate, there was significant skepticism with respect to the efficacy of amphibious attacks. Consequently, the Marines were going to have to prevail over significant professional inertia to have their amphibious warfare doctrine accepted by mainstream Naval and Army leaders (Parker, 1970).

Clearly, the prevailing belief among military professionals in 1914 was that the prospects of a successful landing against a fortified beach were to be regarded with much skepticism (Brown, 1992). There were many historical examples to back up Captain Walker's assessment (Weigley, 1973). The professional conviction was that the accuracy, power and rapid-fire capability of modern weapons simply made killing too efficient for an amphibious assault to succeed. These views were cemented after the First World War as the Allies digested the catastrophe at Gallipoli. Without question, this operation largely validated the prevailing point-of-view that amphibious assaults against modern firepower were positively hopeless (Addington, 1984).

Although the idea behind the Gallipoli campaign was brilliant from a strategic perspective, it ended as a dismal failure. The fighting there evolved into a bloody standoff; and ultimately cost the Australian and New Zealand Army Corps (ANZACs) some 252,000 casualties out of 480,000 troops committed to the operation (Brown, 1992). Even though the initial beach assaults were moderately successful, the ANZACs were quickly pinned-down on the beaches and a protracted war of attrition continued between March 1915 and January 1916. After the war, professional publications in Britain summed up the prevailing attitude toward amphibious assaults by declaring them obsolete. Professional literature in the United States indicated that most Naval and Army officers were doubtful of the future possibility of beach assaults in view of the Gallipoli failure as well (Weigley, 1973).

In the early 1920s, however, the Marines were looking for a definitive mission and they took a critical look at the Gallipoli operation. As a standing force they were on the brink of fiscal extinction as Congress looked to make deep cuts in the military budget following the First World War (Brown, 1992). Although not a formal mission, the Marines always assumed the task of securing forward bases for the fleet. Marine leadership—specifically the Commandant, Major General John Archer Legeune—saw this mission as their new reason for existence in consideration of the perceived threat in the Pacific, and began to review critically the failure at Gallipoli.

The Gallipoli study was very successful. Numerous Marine publications accurately demonstrated that Gallipoli was too badly handled from a command, control and leadership perspective to be used as an object lesson, except how **not** to conduct an amphibious assault (Weigley, 1973). A progressive group of Marines, led by Major Carl E. Ellis, considered by most to be the father of our modern amphibious doctrine, began writing and lecturing on the subject of amphibious assaults during the first part of the 1920s. Their labors benefited from the spreading awareness that Japan was a dangerous potential enemy, and a strategy was needed to cope with a future conflict in the Pacific. Ellis proposed that, "... *it will be necessary for us to project our fleet and landing force across* *the Pacific and wage war in Japanese waters"* (Parker, 1970, p. 46). This view received wider acceptance in the Navy as their leadership reached similar conclusions.

Between the Wars: Plan Orange

The Marine Corps was retrieved from post-WWI oblivion in large measure through the efforts of prescient thinkers such as Major Ellis, MG Legeune and our own strategic evaluation of a future war against Japan. After World War I, the Joint Army-Navy Board resumed work on the color-coded war plans conceived to deal with potential adversaries. The Orange Plan was being prepared to outline a war against Japan, by far our most likely enemy after the defeat of Germany. As the plan evolved, it became increasingly evident that the geography of the Pacific was a central factor in our war plans. The seemingly innumerable islands and archipelagoes that dotted that vast ocean were key terrain and would have to be seized and secured if we were going to prevail (Brown, 1992).

In its original form the Orange Plan envisioned the war against Japan being fought in two phases: 1) a desperate holding action by American garrison forces in our most distant island territories; and 2) the main battle fleet fights its way across the Pacific to relieve the beleaguered forces, retake lost islands and culminate the war in a main fleet action against the combined Japanese Navy (Weigley, 1973). However, the geographic realities of a war in the Pacific began to alter the thinking of strategic planners.

There were no bases between Hawaii and Manila to support the westward advance of the fleet, and the operational distances of the original Orange Plan were vast. The distance between San Francisco and Pearl Harbor is 2,000 miles, and Pearl Harbor to Tokyo is 3,400 miles (see Map 1). Without logistics bases and anchorages along the way, the Navy could not contemplate fighting across such distances (Weigley, 1973). Furthermore, the Japanese held a number of islands and archipelagoes along the intended axis of advance. The concept of using Marines to capture enemy held islands began to come sharply into focus as General Legeune summed up the geographic dilemma in 1921, "*. . . on both flanks of a fleet crossing the Pacific are numerous islands suitable for utilization by the enemy for radio stations, aviation, submarine or destroyer bases . . .*" (Parker, 1970, p. 47).

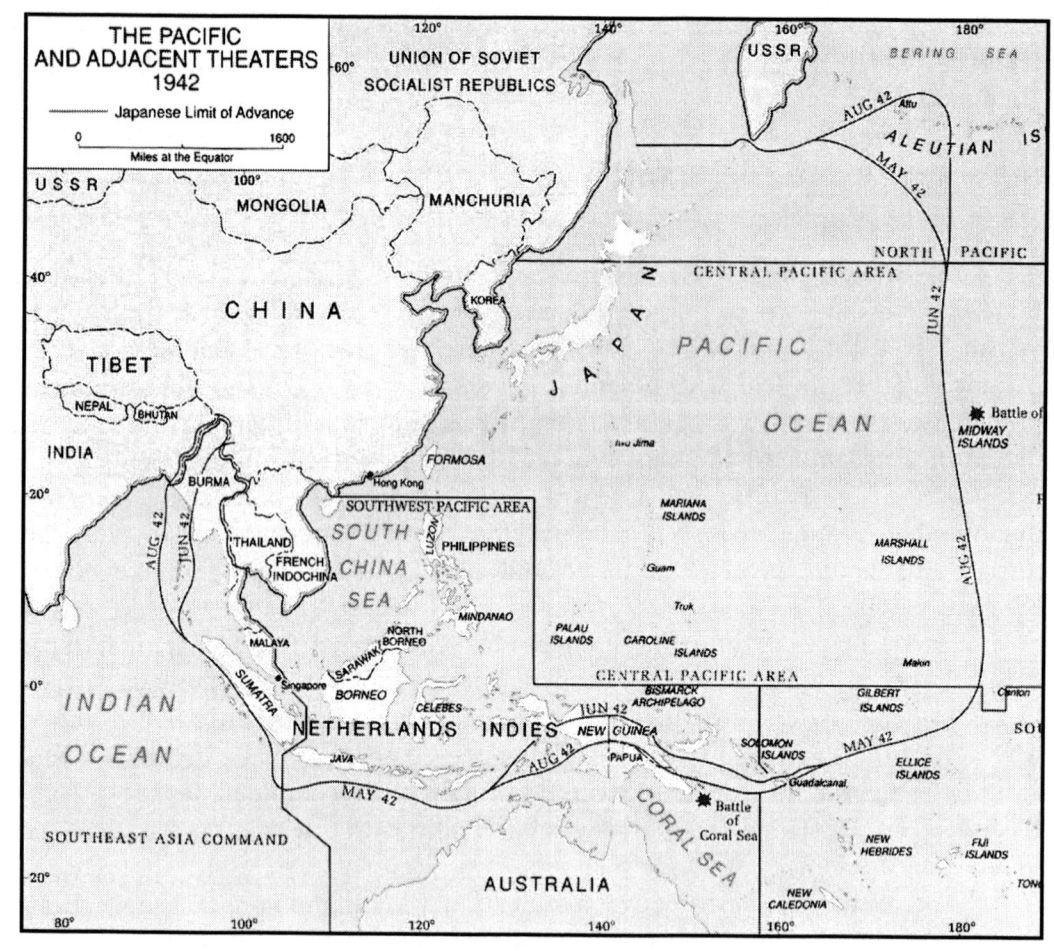

Map 1. The Pacific Theater of Operations.
Source: United States Army.

    War games designed to evaluate the Orange Plan abruptly changed our strategic outlook in the early 1920s. They demonstrated that there was little hope that remote island garrisons in the Pacific could hold out for an extended time should the Japanese decide to attack—most were closer to Japan than the U.S. Hence, planning quickly shifted to the idea that the Japanese would seize and hold these islands, and they would have to be re-taken by assault if we meant to advance our fleet into Japanese waters (Weigley, 1973). The war games also indicated that fleet losses would be unacceptable if we tried to retake the western Pacific without first securing enemy-held islands along the way and establishing advanced bases. The post exercise report summed up the problem, *"Our own greatest weakness in the Far East . . . is the extreme length of our lines of communications . . ."* (Weigley, 1973, p. 254).

Plan Orange began to evolve. The massive naval battle that was expected to decide the war in a single fleet action was discarded. Instead planners adopted a new approach, *". . . the military and naval approach to the Far East should be made in a step-by-step mopping-up process by which all the islands enroute would be taken and occupied in passing."* (Weigley, 1973, p. 254). The new Orange Plan called for an island-hopping strategy to secure or take islands by assault and establish airfields, logistics bases and anchorages to bring the Navy (and the bomber fleet) to within striking distance of Japan. The mission to execute this island-hopping campaign was given to the Marines because they appeared to be particularly suited to the mission and were already thinking it that direction as well (Parker, 1970).

Major Ellis urged Marine commanders to embrace the mission and make amphibious warfare the Corps' specialty. With the endorsement of General Legeune, Ellis began a series of lectures and studies addressing the efficacy of taking a beach by assault in the face of determined enemy resistance. His focus was certainly in the Pacific where the geographic realities of the different types of islands presented the Marines with unique subsets of problems. Many islands were so small that there could be little or no deception as to where the main attack would fall, necessitating the development of specialized tactics, weapons, and landing craft to make a direct assault successful (Weigley, 1970). Further, assaults had to be contemplated against atolls (e.g., Tarawa), volcanic islands in island arc systems (e.g., Saipan), and mainland beaches on larger islands such as Okinawa and Japan itself. Each island type presented distinct surf zone and terrain challenges to planners.

Ellis proved to be a tireless and particularly foresighted researcher and author. Throughout the early 1920s he toured Pacific islands, assembled terrain and hydrographic intelligence and developed new theories on amphibious assaults. He established a solid foundation upon which amphibious assault doctrine would be developed (Brown, 1992). In 1921, he published a paper on how to take back Pacific islands. He was directed by General Legeune to undertake a detailed study to evaluate doctrine, tactics and equipment needed to prosecute the war called for in the Orange Plan (Parker, 1970). Unfortunately, Ellis met with a somewhat suspicious death in 1923 while touring the Japanese-controlled island of Palau. However, his writings up to that point proved to be an important point of departure for a 13-year study that would result in the development of the doctrine that would ultimately win the Pacific war (Costello, 1981; King, 1946)

In the decade prior to World War II, research, evaluations, equipment tests and a number of exercises were conducted by the Marines in a variety of geographic settings to validate the doctrine established in the *"Tentative Manual of Landing Operations"*. Consequently, at the start of the Second World War the United States possessed a well-developed doctrine, as well as specialized landing craft and equipment (Parker, 1970). Certainly the doctrine and equipment would be improved on the basis of hard experience at places such as Guadalcanal, Tarawa and Kwajalien. A new manual titled *"Staff Officer's Manual for Amphibious Operations"* (Fleet Marine Force, 1944) was issued in 1944. Nonetheless, the fundamental tenets of our amphibious doctrine as proposed by Major Ellis in 1921 remained essentially unchanged.

Beach Selection Criteria: A Geographic Perspective

A beach is the geographic area that exists between the outermost breaking waves and the dune line. Only ten percent of a beach is subaerial, the remaining ninety percent is located beneath the waves or is alternately exposed and submerged by tidal fluctuations and is therefore an inherently dynamic environment (see Figure 1). Wave energy is the most obvious manifestation of the dynamic interface between land and sea. Waves typically break at the bar and the onrushing water extends up the foreshore to the berm (see Figure 1). In some instances, such as in a storm or during an exceptionally high tide, wave energy may be transmitted across the backshore to the dune line (Bascom, 1964). However, the geographic considerations that influence amphibious operations are not limited to waves. Beach selection criteria can be broken down into nine separate categories: 1) wave energy; 2) tidal range; 3) bathymetry; 4) sea floor material; 5) beach slope; 6) beach material; 7) currents; 8) inland topography; and 9) weather. Unquestionably, the relative importance of each of these factors will vary on a given segment of coastline based on geographic location and physical conditions.

Wave Energy

Waves form in the open ocean as wind blows across the water surface. Open-ocean waves can be very large indeed, but wave heights near the shore are limited in size by fetch and depth. Waves propagate outward from a formation area and sometimes travel across thousands of miles of ocean before breaking on the shore. The formation of ocean waves is predicated on three factors: wind velocity, duration and fetch. There is a direct relationship between wind speed and wave height (Bascom, 1964). However, waves are limited in size by the fetch; the distance of water over which the wind can blow. Hence, waves are usually small in narrower, enclosed bodies of water such as bays, and estuaries because the wind can act only on a relatively short linear distance of water regardless of its speed.

There is considerable geographic diversity in wave climate (i.e., height, length and direction) driven by prevailing winds or storm events. In large oceans such as the Pacific, large waves can form from anticyclonic activity or storms. However, wave height is further limited by water depth, which is a function

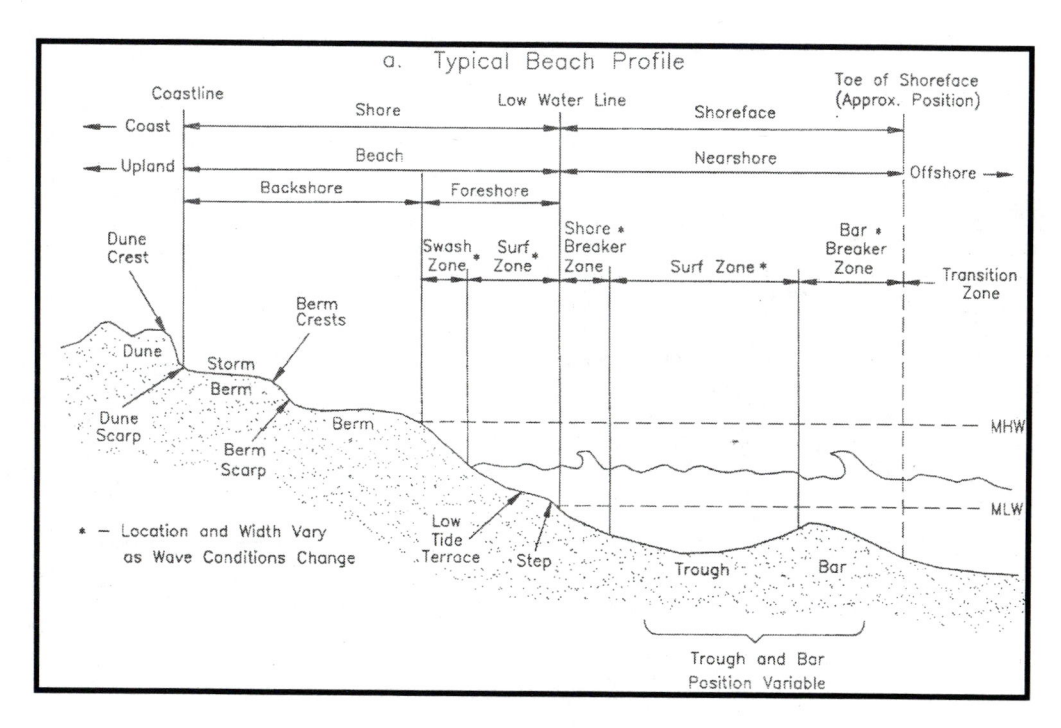

Figure 1. Cross-section of an idealized beach profile.
Source: U.S. Army Corps of Engineers (1984).

of offshore bathymetry. Therefore, at locations where there is a shallow foreshore (see Figure 1), wave heights in the surf zone will be attenuated by the water depth (U.S. Army Corps of Engineers, 1984). Waves travel through the open ocean uninhibited and their height and length will remain relatively constant (see Figure 2). As a wave enters shallower water and approaches the shore, the wave characteristics change as it begins to "feel bottom." Wave height will become larger, wavelength will become shorter and the wave will become translational as the bottom slows through friction and the top portion retains its speed. At a depth, typically defined as ½ the wavelength, the wave becomes unstable and breaks (see Figure 2). Therefore, locations with shallower foreshores will typically manifest much smaller breaker heights. Since wave power is a function of the square of the wave height, the critical selection criteria for an amphibious assault would be at a location with a shallow beach and smaller breaker heights.

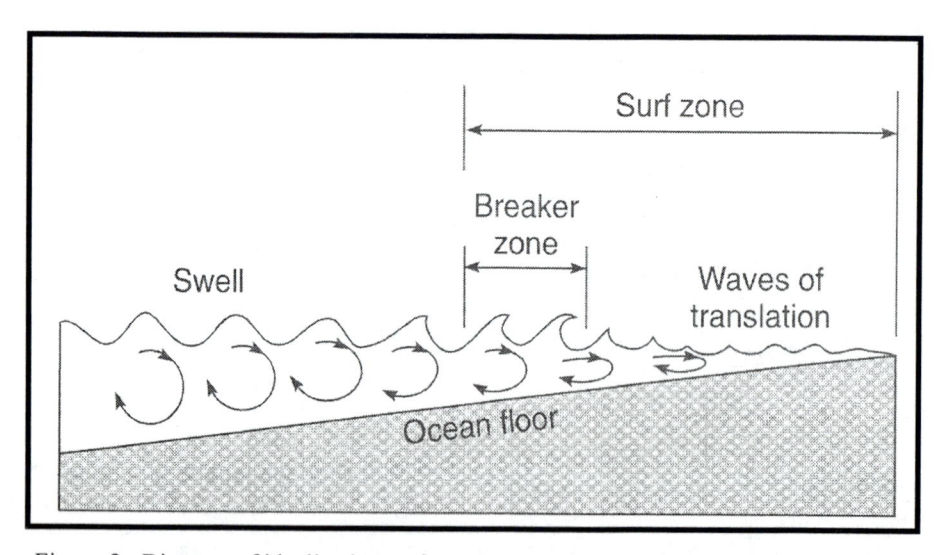

Figure 2. Diagram of idealized waveforms traveling though deep and shallow water. Source: United States Army Corps of Engineers (1984).

Tidal Range

The tides are caused by the gravitational influence of the sun and moon on the ocean surface, which creates large bulges of water on opposite sides of the planet (see Figure 3). Most locations experience semi-diurnal tides (See Figure 4), or two high and low tides per day separated by 12 hours and 25 minutes

(Bascom, 1964). Further, the magnitude of the tides fluctuates on a monthly basis depending on the relative location and alignment of the sun and moon. Therefore, amphibious planners must account for daily high and low water events as well as two spring tides (higher than mean) and two neap tides (lower than mean) per month. These daily and monthly oscillations in tidal-driven water levels produce a significant planning parameter called the tidal range.

Tidal range (T_R) is the vertical distance between the water level at high tide and low tide (see Figure 4). Differences in bathymetry and coastal geometry bring about a wide disparity in tidal ranges. For example, most of the U.S. East Coast experiences a tidal range between 1-3 meters. However, locations within the Gulf of Mexico have tidal ranges of about 5-6 meters. The Normandy beaches were chosen in part for their large tidal range (≈ 7 meters) which permitted the exposure of German beach obstacles at low tide (Winters, 1998).

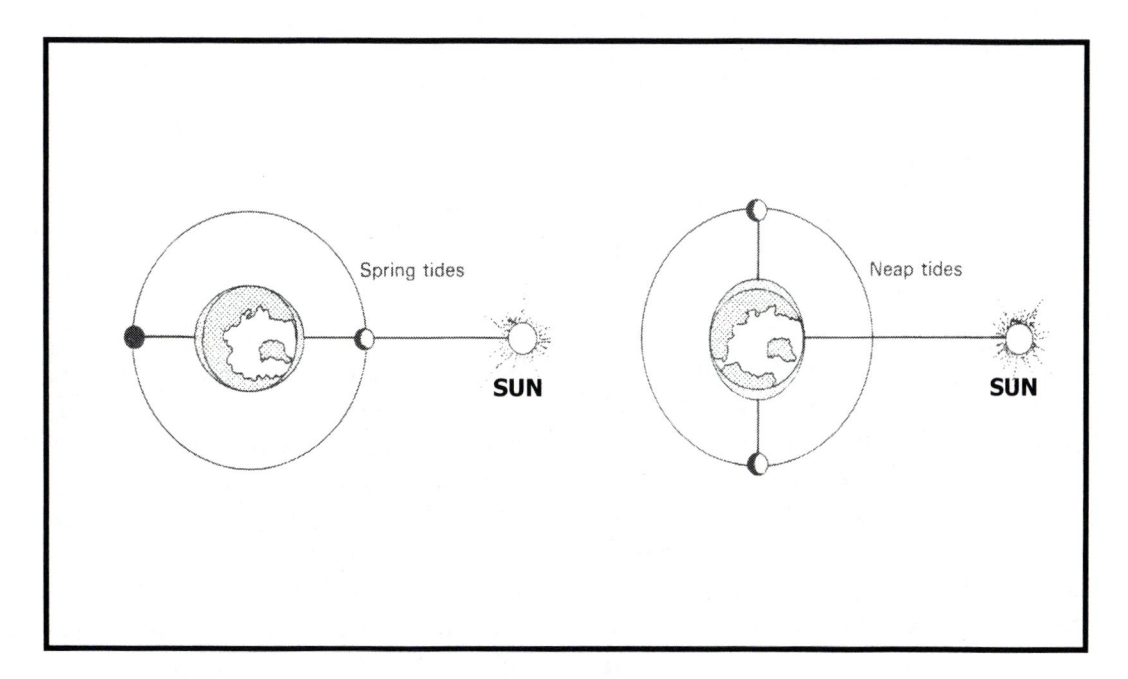

Figure 3. The position of the sun and moon create the daily and monthly tidal cycle.
Source: United States Army Corps of Engineers (1984).

Tidal range is a multi-faceted problem and there is no single "cookbook" planning solution. First, as the tide ebbs, it will expose a portion of the beach face, which may be desirable in some instances (e.g., Normandy). But, the ebbing tide will also generate a seaward current that can become very strong, especially in confined waters (e.g., Inchon). This current may be faster than the forward speed of the landing craft (Fleet Marine Force, 1944). Likewise, the flood, or rising tide will produce a current that can be used by landing craft to generate more forward speed—it was common practice for U.S. assaults to take place during the rising tide for this reason (Fleet Marine Force, 1944). Moreover, the high tide may allow landing craft to land dozens of yards closer to the dune or provide the necessary draft for landing craft to cross a bar or reef (e.g., Tarawa). In addition, when landing at high tide the infantry has less exposed beach to cross (Fleet Marine Force, 1944).

Tidal range is linked to wave energy. Larger tidal ranges are indicative of flatter beach profiles and hence, smaller wave heights. For example, a beach with a tidal range of 1 meter normally will have average, fair weather wave heights of 1-1.5 meters. A beach with a tidal range of 2-2.5 meters may have maximum fair weather wave heights of only 0.5 meters (U.S. Army Corps of Engineers, 1984). Therefore, the desirable planning characteristic is a medium sloped beach with a moderate tidal range that will not generate an abnormal ebb current and manifests relatively low wave energy conditions.

Bathymetry

Bathymetry is the underwater topography. Channels, bars and reefs cause the most common variations in nearshore bathymetry. A bar (See Figure 1) is a submerged or emerged embankment of sand, gravel, or other unconsolidated material built on the sea floor in shallow water by waves and currents. Bars are very common along most sandy beaches and typically pose few problems to landing craft. A reef is a strip or ridge of rocks, sand or coral that rises near the surface of the water. Reefs and bars cause waves to break, thus are high-energy locations. Reefs are more problematic in terms of navigation and can damage ships and stop landing craft if their draft is deeper than the water above the reef's upper surface. Coral reefs were a problem particular to the Pacific Theater, but were limited to atolls in tropical waters, generally 10-15° north and south of the Equator (Fleet Marine Force, 1944).

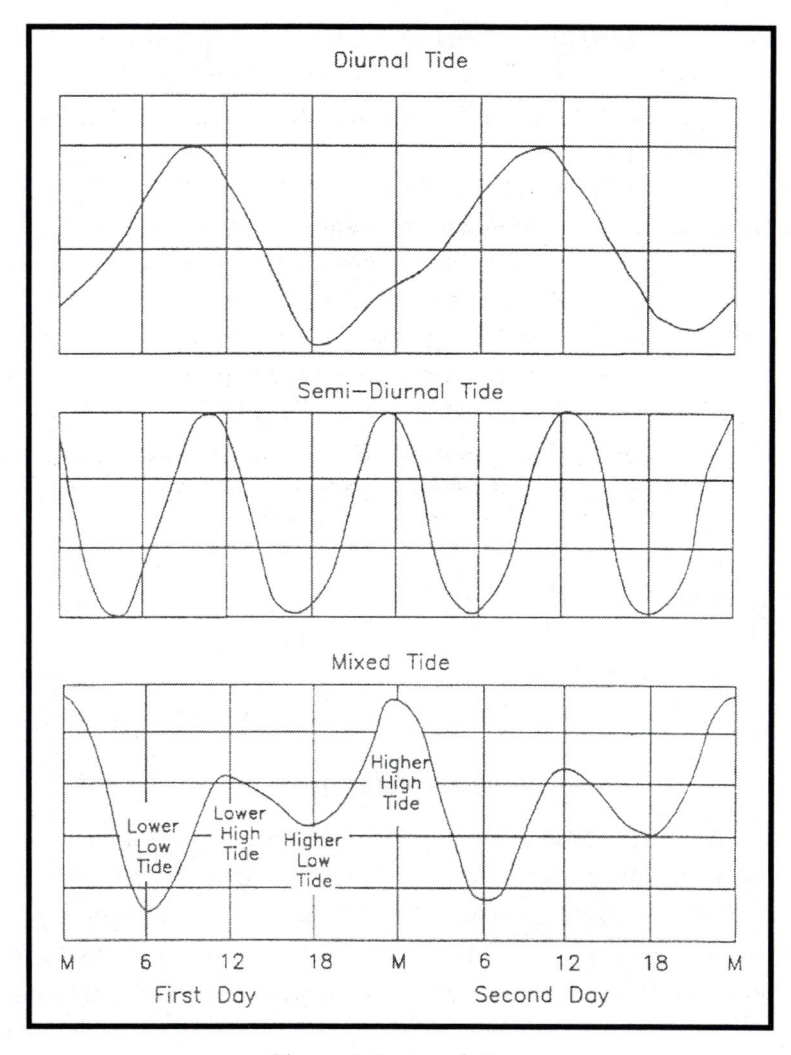

Figure 4. Types of tides.
Source: United States Army Corps of Engineers, 1984.

Seafloor Material

Since 90% of the beach profile is sub-aqueous and a considerable segment of this profile can be exposed at low tide, its makeup is somewhat consequential. Ordinarily, most beaches where a landing would be contemplated are composed of sand which has fair trafficability when wet. Hence, in many scenarios the seafloor material is not problematical. Yet in some instances the foreshore and offshore profile are composed of large rocks or sometimes, thick gooey mud, which in either case will cause significant complications for the landing force. The ideal planning parameter would certainly be a foreshore that is composed of medium to fine-grained sand which will present few problems and support onshore movement of soldiers and sustain limited vehicle traffic during the follow-on phases of the landing (Fleet Marine Force, 1944).

Beach Slope

There is a direct link between beach slope, tidal range, beach material and wave energy. Steeper beaches will have larger wave heights and smaller tidal ranges. Steeper beach slopes will also give rise to larger sediment sizes as well. In most cases an amphibious planner will seek a beach with a relatively moderate or flat slope to take advantage of smaller waves, a larger tidal range and medium to fine-grained sand (Fleet Marine Force, 1944). For example, the assault on Betio Island in Tarawa Atoll was planned for the beaches inside the lagoon in part because the steep seaward beaches were exposed to larger, open ocean waves (Russ, 1974). Similarly, Normandy was considered to be an ideal beach, because its shallow slope meant small waves and fine-grained sand, which eased the onshore movement of soldiers and vehicles (Harrison, 1993).

Beach Material

A beach is composed of unconsolidated material deposited by wave energy. The type and size of that material is dependent upon the available sediment and wave energy, as well as the beach slope. Although many beaches are composed of white quartzite sand grains, the term sand refers to a size of sediment—between 0.2 to 2.0 mm. So, beach sand is not limited to our everyday conception of white sandy beaches, and the type of sediment is dependent on the form of material that is exposed to wave energy. For example, the black beaches commonly seen in Hawaii are composed of sand-sized particles of eroded volcanic material. Likewise the pink beaches of the Bahamas and other tropical locations are composed of sand-sized grains derived from the breakdown of coral reefs and are composed of calcium carbonate. Furthermore, sediment size is strongly correlated to wave energy conditions. In ideal conditions, amphibious operations should be planned in areas with medium to fine-grained sand, which is best for trafficability. For example, when Marines landed on the high energy beaches of Iwo Jima, they were confronted with ball bearing-sized sediment of volcanic origin, which made walking difficult and virtually halted vehicular movement across the beach until a substantial engineer effort created semi-improved roads (Morrison, 1963).

Currents

Beaches are typically subjected to currents along two axes. On- and offshore currents are generated by the rising and falling tide as well as the backwash of wave-deposited water on the foreshore. The tidal current can be considerable, especially in confined waters and can extend many miles out to sea. The current generated by the backwash from waves stems from the downslope movement of water along the foreshore driven by gravity, and can be considerable—especially in rip conditions—from the shoreline to the bar (U.S. Army Corps of Engineers, 1984). The most important current in the surf zone is the longshore current with is more-or-less parallel to the shore. The longshore current is generated by waves that approach and break at an angle to the beach. Most beaches manifest a dominant longshore current direction and velocity. However, very small changes in wave height and angle of approach will cause significant changes in the longshore current (Bascom, 1964). A landing operation has to account for the longshore current in order for the soldiers to be placed ashore on the proper beach. Typically, UDT teams were sent into the surf zone hours before a landing to obtain a final measurement of the longshore current (Costello, 1981). During the D-Day landings at UTAH Beach, an unexpected change in the longshore current caused the assault elements of the 4th Infantry Division to land on the wrong beach. Fortunately, the segment of beach on which they landed was lightly defended and the assault was successful (Harrison, 1983).

Inland Topography

Once a landing force has fought its way onto a beach, it must have suitable egress points to gain access to the mainland areas and carry the fight inland. Beaches are typically backed by a variety of natural landscapes such as marshes, dunes, tidal flats, sounds, cliffs and bluffs. Sometimes beaches are backed by significant cultural landscapes as well. These may include urban areas, sea walls, agricultural areas or polder. Unquestionably, this is an important geographic consideration because a successful landing can be easily stymied and bottled-up if the invasion force cannot move inland. One of the principal causes of the failure of the Gallipoli operation was that the very narrow coastal plain was backed by steep mountains. The ANZACs failed to drive inland and secure the heights, hence the Turks were able to keep them pinned on the beaches (Doyle and Bennett, 1999).

Weather

Storms and foul weather can cause changes in wave energy by an order of magnitude, and threaten the very safety of the landing force. Furthermore, storm conditions will break up naval formations, ground aircraft and impede naval gunfire support. Fog and heavy rain limit visibility and degrade visual coordination, which is a vital component of the assault force. There are a number of excellent examples of the significant influence that weather can have on an amphibious operation. Two of Kublai Khan's invasion fleets were wrecked by large maritime storms, effectively ending his plans to conqueror Japan (Winters, 1998). The interplay between the weather conditions and General Eisenhower's decision to delay the Normandy invasion has been widely published as well (Harrison, 1993).

Case Study: Operation Chromite, September 1950

The amphibious landing at Inchon, Korea was conducted on 15 September 1950 by the United States X Corps, which was composed of the 1st Marine Division and the 7th Infantry Division. The Inchon landing was a turning movement that turned the tide of battle when things looked bleak for the United Nations' forces that were fighting for their lives on the Korean peninsula. Prior to the landing, North Korean forces had control of nearly the entire Korean peninsula except for a small 140-mile semicircular perimeter surrounding the city of Pusan known as the 'Pusan Perimeter' (See Map 2). With the successful landing at Inchon, General MacArthur severed North Korean supply lines, drove the invaders north and perhaps saved the U.N. forces from annihilation (Appleman, 1992).

General MacArthur described the Inchon landing as having a "5,000 to 1" chance of succeeding. Lt. Cmdr. Arlie Capps, a planning officer for the Inchon amphibious assault stated, "We drew up a list of every conceivable natural handicap and Inchon had 'em all." (Utz, 1994, p. 16) The most difficult challenges were Inchon's physical aspects. Most problematic was Inchon's 32-foot tidal range. This was compounded by a sinuous channel with tricky currents, wide mud flats, very poor beaches, and very large seawalls. On MacArthur's side was a very capable amphibious planning team, total air and sea supremacy, and absolute surprise. The amphibious assault led to the re-capture of Seoul and severed completely North Korean lines of communication, smashing their offensive in the south. The turning movement at Inchon crippled the North Korean Army and it never recovered; only Chinese intervention saved it as an entity (Center of Military History, 1989).

Operation Bluehearts: Early Planning

MacArthur directed planning for an amphibious attack against the North Koreans during the first weeks of their offensive. As he saw it, an amphibious attack was the most powerful tactical tool at the United Nations' disposal (Appleman, 1992). His original plan, Operation Bluehearts, called for an amphibious landing in the rear of the North Korean advance on 22 July 1950, but this operation was scrapped as the pace of their offensive quickened and more U.S. units were committed to the peninsula. Undeterred, MacArthur remained focused on the possibilities of a deep amphibious turning movement to rupture North Korean lines of communication and disjoint their offensive. Under his orders, COL D. H. Galloway assembled an unusually talented group of planners to prepare contingency plans. During the first weeks of July 1950, they developed a series of detailed plans (Utz, 1994). The operation was re-designated Operation Chromite.

Chromite called for an amphibious landing during September, chiefly to take advantage of tidal conditions and calmer sea states (Center of Military History, 1989). Galloway's team furnished three different plans. Plan 100B called for a landing at Inchon combined with a simultaneous counterattack by the 8th Army north from the Pusan Perimeter. Plan 100C proposed a landing at Kunsan on the western coast of the peninsula, much closer to the Pusan Perimeter. Plan 100D called for a landing on the eastern side of the Korean peninsula at Chumunjin (See Map 2). The planners knew that MacArthur favored a landing at Inchon and this was the most well-developed of the three (Weintraub, 2000). On 20 July 1950, MacArthur selected Plan 100B. In his estimation, Inchon was the most decisive location for a landing.

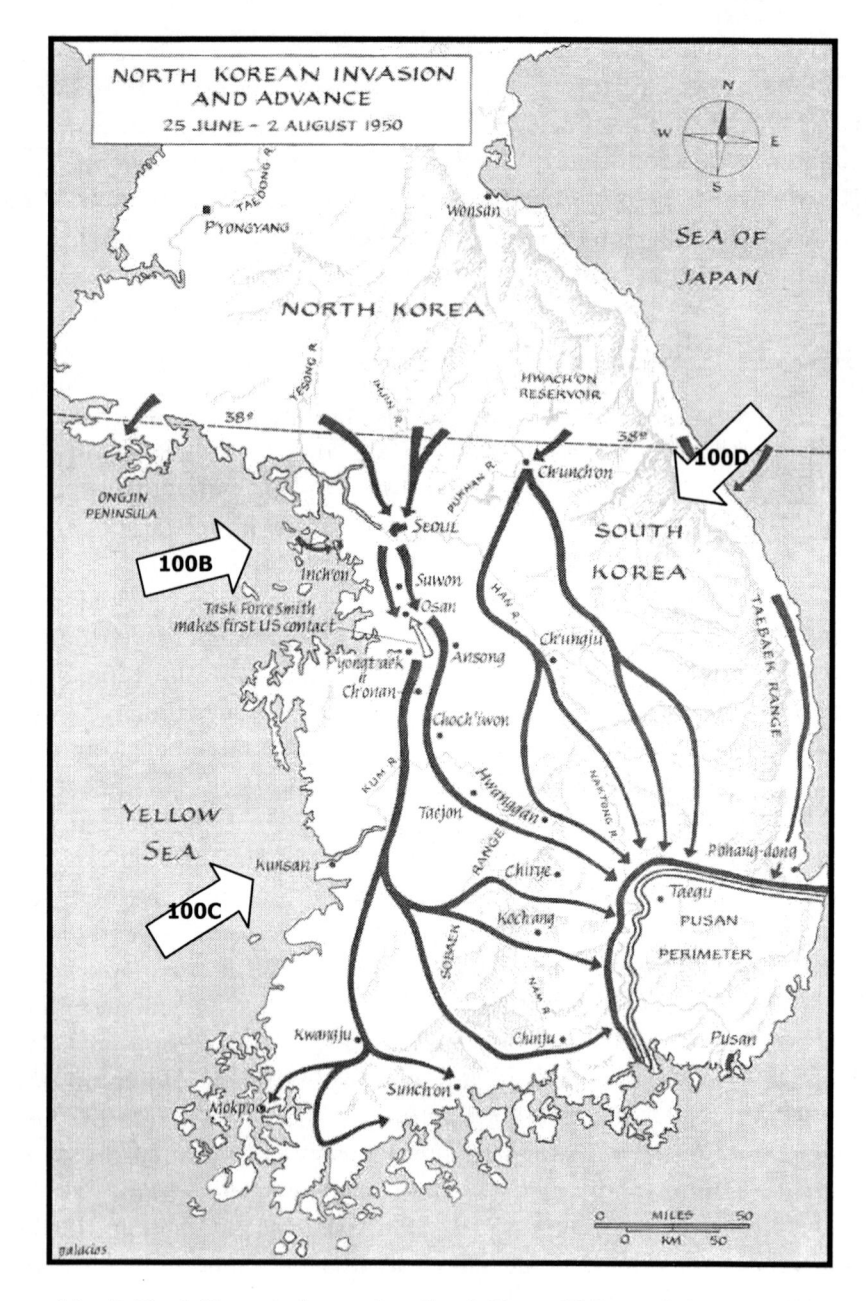

Map 2. North Korean advance into South Korea 25 June - 2 August, 1950.
Source: Adapted from Knox, 1985.

The unexpected success of the North Korean attack, combined with a paucity of combat-ready American units dimmed Chromite's chances of ever coming to fruition. MacArthur originally designated the 5th Marine Regiment and 2nd Infantry Division as the assault units for Operation Chromite. However, reverses on the peninsula necessitated their commitment to the front lines in late July. Similarly, the 1st Cavalry Division was likewise committed to the peninsula as the North Koreans continued driving south (Appleman, 1992). Notwithstanding these diversions, MacArthur continued to direct planning for the invasion.

The X Corps was assembled in Japan as the operational headquarters for Operation Chromite in late July. The corps staff was created from the best amphibious planners in the Far East Command. MacArthur reluctantly turned to the 7th Infantry Division; his last remaining American unit not committed to the fighting in Korea as one of the two tactical units for the landing. The 7th Infantry Division was a division in name only in July 1950. This unit had performed garrison duty in Japan and was not well trained. Further, it was only at half strength because many of its units were stripped of soldiers to serve as replacements. MacArthur had the division rebuilt with nearly 8,600 Korean augmentees and individual replacements culled from American units in the States. Chromite's other division was the hastily assembled 1st Marine Division (Appleman, 1992). MacArthur's logistics and naval staff likewise performed a herculean effort and assembled supplies and nearly 230 landing craft and support ships to transport the invasion force. It at last appeared that the landing was going to happen.

The Geography of Inchon

The amphibious assault at Inchon had one overriding strategic geographic consideration and it was rooted in Inchon's relative location. General MacArthur preferred Inchon as the objective for Operation Chromite for one paramount reason; it was the port city of Seoul, which was only 18 miles inland. Not only was Seoul the capital city of South Korea, but it served as the logistical and administrative hub for the North Korean Army. Although the North Korean Army had nearly over-run South Korea and was laying siege to the American 8th Army in the Pusan Perimeter, it was dangling at the end of a very long, vulnerable logistical rope. The North Koreans concentrated nearly all of their combat formations along the Naktong River (see Map 2) and neglected their rear.

A successful assault at Inchon followed by a rapid capture of Seoul would have far-reaching strategic repercussions in his view. First, it would place a U.S. corps astride the North Korean lines of communication with no significant reserves to retrieve the situation. Second, MacArthur was convinced that the psychological and political ramifications of retaking Seoul were powerful indeed. He argued that this would capture the imagination of Asia and win support for the United Nations' cause (Appleman, 1992). General MacArthur summed up Inchon's strategic geographic importance in his summary to the Joint Chiefs of Staff on 8 September 1950:

> *"The seizure of the heart of the enemy distributing system in the Seoul area will completely dislocate the logistical supply of his forces now operating in South Korea and therefore will ultimately result in their disintegration. This indeed, is the primary purpose of the movement . . . the enemy cannot fail to be shattered."* (Appleman, 1992, p. 495).

Despite Inchon's appeal from a strategic standpoint, at the tactical level it was a dreadful collection of everything wrong from an amphibious perspective. Inchon is located in the estuary of the Yom-ha River. The estuary is a low-lying tidal basin and the tidal range at here is approximately 32 feet, second only to the 50-foot tidal range in the Bay of Fundy. At low tide, all of the water leaves the harbor exposing acres of thick, gooey mud. The harbor was connected to the Yellow Sea by Flying Fish Channel; a shallow, narrow, sinuous affair through which an invasion force of some 230 ships would have to pass. There were no real beaches at Inchon; instead, planners were faced with acres of mud flats exposed at low tide and a huge seawall that surrounded the city to protect it from the tides. The landing "beaches" were backed by an urban area with a population of about 250,000 (Utz, 1994). Finally, the harbor was split into inner and outer sections by Wolmi-Do (note: "Do" is the Korean term for island). This island was occupied by North Korean troops and they did have some large-caliber guns (Appleman, 1992). Photograph 2 illustrates the location of Wolmi-Do, the bisected harbor and the city of Inchon, directly behind the invasion beaches.

Unquestionably, the tides were the most difficult problem to surmount at Inchon, especially combined with the physical constraints of Flying Fish Channel. The draft of the landing craft and tidal range meant that the assaulting forces had only two hours to enter and leave the harbor. The largest of the landing craft, the Landing Ship Tank (LST), required a minimum of 29 feet of water; a condition that existed only one time a month and only for four days. The 15th of September was the only day with a maximum water depth of 31.2 feet over the mud flats, which satisfied the Navy's ideal conditions (Appleman,

1992). The 27th of September was marginal, but once this narrow windowed closed the invasion would have to be delayed until 11-13 October when the spring tide again created a water depth of 30 feet (Utz, 1994).

A tidal range this large also means that a great deal of water must be exchanged as the tide goes in and out over the 12 hour, 25 minute tidal cycle. This meant that the ebb current in the main boat channel exceeded the forward speed of the landing craft, hence operations would have to be suspended as the tide ebbed. The large tidal range also exposed a vast mud flat at low tide stretching three miles out to sea and changing every channel and inlet to coils of twisting mud. The only channel deep enough to allow the assaulting forces into Inchon Harbor was Flying Fish Channel which has a five-knot current and is littered with rocks, shoals, reefs, and islands. The combination of tidal currents and the difficult channel called for a daylight approach and a main landing during the late afternoon high tide, which meant an assault at 5:30 p.m. (Center of Military History, 1989).

Wolmi-Do made this problematical because of its fortifications and by its prominent location in the harbor (See Photograph 2). Invasion planners did not think it feasible or practical that the invasion force could land on the main beaches without first securing Wolmi-Do (Weintraub, 2000). However, the tidal and channel conditions did not allow a simultaneous landing on Wolmi-Do and the main beaches. Hence, a landing was planned on Wolmi-Do by the 3d Battalion, 5th Marines at 6:30 a.m. coincident with the morning high tide with a very tricky approach in Flying Fish Channel during hours of darkness. During the intervening 12 hours, the battalion was essentially on its own until the fleet could return with the late afternoon high tide (Appleman, 1992).

Operation Chromite faced other difficult challenges as well. The 16-foot seawall that surrounded the city required planners to equip the assault forces with ladders (see Photograph 3). So, the main invasion "beaches" along Inchon's waterfront were not really beaches at all. The Marines in the assault force would have to scale a difficult obstacle from bobbing landing craft, perhaps in the face of enemy fire (Utz, 1994). Further, once ashore, the assault troops were involved immediately in an urban setting (see Photograph 4). One of Chromite's important assumptions was that the X Corps could move quickly inland and seize Seoul; severing the North Korean's lines of communications before they could recover (Appleman, 1992). An urban battle

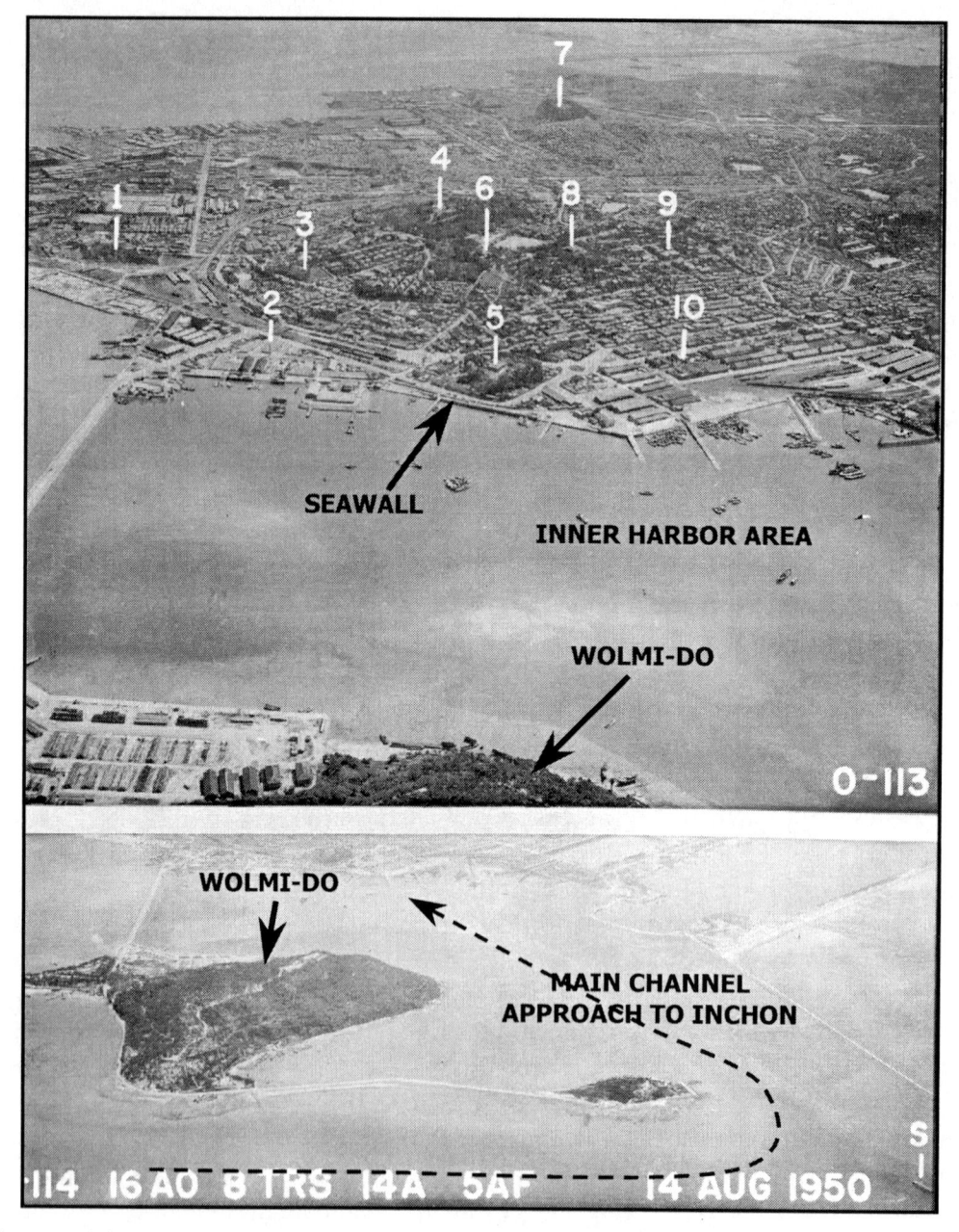

Photograph 2. Reconnaissance photographs (14 August 1950) of Inchon and its harbor. The upper photograph illustrates the built up nature of the beaches and the seawall is evident. The lower photograph demonstrates how Wolmi-Do dominates the approaches to the inner harbor.
Source: United States Military Academy.

Operation Chromite: Turning Movement at Inchon 101

Photograph 3. Marines scale the seawall at Inchon during the second phase of the assault.
Source: United States Marine Corps photograph.

could clearly upset that timetable. MacArthur paid little mind to the risk because his intelligence staff assured him that there was only a maximum of 6,500 mostly poorly trained North Korean soldiers in the entire Seoul region (Weintraub, 2000).

A September landing at this latitude has implications in terms of weather effects. First, September is the transition season in the regional monsoon. Seas are typically low during the summer period (May through August). Higher seas dominate between October and March (Appleman, 1992). Although episodic events could lead to difficult sea states, Naval planners thought that conditions would remain adequate during the projected invasion time. However, a consideration of major importance is that September is the height of the typhoon season in this part of Asia. A typhoon of any size could delay the invasion til

October, which everyone deemed too late (Utz, 1994). As it turned out, the invasion had two brushes with typhoons. The first, Typhoon Jane, struck Japan on 3 September while the invasion force was loading out. It caused some damage to some ships, but not enough to suspend the operation (Center of Military History, 1989). The second, Typhoon Kezia, threatened the region on the 13th of September, but thankfully shifted direction at the last possible moment permitting the landing force to sail from Japan (Appleman, 1992).

The Landing Controversy

The Joint Chiefs of Staff knew that MacArthur wanted to land at Inchon; however, they (specifically senior Naval officers) were concerned about the difficult physical challenges. While they did not disapprove of his plan, they did not formally accept it either and promoted forcefully the Kunsan option (Weintraub, 2000). The Navy's opposition to the plan centered largely on the difficult tidal conditions and the intricate timing required for the operation. The Marines were uneasy over the plan to leave a battalion isolated on Wolmi-Do for 12 hours and difficult nature of the landing beaches in general. Both the Navy and Marines favored Kunsan as the landing objective for Operation Chromite. The Army Chief of Staff, General Collins and the Chief of Naval Operations, Admiral Sherman departed for Japan to discuss the matter with General MacArthur (Appleman, 1992).

On 23 July, Collins and Sherman met with MacArthur and his staff, along with Rear Admiral J.H. Doyle, commander of the invasion fleet. Admiral Doyle spoke for one hour and presented the landings naval considerations. His mood was pessimistic as he enumerated the formidable physical obstacles and the intricate timing needed for the two-phase landing. Collins and Sherman asked many pointed questions and the efficacy of the landing was debated hotly by MacArthur's staff and the Pentagon representatives. Finally, the Chief of Naval Operations asked Admiral Doyle directly for his appraisal of the operation. Doyle replied that, *"The operation is not impossible, but I do not recommend it."* (Appleman, 1992, p. 493).

The fate of Operation Chromite was uncertain. Doyle, a veteran of innumerable amphibious operations in the Pacific did not exactly give it a ringing endorsement. At that point, MacArthur stood and gave Collins, Sherman and all in attendance a 45-minute soliloquy on the strategic and tactical advantages of the Inchon plan. He reasoned that Kunsan was a good idea, but the wrong location. It was too close to the front lines and afforded the North Korean's interior lines (see Map 2). Furthermore, the turning movement was not deep enough to cause the North Korean front line units to collapse (Weintraub, 2000). He pointed out that the enemy had neglected his rear area and was at the end of a long, exposed logistical tail. North Korea committed practically all of their combat ready formations against the Pusan Perimeter and no trained reserves were available. Hence, the ability for the enemy to disrupt the invasion was limited (Appleman, 1992).

Photograph 4. Vertical aerial photograph taken of Inchon on 16 August 1950. The photograph illustrates the extent of the urban area behind the "beaches".
Source: United States Military Academy.

From the beginning, MacArthur understood the strategic impact of severing the North Korean lines of communications and the immediate influence it would have on the course of battle. He pointed out that the amphibious operation was the most compelling strategic instrument available and by using it decisively at Inchon, he could reverse the tide of the war. Clearly, as he saw it, the proper use of the amphibious attack was to strike deep and hard into the enemy rear. Kunsan was certainly low risk, but it also promised a much smaller payoff, and could not guarantee the precipitation of a North Korean retreat (Appleman, 1992). MacArthur called attention to the fact that everything indicated that the North Koreans considered a landing at Inchon to be geographically impossible and had devoted only minimum efforts to securing the harbor. Hence, surprise would be complete. The most important advantages of an amphibious operation were his to enjoy; mobility, surprise and absolute, overwhelming firepower (Utz, 1994).

Finally, MacArthur recalled his operations in World War II. He praised the Navy, which made his historic island hopping campaign possible. He summed everything up by declaring his unequaled conviction in the Inchon operation and that the Navy would be able to pull it off. He finished by saying that, *"The Navy has never turned me down yet, and I know it will not now."* (Appleman, 1992, p. 494). He evidently convinced Chromite's doubters and won approval from the Joint Chiefs of Staff. It is important to note that under ordinary circumstances, Inchon would likely not have happened. Who but MacArthur could have swayed the Joint Chiefs of Staff, who were by the way all his military juniors? However, we cannot overlook the fact that Operation Chromite was skillfully designed and MacArthur was able to maintain an absolute clear vision of its tactical requirements and strategic goals and possibilities (Weintraub, 2000).

Epilogue: Victory at Inchon

"The Navy and Marines had never shone more brightly than this morning."
                                                General Douglas MacArthur

The preliminary naval and air bombardment of Wolmi-Do and the Inchon beaches began in earnest on 13 September. The X Corps arrived offshore early on 15 September with nearly 70,000 soldiers and Marines. The landing group took position off Wolmi-Do at 0530 and the Marines began loading the landing craft for the first of the two-phase landings. During the pre-dawn light of 15 September 1950, the 3d Battalion, 5th Marines landed at Green Beach on Wolmi-Do (see Map 5). The landing was preceded by a short naval bombardment. The assault was a spectacular success—it was the first American amphibious assault since Easter Sunday, 1 April 1945, at Okinawa (Appleman, 1992). The fist two waves of assault troops were ashore by 0645 and encountered very little resistance. The reduction of the island was complete and it was declared secured at 0750. The North Koreans were bewildered by the size and swiftness of the assault. The Marines captured nearly 400 prison with the loss of only a handful of the assault troops (Appleman, 1992).

Following the relatively effortless capture of Wolmi-Do came the long anxious wait during the period of ebbing tide when all activity was effectively suspended. The remainder of the 1st Marine Division boarded their landing craft at 3:30 p.m. Another short, violent Naval bombardment blasted Red and Blue Beaches (See Map 5) and the 5th and 1st Marines breasted the seawall at 5:33 p.m. (Appleman, 1992). The North Koreans were fully alerted by now and the fighting once ashore was much heavier. Nonetheless, the Marines pushed inland quickly and their biggest obstacle after the seawall was the approaching night. The timing of the complex plan worked nearly to perfection (Utz, 1994).

Conclusion

The North Korean invasion was defeated by a single devastating stroke. Why was Inchon chosen? General MacArthur made the decision to use Inchon because; "*the very arguments you have made as to the impracticalities involved will tend to ensure for me the element of surprise. For the enemy commander will reason that no one would be so brash as to make such an attempt.*" (Utz, 1994, p. 23). The landing at Inchon was a success and the course of the Korean Conflict dramatically changed.

The geography of Korea and specifically that of Inchon drove the operational and tactical decision making during the planning phase of the invasion. Inchon's relative location made it the single decisive place for a turning movement. Likewise, Inchon's hydrographic characteristics made a landing there so unlikely, that the North Korean's guarded the area with only weak forces. MacArthur was able to clearly "see the terrain' so to speak, and employ the inherent advantages of amphibious operations in a single, devastating stroke against a weakly defended critical point.

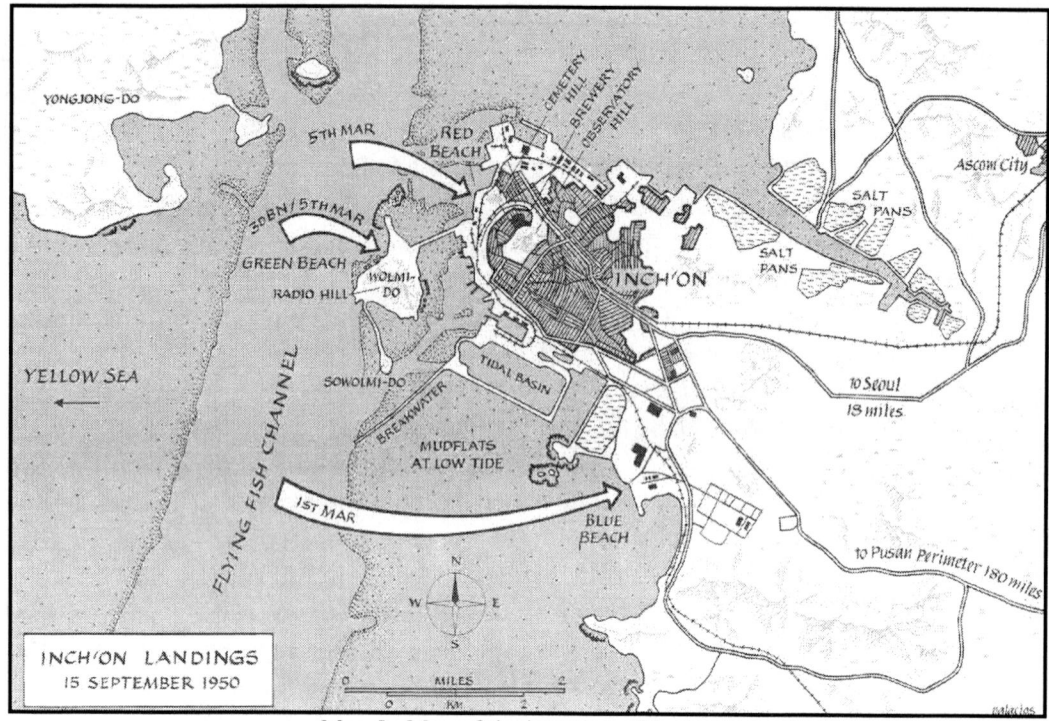

Map 5. Map of the invasion of Inchon.
Source: Adapted from Knox, 1985.

Amphibious operations are perhaps the most difficult to plan and undertake because they take place in a very dynamic environment where three spheres (i.e., land, ocean, and atmosphere) converge. Small changes in weather patterns can alter wave height and direction, currents can rapidly shift, and tides must be dealt with twice a day during the operation. The physical challenges are compounded by the fact that there are a finite number of suitable landing beaches and the enemy typically knows where they are as well. Consequently, military planners are challenged by daunting and complex physical parameters such as waves and tides, and the intricate tactical challenge of outwitting the enemy to insure that the invasion force has a reasonable chance of success.

☙ ❧

Bibliography

Addington, Larry H. 1984. *The Patterns of War Since the Eighteenth Century.* Bloomington, Indiana: University of Indian Press.

Appleman, Roy E. 1992. *South to the Naktong, North to the Yalu.* History of the United States Army in the Korean War, Center of Military History, Washington, D.C.: United States Government Printing Office.

Barber, Daniel E. 1969. *MacArthur's Amphibious Navy: Seventh Amphibious Force Operations, 1943-1945*, Annapolis, Maryland: U.S. Naval Institute Press.

Bascom, W. 1964. *Waves and Beaches.* New York, New York: Doubleday Academic Press.

Brown, Jerold, E. 1992. "Tarawa: The Testing of an Amphibious Doctrine". In Spiller, R.J. (ed.), *Combined Arms in Battle Since 1939.* Fort Leavenworth, Kansas: U.S. Army Command and General Staff College Press.

Center of Military History. 1989. *Korea-1950. Publication 21-1*, Washington D.C.: United States Government Printing Office.

Costello, John, 1981. *The Pacific War.* New York, New York: Rawson, Wade Inc.

Department of the Army. 1989. *FM 100-15, Corps Operations.* Washington, D.C.: United States Government Printing Office.

Department of the Army. 1998. *FM 101-5-1, Operational Terms and Graphics.* Washington, D.C.: United States Government Printing Office.

Department of Defense. 1992. *Conduct of the Persian Gulf War.* Washington, D.C.: United States Government Printing Office.

D'Este, Carlo. 1991. *Fatal Decision, Anzio and the Battle for Rome.* New York, New York: Harper Collins Publishers.

Doyle, Peter, and Bennett, Matthew R. 1999. "Military Geography: The Influence of Terrain in the Outcome of the Gallipoli Campaign." 1915. *The Geographical Journal*, 165 (1): 12-36.

Fehrenbach, T.R. 1963. *This Kind of War*. Washington, D.C.: Brasseys.

Fleet Marine Force. 1934. *Tentative Manual of Landing Operations*. Washington, D.C.: United States Government Printing Office.

Fleet Marine Force, Pacific. 1944. *Staff Officer's Field Manual for Amphibious Operations*. Washington, D.C.: United States Government Printing Office.

Galgano, Francis A.. 1993. "The Landings at Anzio." *Military Review*, 74 (1): 69-73.

Harrison, Gordon A. 1993. *Cross-Channel Attack. United States Army in World War II, The European Theater of Operations*. U.S. Army Center of Military History, Washington, D.C.: United States Government Printing Office.

Hastings, Max. 1987. *The Korean War*. New York, New York: Simon & Schuster.

King, Ernest J. 1946. "The United States Navy at War, 1941-1945." *Official Reports to the Secretary of the Navy*, Washington, D.C.: United States Government Printing Office.

Knox, Donald. 1985. *The Korean War, An Oral History, Pusan to the Chosin*. New York, New York: Harcourt, Brace and Jovanovich.

Miller, John. 1989. *Guadalcanal: The First Offensive. United States Army in World War II, The War in the Pacific*. U.S. Army Center of Military History, Washington, D.C.: United States Government Printing Office.

Miller, John. 1990. *Cartwheel: The Reduction of Rabaul. United States Army in World War II, The War in the Pacific*. U.S. Army Center of Military History, Washington, D.C.: United States Government Printing Office.

Morrison, Samuel E. 1960. *Volume II: Operations in North African Waters, October 1942-June 1943, History of United States Naval Operations in*

World War I., Washington, D.C.: United States Government Printing Office.

Morrison, Samuel E. 1961a. *Volume III: The Rising Sun in the Pacific, 1931-April 1942, History of United States Naval Operations in World War II.* Washington, D.C.: United States Government Printing Office.

Morrison, Samuel E. 1961b. *Aleutians, Gilberts and Marshalls, June 1942-April 1944, History of United States Naval Operations in World War II.* Washington, D.C.: United States Government Printing Office.

Morrison, Samuel E. 1963. *The Two Ocean War: A Short History of the United States Navy in the Second World War*, Boston, Massachusetts: Little, Brown Inc.

Parker, William D. 1970. "Advanced Base Expeditionary Era, 1916-1941." In *A Concise History of the United States Marine Corps, 1775-1969.* U.S. Marine Corps Historical Division, Washington, D.C.: United States Government Printing Office, pp. 45-58.

Polmar, Norman and Mersky, Peter B. 1988. *Amphibious Warfare: An Illustrated History.* New York, New York: Blanford Press.

Reynolds, Clark G. 1990. *War in the Pacific.* New York, New York: Military Press.

Robertson, William G. 1985. *Counterattack on the Naktong, 1950.* Leavenworth Papers, No. 13. Fort Leavenworth, Kansas: United States Army Command and General Staff College Press.

Russ, Martin, 1975. *Line of Departure: Tarawa.* Garden City, New York: Doubleday & Company, Inc.

STANAG 1149, 1984. *Doctrine for Amphibious Operations.*

Unites States Army Corps of Engineers. 1984. *Shore Protection Manual (3 Volumes).* Vicksburg, Mississippi: Waterways Experimental Station.

Utz, Curtis A., 1994. *Assault from the Sea, The Amphibious Landing at Inchon.* The U.S. Navy in the Modern World, Publication No. 2, Naval Historical

Center, Washington, Navy Yard, Washington, D.C.: United States Government Printing Office.

Webb, William J. 1995, *The Korean War: The Outbreak, 27 June-15 September 1950.* United States Army Center of Military History, Washington, D.C.: United States Government Printing Office.

Weigley, Russell, F. 1973. *The American Way of War, A History of the United States Military Strategy and Policy.* Bloomington, Indiana: Indiana University Press.

Wientraub, Stanley. 2000. *MacArthur's War, Korea and the Undoing of an American Hero.* New York, New York: The Free Press.

Winters, Harold A. 1998. *Battling the Elements.* Baltimore, Maryland: The Johns Hopkins University Press.

Peleliu:
GEOGRAPHIC COMPLICATIONS OF OPERATION STALEMATE

By

Jon C. Malinowski

Introduction

Eight hundred miles east of Mindanao lies the tiny island of Peleliu, one of several islands in the Palau archipelago. Only about six miles long and two miles at its widest, this small coral island was soaked with the blood of over 12,000 United States and Japanese soldiers and marines killed between September and December of 1944. Perhaps more than any other island battle of World War II, soldiers on Peleliu fought not only against their enemy, but also against terrain and climate. Even before the battle, the physiography of the island was severely underestimated by U.S. commanders while Japanese forces learned to use terrain to prolong the suffering of their attackers. Heat, lack of water, and coral terrain all contributed to one of the costliest—and as it turned out, needless—battles of the war.

Strategic Importance

The strategic importance of the Peleliu campaign is greatly debated to this day. Originally, the Japanese-occupied Palau Islands, including Peleliu, were considered necessary to provide flank protection for MacArthur's invasion of the Philippines. Morotai in the south, and Peleliu in the north, would serve to guard the left and right flanks during the Mindanao invasion. In particular, invasion planners were concerned with the Japanese air threat from these islands and the Navy insisted on the elimination of this perceived threat. Thus, the initial focus on the Palau chain was rooted in its geographic location. The invasion was set for September of 1944.

Just before the invasion, air raids on Mindanao indicated that the southern Philippines were only lightly defended. There were few front-line air units nor was heavy ground resistance expected. In addition, an interview with a rescued fighter pilot shot down on Leyte indicated that the Japanese were not heavily defending that island, making it a great first step for a Philippine invasion. Admiral William F. Halsey warned in June of 1944 that any Palau invasion would be prohibitively costly. Furthermore, he sent last minute messages just two days before D-Day on Peleliu, indicating that the Palaus were no longer needed (Alexander, 1998). However, Admiral Nimitz (Commander in Chief of the Pacific Fleet) rejected the suggestion that the Peleliu invasion be scrapped and the invasion went forward. This decision remains controversial, magnified through hindsight by the carnage that followed.

Peleliu itself was not the most heavily defended island in the Palaus. Babelthaup and Koror islands to the north were thought by the Japanese to be the likely targets for an invasion, and as many as 25,000 troops had been stationed on Babelthaup alone. Peleliu was thus attractive to U.S. planners because it would be less heavily defended, at least in terms of troop strength, and it had an important airfield that could be used as a staging point for operations to the west.

For the Japanese, the Palaus were an important control point. Seized during World War I from the Germans and confirmed by League of Nations mandates, thousands of Japanese settled the islands to improve political and cultural control. Roads and ports were built or improved, and an airfield was added on Peleliu. During the early years of the war, the Palaus were not seen by the Japanese as vital to their defense strategy and served primarily as supply depots; but, by 1943 defeats elsewhere in the Pacific reinforced the need to strengthen forces in the western Pacific (Gailey, 1983). At roughly the same time, Japanese leaders shifted their attitude towards holding Pacific territories. Until early 1944 naval forces were in charge of holding islands in the region. A fierce U.S. bombing raid in March of 1944 destroyed numerous Japanese ships in the area. Now the Japanese high command shifted control of the islands to Army units. These newly assigned forces were given orders to destroy attacking forces on the beaches, then defend from entrenched positions inland in order to inflict as many casualties as possible on attackers. It was hoped that this technique would delay any movement towards Japan (Gailey, 1983).

Thus, in planning for the Peleliu battle, both sides had a clear purpose. The Americans wanted to seize the airfield and neutralize the enemy presence on the island. The Japanese sought to make any attack costly and time consuming. It would seem that both sides were in the end, victorious.

Physical Geography

Peleliu's physical geography is forbidding. The island is largely made of limestone coral, formed centuries earlier from the remains of aquatic life. At some point, parts of the island were uplifted to elevations of several hundred feet. The erosive effects of water and vegetation subsequently dissected these areas of higher elevation. The resulting karst topography is one of steep cliffs, parallel ridges, and odd-shaped towers of coral. As indicated on Map 1, the areas north of the Japanese barracks formed a long, complicated ridge of rough, rocky terrain including features such Umurbrogol Mountain. Marines would name this area "Bloody Nose Ridge" soon after the battle began.

The southern part of the island is relatively flat and it is here that the Japanese maintained an airbase. The underlying ground in this area is coral as well. In several parts of the island mangrove swamps dominate. Before the U.S. naval bombardment started, dense tropical vegetation covered nearly all of the island, but much of this was destroyed by explosions. Temperatures on Peleliu average in the 80s (°F) during all months, and between 6 and 17 inches of rain are expected per month. These conditions support an abundance of insects, crabs, and small mammals. Surrounding the island is a series of reefs.

Fresh water is a problem on coral islands. Because of the permeability of limestone, water tends to seep directly into the ground rather than forming streams or ponds. A few small bodies of water did exist on the island at the time of the battle, but generally the water supply was unreliable. Sitting water quickly becomes stagnant or brackish, and coral dust creates a milky appearance.

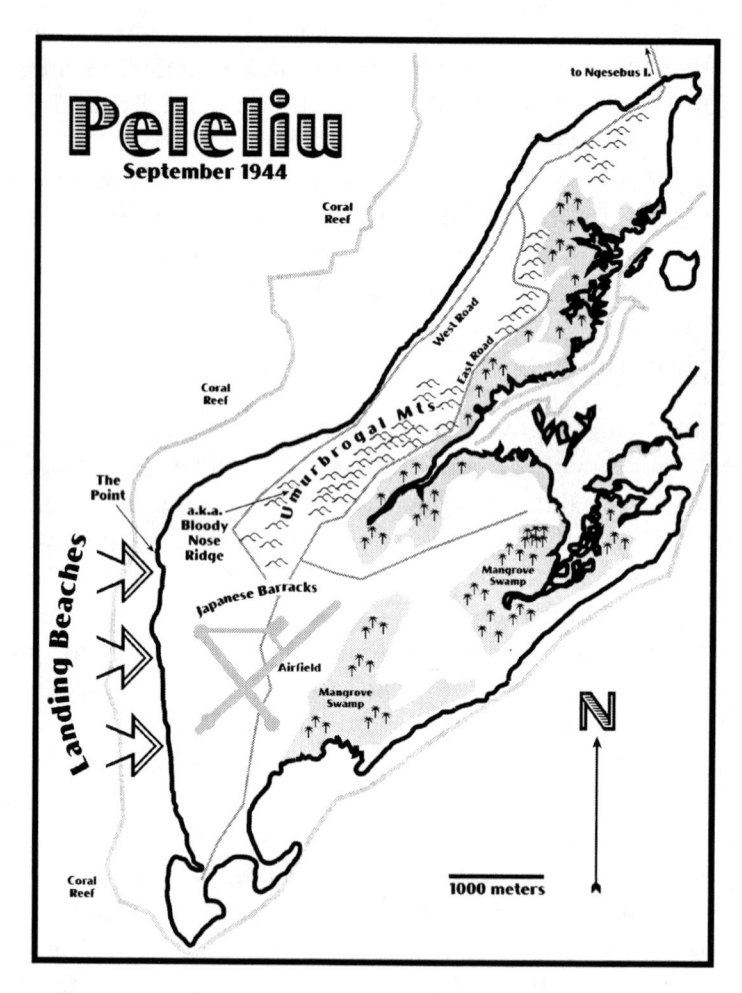

Map 1: Peleliu on D-Day.

Preparations

U.S. planning for Peleliu remains controversial because of conflicting evidence about the extent of accurate information known to planners. Some sources indicate that many planners believed that Peleliu was "low and flat" (Gailey, 1983). Yet military maps from July 1944 clearly show significant topography on much of the island. A July 1944 military map clearly shows elevations of over 200 feet in areas north of the airfield ("Sheet 37", 1944). This same map also shows a "phosphate refinery" which should have clued intelligence personnel that the island could have mines or caves. Phosphates are formed from a mix of guano and coral that hardens over time. It is generally surface mined and can produce steep cliffs and rugged terrain. There is little evidence that Peleliu was substantially mined, and the plant may have refined phosphates from nearby Anguar Island, but it is a clue to the possible nature of the terrain. Furthermore, a Joint Army-Navy Intelligence Study (JANIS) of the islands, dated April 1944, says of Peleliu that *"rough areas in the northwestern part ... and mangrove swamps along the northeast coasts"* are the only features that would prevent cross-country movement (Joint, 1944). This description could be interpreted as both warning of high ground and as an assurance that few problems would be encountered.

Part of the problem may have arisen from the use of composite aerial photos. These photomaps were pieced together from available images, in this case from six different sorties (Gailey, 1983). Additional ambiguity stems from the tropical canopy that grows on Peleliu. Rough terrain tends to be smoothed out by dense forest cover, turning sharp edges into smooth, rounded formations. A photomap dated August 1944 has mountains labeled on it, but is fairly difficult to distinguish these areas as significantly higher than surrounding areas, although some parallel ridges can be discerned (Photomap, 1944). The lingering idea from studying the evidence is that some planners knew that Peleliu was not "low and flat", but this information was not adequately distributed to all of the personnel who could have benefited from this information. The same JANIS report does provide accurate information on heat and water availability on Peleliu. The report describes the climate as *"humid tropical"* and as *"hot, rainy, and cloudy at all seasons"* (Joint, 1944, p. I-5). Water supply problems are described as *"acute, particularly on the low coralline limestone islands and islets"* such as Peleliu. The report notes that rainwater is collected and used for drinking, and that some wells are dug, but the water is warned to be impure (Joint, 1944, p. I-9). Thus, some plans were put in place for the provision of water during the invasion, including the shipping of water in steam-cleaned, 50-gallon diesel drums.

The 1st Marine Division, under the command of General William H. Rupertus, decided to land on the southwest coast, designated White and Orange Beaches. The southwestern coast (Purple Beach), although an easier place to land, was rejected because Marines would be forced to cross mangrove swamps to get to the airfield, or would get caught in a bottleneck on the narrow strip of land circumventing the swamps. Japanese defenses also seemed heavier in this area. Likewise, the southern tip of the island (Scarlet Beach) was also rejected because of the narrowness of the land and because of coral cliffs that would impede landing troops. Northern portions of the island were rejected because of supporting positions on nearby Ngesebus Island. Thus, the southwest beaches became the prime landing spot.

Photograph 1: Japanese cave on Peleliu.
Source: U.S. Marine Corps (U.S. Naval Institute).

The basic attack plan called for landing three regiments of the 1st Marine Division on the southwest coast of the islands. Troops would come ashore in amphibious tractors that would then serve as supply vessels to ferry supplies and personnel from transport ships to the beaches and wounded from the battlefield to hospital ships. The 1st Regiment would secure the left flank of the invasion area, the 5h Regiment would move straight across and secure the airfield while the 7h Regiment would secure the right flank at the southern part of the island.

Once the airfield was secure, troops would turn north to attack remaining Japanese soldiers in the northern hills. Despite considerable skepticism on the part of some officers, the senior officer, General Rupertus, predicted a tough but quick fight of 2-3 days.

Japanese troops in Peleliu, numbering about 10,000, were under the command of Colonel Kunio Nakagawa. Nakagawa, given his orders to make any invasion as protracted as possible, understood and took advantage of the available terrain. Building on work started by earlier navy personnel, Nakagawa set about to reinforce and enlarge the existing series of caves on the island. The cave system became so elaborate some could hold hundreds of men and were complete with fresh-water springs and cooking areas (see Photograph 1). Militarily, Nakagawa ensured the cave system was designed in such a way that troops attempting to attack one cave would come under fire from another. Many caves had several access holes to allow Japanese soldiers to move around or escape. Many had openings of only a few inches through which to fire, making all but a point-blank attack useless. Cave shapes, such as in the shape of the letters "E" or "H", allowed personnel to escape direct fire or concussions coming through an opening. Overall, the Japanese maintained over 500 caves of differing sizes and complexities.

In other areas, Nakagawa's troops built or reinforced pillboxes closer to the beach areas (see Photograph 2). On "The Point", a 130 foot high piece of coral jutting out into the ocean at the left flank of White Beach, the Japanese blasted out the coral at high and low elevations and built concrete pillboxes. This would prove to be an excellent defensive position during the first day of the invasion. Elsewhere, mortars and other weapons were registered on strategic locations such as the surrounding reefs and the beach areas. Some mines were also laid in likely landing areas. Nakagawa also had some light tanks, and these were positioned near the airfield.

The Invasion

The invasion of Peleliu began on the morning of September 15, 1944. Coming aboard in amphibious vehicles (LVTs) just behind armored tractors (LVTAs) carrying 75mm howitzers, the Marines of the 1st Division hoped that the days of naval bombardment before D-Day had destroyed many of the Japanese defenses. They were soon painfully aware that the bombardment had missed most targets. Because of tree-cover and inadequate distribution of information about the island, naval gunners were unaware of most of the caves and dug-in pillboxes that the Japanese had built. In reality, Japanese defenses were largely unharmed by Navy's big guns. Many LVTs were hit crossing the half-mile coral reef that needed to be crossed before the beaches were reached. Many attacks originated in hidden bunkers flanking the landing beaches, such as those on "The Point." The destruction of LVTs greatly disrupted supply efforts and caused considerable confusion on the beaches.

Photograph 2: Japanese pillbox. Notice the lack of a prominent opening.
Source: U.S. Marine Corps (U.S. Naval Institute).

The Coral

After hitting the beach, the Marines moved quickly inland under heavy enemy fire. Once on the beach, they realized one of the first lessons of combat on a coral island, namely that you can't dig easily in coral. Digging a defensive position proved to be very difficult if not impossible, and the natural cover had generally been destroyed. Some soldiers used existing Japanese anti-tank trenches (Watkins, 1999). Others piled up pieces of wood and coral while still others made use of artillery craters. Bullets and shrapnel easily bounced off the coral, making any explosion even more dangerous. This problem would linger throughout the rest of the battle, and many casualties arose from the simple fact that troops could not get below the ground surface. Later in the campaign, the Army increased the use of sandbags to overcome this problem. Bags were filled in the rear areas and brought as close to the lines as possible, where they were passed hand-to-hand to the front (Gailey, 1983). When the front moved, the sandbags were inched forward as well. The inability to dig in the coral also affected basic sanitation. Human waste and everyday garbage could not be buried, and thus much remained on the top of the ground. The stench from this refuse is described by veterans as a particularly bad memory of battle on Peleliu (Sledge, 1981).

The 1st Regiment, under the command of the legendary COL Lewis "Chesty" Puller, met stiff resistance along the left flank. This area was closest to the high terrain in the center of the island and was under easy observation by Japanese gunners. In effect, the 1st Marines could not move without being observed and fired upon by a well-concealed enemy.

The 5th Regiment drove across the island and captured portions of the airfield while the 7th Regiment fought against tough resistance on the southern end of the island. During the late afternoon of D-Day, the Japanese launched a tank and infantry counterattack from the northern side of the airfield. The Japanese tanks were no match for their American counterparts and naval firepower and were easily destroyed. The first day on Peleliu was difficult and discouraging, and quickly proved the initial assessments for an easy battle wrong. By the end of the first day, Marines had not met their objectives except in the middle of the attack (see Map 2). On both the left and right flanks, stiff resistance, counterattacks, and supporting fire from high ground and dug-in positions slowed the operation. The next days of battle would bring home two other realities of coral island geography.

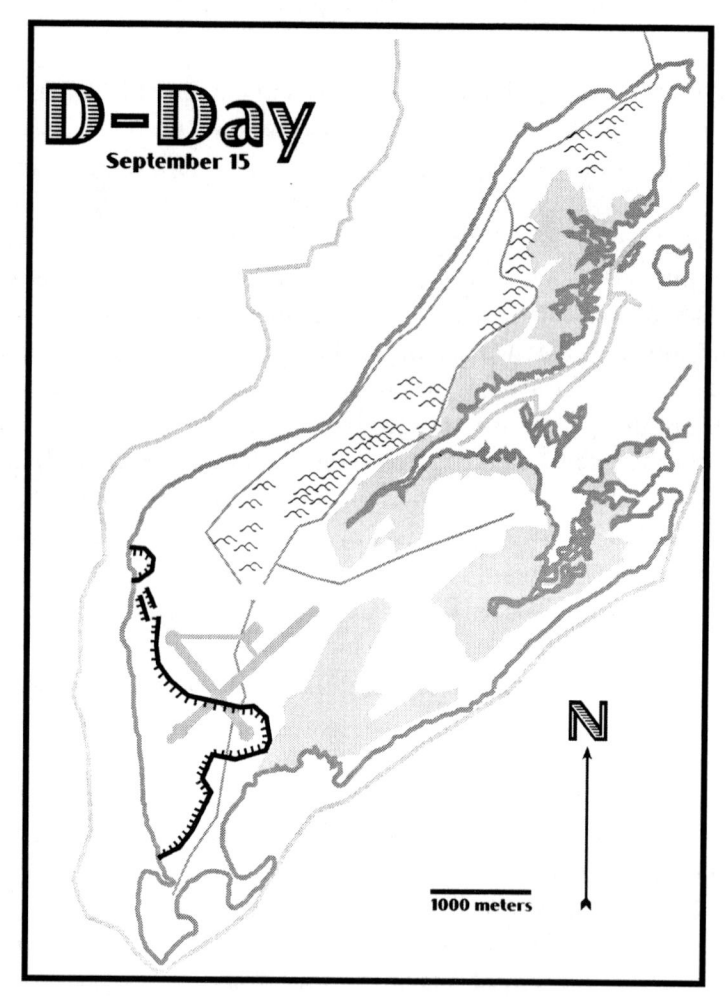

Map 2. D-Day Advances.

Heat

The first reality was the heat, and even on the first day, heat prostration was a problem. Early on D + 1 the temperature rose to over 100 degrees, and there was little shade because heavy bombardment had destroyed most trees in the battle areas. Marines came ashore with two canteens of water, and these quickly became depleted. As mentioned above, much of the water supply was brought in cleaned out oil or diesel drums, but many had not been cleaned properly and the water was tainted with oil, diesel, or rust (Sledge, 1981). Some

Marines resorted to drinking water collected from the bottom of shell holes blasted out the coral and then treated with halazone tablets (Watkins, 1999). Shell holes five to six feet deep would often reach the water table, revealing muddy but drinkable water.

At many times during the Peleliu campaign temperatures would reach well over 100°F. Heat related casualties became common. Soldiers overcome by heat or dehydration would be sent to the rear to recover before being back to combat, reducing the numbers available to fight. One survivor tells of lifting a leg and having water run out of his boots because of so much sweat collected in the bottom (Sledge, 1981). Salt tablets were used, but did not help much for some. Sweat and heat produced fungus infections that lingered on most combat troops until well after they had left the island. The combination of direct sunlight and the reflection off the coral caused severe sunburns and cracked lips. One veteran tells of the mixed pain and pleasure of drinking some grapefruit juice with cracked lips during his time on Peleliu (Watkins, 2000).

The intense heat also played a more gruesome role on the island. Human waste and garbage, as mentioned above, could not be buried, and lingered to rot on the surface. The large number of dead bodies rotted as well. American dead were generally covered and removed to the beach areas at the rear to await burial. The Japanese, however, were unable to bury or collect their dead because they were entrenched in caves and defensive positions. Therefore, as the battle wore on, hundreds and later thousands of Japanese bodies were rotting on the island. The heat contributed to a horrible stench that many U.S. veterans recount. The corpses also attracted flies during the day and land crabs at night, adding to the horror witnessed by all involved. The flies got so bloated and used to humans that they had to be picked off food because they were so fearless (Sledge, 1981). Many could barely fly.

The Terrain

The second reality of Peleliu's physical geography was the rugged nature of the coral highlands that ran north from the airfield. Named "Bloody Nose Ridge" by the Marines, this area contained a series of ridges and small, steep hills that defied any logical arrangement. Many had cliff faces on one of more sides (see Photograph 3). Sometimes ridges paralleled each other, resulting in dangerous alleys that were deathtraps for attackers. Nearly every hill and ridge had a series of caves, so when assaulting one hill an attacker was usually under fire from another angle. Because many of the caves had such small openings and were at different elevations, direct, point-black fire was necessary. Flame-throwers also proved somewhat useful. Many of the most heroic stories from the Peleliu invasion center upon Marines or Army soldiers and officers who took, held, or lost a particular hill or ridge.

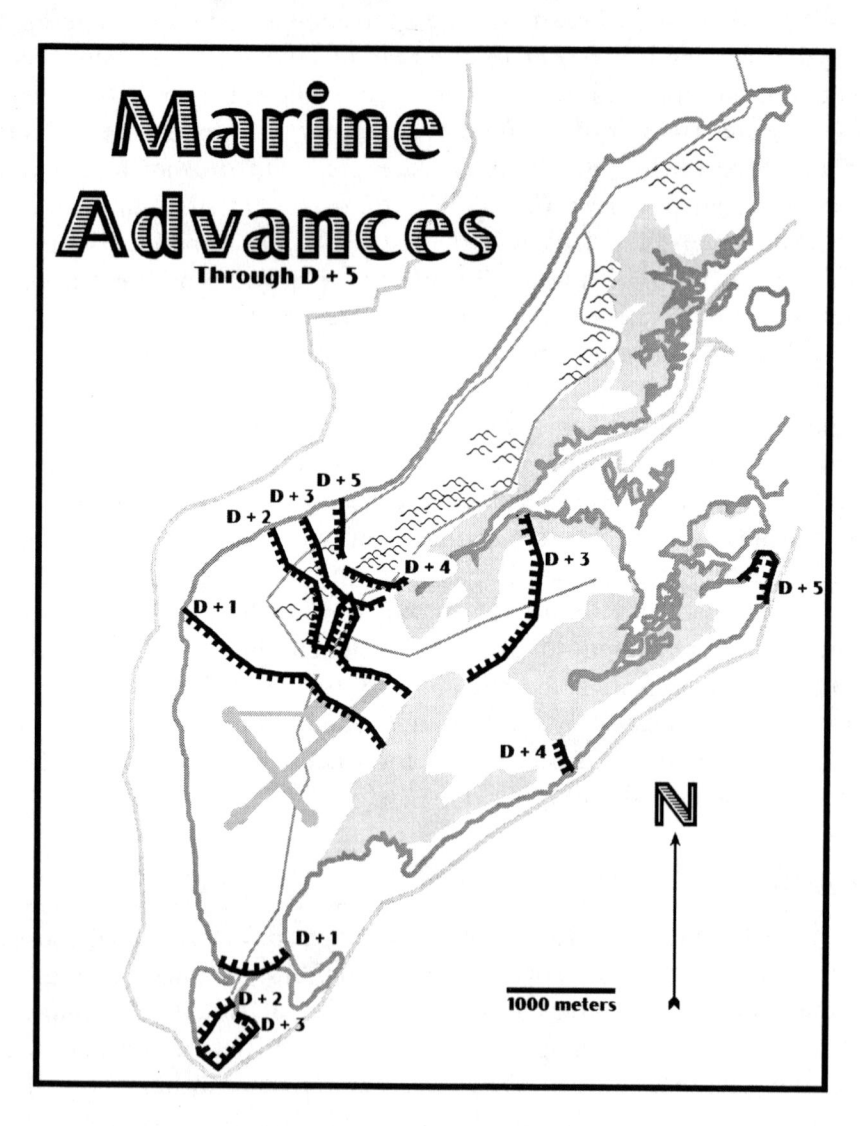

Map 3. Marine advances D+1 through D+5.

Because it was nearly impossible to dig-in, attacking this terrain was especially costly. Marine doctrine was to send wave after wave of men against a position to neutralize the enemy. This proved to be disastrous at Peleliu. Because of leg injuries, many high ranking officers, including Division Commander General Rupertus, and Chesty Puller, were stuck in rear areas away from the front and not as aware of the terrain difficulties as they could have been. This led to frustrated calls to attack nearly indestructible defensive positions that front-line officers knew would result in carnage. Many Peleliu veterans express deep bitterness over the human cost of these orders to attack. While all obeyed out of loyalty and a sense of duty, the price of attacking these hills remains controversial.

One terrain-related problem that U.S. forces faced was the inability to see their Japanese enemies. Dug into caves and bunkers, Japanese troops often fired through openings only inches wide. One Marine Lieutenant remembers:

> *"People who haven't been in combat think that one always sees the enemy. When we were in attack the Japs saw us readily-- but in their caves, trenches, and behind coral outcroppings, we generally only got glimpses of running figures until we found a means of flushing them out. On this occasion, we were getting lots of rifle fire and trying desperately to locate the Japs in the confusion of the battle. I happened to spy one Jap sharpshooter raise his head quickly and fire. I watched him do this at regular intervals about three times and lined my sights up on the spot. The next time his head appeared, I had the trigger half-squeezed. I saw his head jerk back as the bullet struck the center of his helmet."* (Watkins, 1999).

On a micro-scale, the terrain was extremely difficult on the average Marine or soldier. Falling on the coral resulted in numerous cuts, scratches, and bruises. Coral slashed clothing and ruined boots. One veteran recalled that he destroyed a new pair of boots in only six days on Peleliu (Watkins, 2000)! As mentioned before, numerous injuries resulted from explosions kicking up jagged pieces of coral that were often inescapable in shallow foxholes or behind quickly constructed shelters. Coral dust covered everyone, crusting over after periods of rain to create ghostly gray figures.

The first major attacks on the ridges began on D + 2 (Map 3). For weeks to follow, each day approximated the last. Orders were given to take a hill and heavy causalities followed on both sides. By September 23, most of the 1st Regiment had been wiped out, and other units took up the fight, including the 321st Regimental Combat Team of the U.S. Army's 81st Infantry Division ("Wildcat Division"). By the end of October, almost all of the remaining U.S. fighters wore Army patches. The Marines had fought gallantly against tremendous odds and paid dearly. The 1st Marines alone had suffered 1,672 causalities in just eight days. Losses of as high as 71% in some battalions marked the most severe losses of any Marine unit in the Pacific (Gailey, 1983).

Photograph 3: Marines on Peleliu.
Source: U.S. Marine Corps (U.S. Naval Institute).

The Pocket

By the end of September, Marine and Army troops managed to isolate COL Nakagawa and his remaining troops in a section of the Umurbrogol Mountains known as "The Pocket" (see Map 4). The Americans clearly believed that the end was in sight, but few realized at the time that they would be fighting for another two months. Within "The Pocket" lay the most forbidding of Peleliu's terrain—a complex maze of hills and ridges that would take on names such as the Five Sisters, Baldy Hill, the Five Brothers, and the China Wall. Only 900 by 400 yards long, the Pocket with riddled with hundreds of Japanese caves, each expertly hidden and fortified. Between the hills were found long valleys such as Wildcat Bowl and the Horseshoe (see Photograph 4). These were especially difficult to attack because walls surrounded the valleys on three sides, allowing excellent crossfire by Japanese holdouts.

American plans to attack in "The Pocket" did not initially consider the terrain adequately. Commanders established linear, time-based goals thinking that steady and consistent gains could be made. The decision to push troops into "The Pocket" remains controversial, for by this time the airfield had been secured and the Japanese positions that threatened the southern part of the island were destroyed. This portion of the invasion highlights doctrinal differences between the contemporary Marine doctrine of to attack with infantry, maintain momentum, and break the enemy; and the Army's tenet to blast enemy positions with artillery for as long as possible before committing soldiers. Not all terrain can be taken with the same strategy.

In "The Pocket," the terrain became difficult to control. Valleys and canyons offered natural pathways for movement, but these corridors were well defended by mutually supporting Japanese cave positions. Attacking one side of the valley resulted in fire from one or two other sides. These caves and spider holes either had to be blasted shut at close range individually, or their occupants seared to death with flame-throwers (mounted on LVTs), or close range rifle fire. Engineer bulldozers were used to clear paths for tanks and LVTs to penetrate into the canyons. Close range tank and light artillery fire actually demolished several hills and the rubble used to build ramps to other firing positions. Here sandbagging was increasingly used to protect 75mm howitzers that sometimes had to be disassembled, carried up the cliffs, and then reassembled (Gayle, 1996). As seen in Photograph 5, Corsairs from Marine Aircraft Group 11 also proved useful by dropping napalm on cave positions, often very close to advancing troops. Because the airfield was so close, bombers did not even bother to put their landing gear up. Targets were only a few hundred yards away at most.

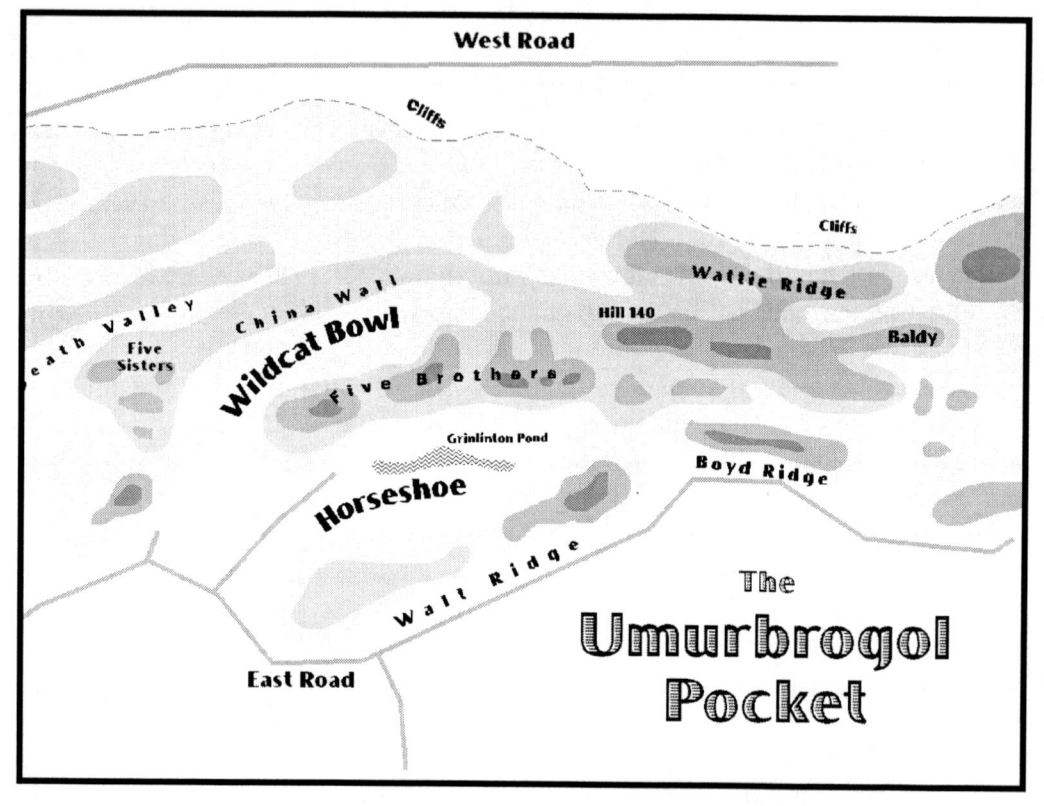

Map 4: The Umurbrogol Pocket.

Steep ridges and high ground presented their own problems. First, troops needed to scramble up extremely steep slopes, often under fire. This difficult task completed, Marines and Army soldiers often came under attack from facing ridges and caves. Many hills and ridges were taken several times only to be surrendered time and time again. Japanese soldiers used the cover of darkness to attack U.S. forces, to get water where available, and to communicate with their comrades, only to disappear again into their caves. These night attacks were useful as psychological warfare against a tired and frustrated enemy.

Photograph 4: Advancing up the Horseshoe.
Source: U.S. Marine Corps (U.S. Naval Institute).

On October 20 General Mueller of the 81st Infantry took over the job of reducing the remaining resistance on Peleliu from General Rupertus and the Marine Corps. Mueller treated the Pocket differently, adopting more of a siege mentality. The Army was content to use tanks, napalm, artillery, and flame-throwers to pulverize Japanese positions before sending in soldiers. Oil pipelines were improvised to bring oil to the front, a distance of over 300 yards. Under pressure, the oil was sprayed into caves and defensive positions and then lit on fire with incendiary grenades to burn out the Japanese. Although the Japanese remained in highly defensible positions, they were now cut off from resupply and, in many cases, from water which they had previously collected from a freshwater lake in the Horseshoe known as Grinlinton Pond. Floodlights were actually set up on this small pond knowing that desperate Japanese would try to get the water (Gailey, 1983). Thus, deprived of basic life-support, Mueller knew that there was no need to move in too quickly.

Photograph 5: Navy bombers drop napalm. Notice that the wheels are still up.
Source: Marine Corps Photograph (U.S. Naval Institute).

Rains in late October and November slowed the final mop up efforts. The last week of October included heavy rain and fog and was followed by a small typhoon between 4-10 November (Gailey, 1983). This time was used to reinforce positions, bring in more sandbags, and add artillery pieces. Although artillery and napalm attacks continued, the general pace slowed. The Japanese continued sporadic night attacks, further weakening and reducing the number of soldiers Nakagawa had available. The rains did help the Japanese water supplies however.

In the end, the Army continued to pound positions and inch forward with infantry soldiers (Map 5). Most advances made were measured in yards. Eventually Army troops were able to take charge of parts of the China wall, dividing the remaining Japanese troops. This was completed on the 25th of November. By this time, the end was at hand for the remaining brave Japanese soldiers. A day earlier, unbeknownst to U.S. forces, Nakagawa had sent a final message to his superiors and committed *seppuku,* ritual suicide.

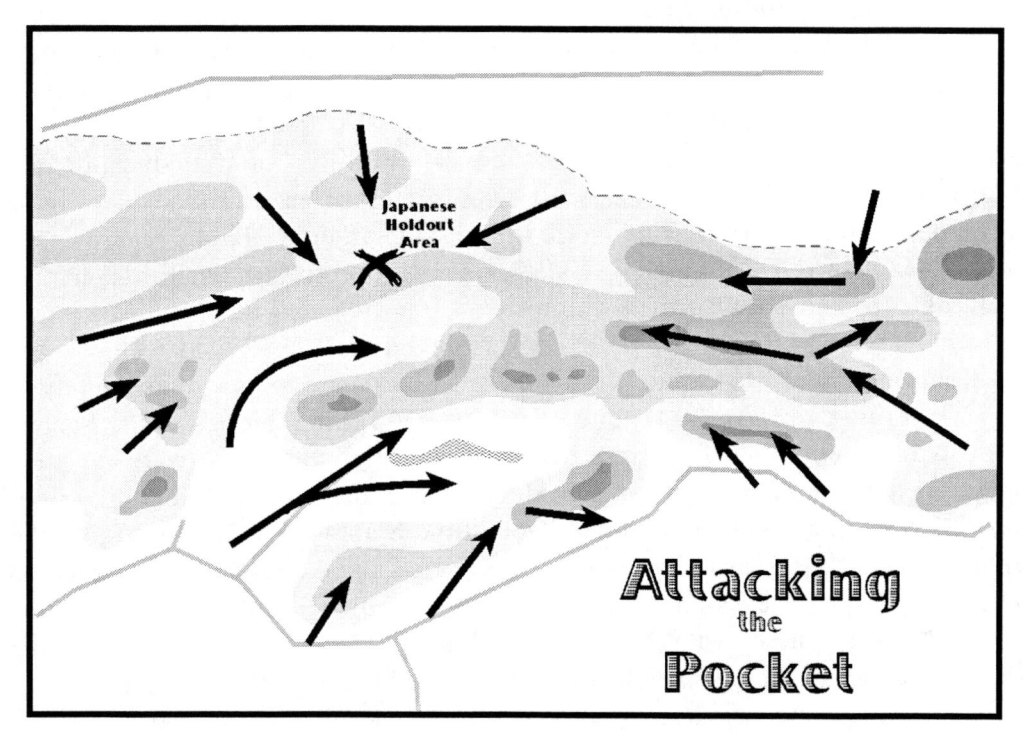

Map 5: Attacking the Pocket.

Aftermath and Lessons

Although Operation Stalemate was essentially finished, Army forces took sporadic causalities well into December from Japanese holdouts. In the end, Marine and Navy casualties totaled 6,526, including 1,252 killed. Army causalities surpassed 3,000, including 404 dead. Japanese casualties are not known, but estimates are just under 11,000 killed. Estimates are that each Japanese casualty cost 1,589 rounds of heavy and light ammunition and one American casualty (Gayle, 1996).

In February 1945 five Japanese soldiers dug out of a collapsed cave and surrendered to American occupation troops. Two years later, in April of 1947, a Japanese lieutenant and 26 of his men surrendered. In 1949 a lone Japanese survivor was found on the island, and nearly a decade after the battle a Korean in Japanese service gave himself up (Gailey, 1983). The terrain had truly proven to be an excellent ally for the Japanese.

The lessons of Peleliu are numerous. U.S. forces were not adequately prepared for fighting in this type of environment. Lingering uncertainties about pre-operation intelligence raise serious questions about preparation. Water and heat related problems resulted in numerous causalities and unquestionably, American troops suffered unnecessarily from a lack of preparation. The rough terrain defeated standard Marine tactic of constant infantry attacks regardless of success. On the other side, COL Nakagawa was able to last much longer than would be expected because of his significant improvement and modification of the island's terrain. In reality, both sides achieved their objectives. The United States controlled the island, but the Japanese had forced a long and costly battle.

Peleliu is, unfortunately, one of the often forgotten battles of the Second World War. The bloody fight was overshadowed by events that were far more momentous. Events in Europe and MacArthur's invasion of the Philippines overshadowed the ironically named Operation Stalemate. In hindsight, the genuine need for American forces to seize Peleliu can be questioned and debated. Nevertheless, this discussion should not be used to question the heroism of any, American or Japanese, who fought on Peleliu. Their blood and sweat will forever be a part of the history of this small, jagged piece of Pacific coral.

Bibliography

Alexander, J. H. 1998. What was Nimitz Thinking? *United States Naval Institute. Proceedings.*

Gailey, H. A. 1983. *Peleliu:1944.* Annapolis, MD: The Nautical & Aviation Publishing Company of America.

Gayle, G. D. 1996. *Bloody Beaches: The Marines at Peleliu.* Washington, DC: Marine Corps Historical Foundation.

Hallas, J. H. 1994. *The Devil's Anvil: The Assault on Peleliu*. Westport, CT: Praeger.

Hunt, G. P. 1946. *Coral Comes High.* New York: Harper & Brothers.

Joint Intelligence Study Publishing Board, 1944. *Joint Army-Navy Intelligence Study of the Palau Islands, Chapter 1: Brief.*

"Photomap of Peleliu" 1944. Department of Defense Map.

Ross, B. D. 1991. *Peleliu: Tragic Triumph.* New York: Random House.

"Sheet 37 of 40: Peleliu" 1944. Department of Defense Map.

Sledge, E. B. 1981. *With the Old Breed at Peleliu and Okinawa*. New York: Oxford University Press.

Watkins, R. B. 1999. *Brothers in Battle: One Marine's Account of War in the Pacific*. Amston, CT: Published by the Author.

Watkins, R. B. 2000. Interview with Author, April 2000.

Aerial photograph of Peleliu taken in 1946.
Source: National Archives photograph.

6

Protecting the Force:
MEDICAL GEOGRAPHY AND THE BUNA CAMPAIGN, WWII

By

Eugene J. Palka
and Francis Galgano, Jr.

Introduction

Army leaders at all echelons are responsible for conducting risk assessments as an integral part of designing, planning, coordinating, and executing training, regardless of whether it occurs at home station, elsewhere within the continental United States, or overseas. Leaders have the same inherent responsibility when planning and orchestrating operational missions at home and abroad in both peacetime and war. An integral part of the planning process involves addressing the specific environmental health hazards that may be encountered in a specific place.

As a subdiscipline of geography, medical geography dates back to the late 18th century in England. The field gained significant credibility within American academic geography by the late 1940s. Two early founders of medical geography were Jacques May and Ralph Audy. Both developed much of their expertise during the course of their military experience.

Jacques May was educated in Paris and subsequently practiced medicine in Southeast Asia in the 1920s and 30s. He also practiced in Vietnam prior to the war in Indo-China, and was the head French doctor in Hanoi. May was the first to describe the cultural ecology of disease. He recognized that every disease revealed a specific distribution pattern, or geography, and that human behavior was an integral part of any disease-complex. May was a prolific writer and many of his early works laid the foundation for medical geography, earning him the distinction as the "father" of the subfield in the US.

Ralph Audy traveled with the British Army throughout Southeast Asia, India, and Malaysia. He was a medical doctor who discovered the etiology of scrub typhus in Malaysia, and later served as the head of the Institute of Medical Research in that country. Audy later came to the US and served as the head of several international research foundations. His concept of "health" has been adopted throughout the medical geography community and by the World Health Organization.

As Jacques May (1958) concluded, specific environments present specific environmental hazards. All diseases have specific etiologies that can be described using basic physical geographic criteria. Primary consideration must therefore be given to identifying potential disease hazards, and subsequently protecting soldiers from acquiring diseases via implementing a wide range of preventive measures (such as immunizations, information briefings and pamphlets, protective clothing, specific types of personal hygiene, etc.). In the event that a disease is contracted, then proper treatment must be administered by trained medical personnel in a timely fashion, and all efforts must be made to prevent the spread of the disease.

Assessing the health hazards of a particular region from a military perspective is yet another aspect of effective area analysis during operational planning. Medical geography specifically focuses on the interaction between people and their environment and accentuates the activities that expose soldiers and civilians to health risks. This particular subfield of geography is unquestionably relevant to the military personnel from a professional development and practical perspective.

Medical geography uses the concepts and techniques of the discipline to investigate a wide range of health related topics. It explains the distribution of health and disease and identifies efficient ways to intervene and distribute trained personnel and technology to provide effective health care. The subfield of medical geography is closely related to epidemiology, the study of epidemics and epidemic diseases. Epidemiology, however, tends to be focused on one or two scales. By comparison, medical geography is much broader-covering multiple scales that may range from a single house to a culture realm. Moreover, medical geography employs all of the systematic geographies at various scales over time, and as such, is extremely integrative in its approaches. Finally, the scope of medical geography is not restricted to the presence or absence of disease, but considers the entire range of health concerns in a spatial context.

History is replete with examples of battles, campaigns, and wars, where far more casualties were attributed to disease than to enemy fire. This chapter provides an overview of medical geography and recalls the Buna-Gona Campaign of World War II as a means to emphasize the relevance of this particular subfield to military operations and warfighting.

Health Versus Disease

As a point of departure, we will consider the basic notion of health, a fundamental concern of all military units in virtually any context. Health is a state of complete physical, mental and social well-being, and not merely the absence of disease or infirmity (Meade and Earickson, 2000). Health is perhaps best understood as a continuing property that can be measured by an individual's ability to rally from a wide range and considerable amplitude of insults (Audy, 1971). The latter include the impacts stemming from chemical, physical, infectious, psychological, and social stimuli. Insults may be best understood as an array of negative environmental affects that impact on a person and trigger a bodily response. Physical insults could refer to air quality, temperature, humidity, light, sound, atmospheric pressure, and trauma - just to name a few. Chemical insults might include pollen, asbestos, various pollutants, smoke, or even food. Infectious insults would include virus, rickettsia, bacteria, fungi, protozoa, and helminthes. And, psycho/social insults could involve crowding, isolation, fear, anxiety, excitement, humor, and/or alienation. Regardless of location, occupation, or lifestyle, every person is subjected to a continual range of insults from birth until death. Consequently, health is regarded as an ever-present measure of a person's total vitality. It is easy to see that given the range of insults mentioned above, soldiers are routinely exposed to each and all of the possibilities during both peacetime training and war.

Disease refers to the alteration of living cells or tissues in such a way so as to jeopardize their survival in their environment (May, 1961). Medical geographers are concerned with several broad categories of disease. Congenital diseases are those which are present at birth. Chronic diseases are present or recurring over a long period of time. Degenerative diseases involve the impairment of an organ or the deterioration of its cells and the tissues. Finally, infectious diseases result from an invasion of parasites and their multiplication in the body. Given the various categories above, the military is most concerned with infectious diseases, because of the relative ease that they can be contracted then subsequently spread throughout a unit. Moreover, given the routine medical screening that all soldiers undergo early in their career, the first three categories are rarely applicable to career soldiers, or even draftees.

Diseases may also be classified based on the conditions or contexts within which they occur. A deficiency disease comes about when one's diet is not varied enough to satisfy all of the body's demands. Substances important for health may be missing from the person's food intake for cultural or physical reasons. *Kwashiorkor* is a widespread example, and is due to a shortage of protein in the food eaten by some of the poverty-stricken residents of rice growing areas of Africa and Southeast Asia. An occupational disease is caused by conditions at the work place. For example, the chest disease, *silicosis*, has long been related to coal mining and the inhalation of harmful coal dust. In some cases, the disease may become quickly apparent, while at other times, symptoms do not appear for many years after the occupation. Stress diseases occur when environmental pressures are too great for an individual to bear. Overcrowding in an urban area or one of many different types of social pressures could serve as the root of the problem. Senescent diseases are due to old age, rather than to environmental factors. Several types of heart failure would fall into this category. Obviously, the military is not concerned with senescent diseases, however, occupational and stress diseases are prevalent in wartime.

A third classification scheme that is routinely employed in undeveloped or under-developed countries of the world is based on the relationship between the disease and water. Water-borne diseases are ingested. Examples include typhoid, cholera, diphtheria, polio, and bacillary dysentery. Water-washed diseases are those that can be avoided with proper hygiene and other preventive measures. Typhus, plague, and various intestinal worms are examples of potential water-washed diseases that commonly occur under unsanitary conditions. With water-based diseases people do not acquire the disease directly from the water, but the vector requires water. Specific examples of diseases and their associated vectors include malaria (mosquito), river blindness (black fly), and schistosomiasis (snail). Given the global nature of the US Military's mission, each of the above water related disease categories are of a special concern.

Key Terms Related to Disease Ecology

As is the case with any discipline, the lexicon of medical geography includes many basic concepts that must be understood and applied in the appropriate context. The following terms are interrelated and fundamentally important to the medical geographer, as he/she seeks to understand how diseases are acquired and spread to different locations. An *agent* is a disease-causing organism, including animals and viruses, ranging in size from one-celled organisms to parasitic worms and insects. A *host* is the organism infected by the disease agent. After being afflicted by an infectious disease, a host (person, animal, bird, or arthropod) supports the disease organism by providing lodgment or subsistence. A *reservoir* refers to a large number and concentration of hosts in a population, from which a disease may expand or diffuse. The infectious agent normally lives and multiplies within the reservoir. Thus, a reservoir may serve as a continuing source of possible infection for humans. A *vector* is a carrier of a disease and is capable of transferring the latter between hosts. The agent often goes through life-cycle changes in form within the vector. The habitat of the vector may determine the location of the disease. Biological vectors (such as insects or rodents) are alive and provide the habitat for an agent to develop or multiply within prior to becoming infective. By comparison, nonbiological or mechanical vectors such as water, soil, food, or fecal matter, are not essential for the agent's life-cycle but may serve as a vehicle for transmitting the infectious agent.

Distribution and Diffusion of Diseases

In keeping with traditional practices, medical geographers are concerned with the questions of what is where and why?-as it relates to their subfield. Locating specific places through locational analysis, defining areal extents or regions, and identifying spread mechanisms are all integral parts of research in medical geography. Several terms are especially important to understanding the areal extent and diffusion of diseases. Unfortunately for military personnel, however, the operational mission in wartime routinely requires personnel to venture into an infested area.

The term epidemic refers to a sudden and severe outbreak of a disease, leading to a high percentage of afflictions and a substantial number of deaths within a population. A disease of pandemic proportions begins regionally and then spreads worldwide (as in the case of many forms or influenza or AIDS). Endemic refers to the situation where a disease is carried by many hosts in a condition of "near equilibrium," without leading to a rapid and widespread death toll among a population. A disease that is endemic within a population may gradually sap the strength from members or at least render them more vulnerable to other maladies. In either case, the endemic disease significantly decreases life expectancy. Endemic regions are areal extents that are best understood as reservoir areas where a large percentage of a population is host to a particular disease.

Medical geographers place tremendous reliance on the use of maps for conducting locational analysis, defining regions based on specific criteria, plotting incidence of disease, and graphically portraying diffusion. Indeed, the advent of medical geography can probably be traced to medical topography that evolved in Europe during the late 18th century (Meade et al., 1988). The earliest practitioners were medical personnel who recognized the tremendous utility of analyzing the spatial aspects of disease in order to find the root problem. Maps of the yellow-fever and cholera epidemics of the late 18th and early 19th centuries were the first examples of disease maps. Jon Snow's 1854 dot map of cholera around the Broad Street water pump in London was perhaps the most famous of the early maps that convinced investigators of the utility of geography's most enduring tool.

In general, maps are two-dimensional representations of all or part of the earth's surface to scale. Many are designed to show a specific distribution of a particular phenomenon within a given area. Within medical geography, maps can address a wide range of topics ranging from specific diseases (see Map 1), to deficiencies such as malnutrition or under-nutrition, to specific physical environments that provide the optimum habitats for a particular malady. Moreover, maps can be generated at various scales, enabling geographers to examine the areal extent of the disease in a global context, or focus on the particular region of concern. A current edition of *Goode's World Atlas* (2000) provides several examples of maps (specifically cartograms) at the global scale that graphically portray calorie supply, protein consumption, life expectancy, and the number of physicians per populations size. Each of these graphic representations enables one to assess the general health of a population in a particular place. Maps can also provide a revealing historical record of a particular disease. By plotting each occurrence for a given time period within a designated area, specific changes and trends can be observed (see Map 2).

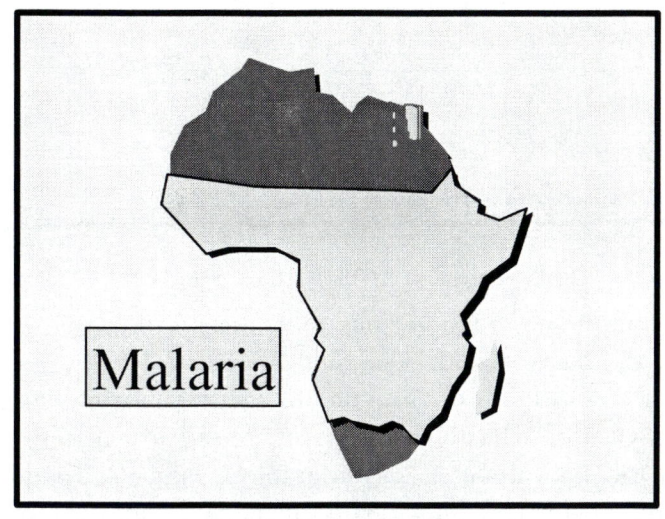

Map 1. Distribution of Malaria in Africa.

Disease Ecology

Medical geographers have a long established pattern of focusing research efforts into two major foci - disease ecology and medical health care. The latter focus is more recent and includes themes such as, health services location, facility utilization, accessibility of health care to various socio-economic groups, the relationship between health and poverty, and a number of different health-care issues in undeveloped countries. Disease ecology has a much stronger tradition, and has sought to understand the relationships between health, disease, cultural behavior, and human/environmental interactions. The spatial ecology of disease and the geographical aspects of the health of populations provide fertile ground for medical geographers,

BY THE NUMBERS

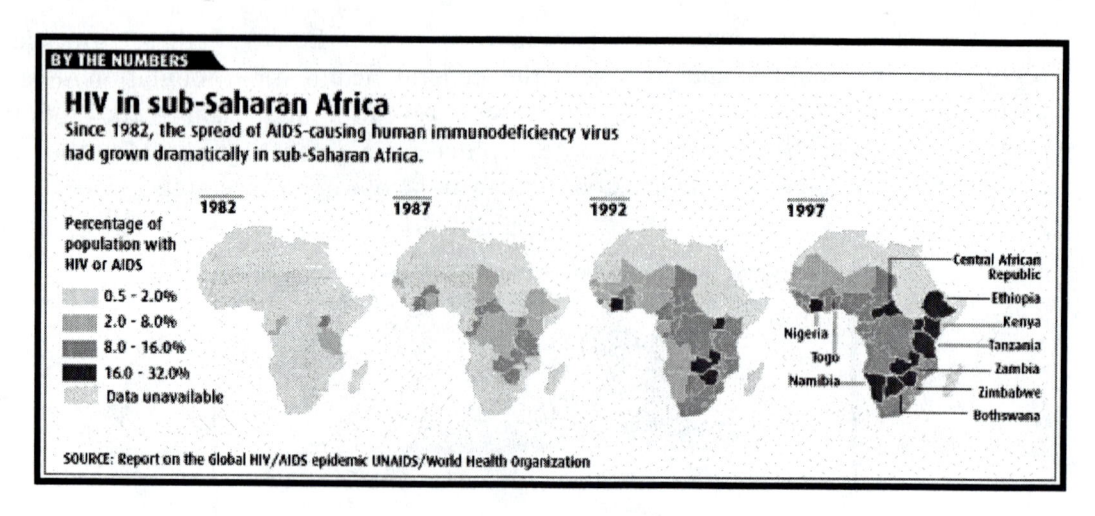

HIV in sub-Saharan Africa
Since 1982, the spread of AIDS-causing human immunodeficiency virus had grown dramatically in sub-Saharan Africa.

1982   1987   1992   1997

Percentage of population with HIV or AIDS
- 0.5 - 2.0%
- 2.0 - 8.0%
- 8.0 - 16.0%
- 16.0 - 32.0%
- Data unavailable

Central African Republic, Ethiopia, Kenya, Nigeria, Tanzania, Togo, Zambia, Namibia, Zimbabwe, Botswana

SOURCE: Report on the Global HIV/AIDS epidemic UNAIDS/World Health Organization

Map 2. Spread of AIDS in Africa from 1982-1997.

ecologists, epidemiologists, and various specialists within medical science. More often than not, the map becomes the fundamental link between the disciplines. The following series of maps enables one to compare various diseases that occur within Africa (see Map 3). Each, however, requires a different environmental niche. Compare the climate, physiography, land-use patterns, population distribution, and vegetative biomes, with the disease patterns. What questions do these patterns raise? What conclusions can you draw?

Given the patterns that emerge in the maps above, it is tempting to arrive at premature conclusions, and to recall various discussions about environmental determinism. Cultural adaptations, however, have a significant impact on whether or not the population of a particular region comes into contact with a particular disease or health hazard. Moreover, cultural attitudes and beliefs have an influence on how various groups prevent or treat different diseases. Cultural buffers refer to behavior or innovations that shield against disease or illness. The examples are too many to list, but consider the following cultural generalizations. Chinese drink tea, not water. They must boil the water first to make tea, so consequently, they experience few water-borne diseases. Europeans derive many similar benefits from drinking moderate amounts of wine instead of poor quality water. Considering culture traits, sandals have proven to be a barrier to hookworms all over the world. Another simple innovation, the handkerchief, prompted the cessation of TB in many regions.

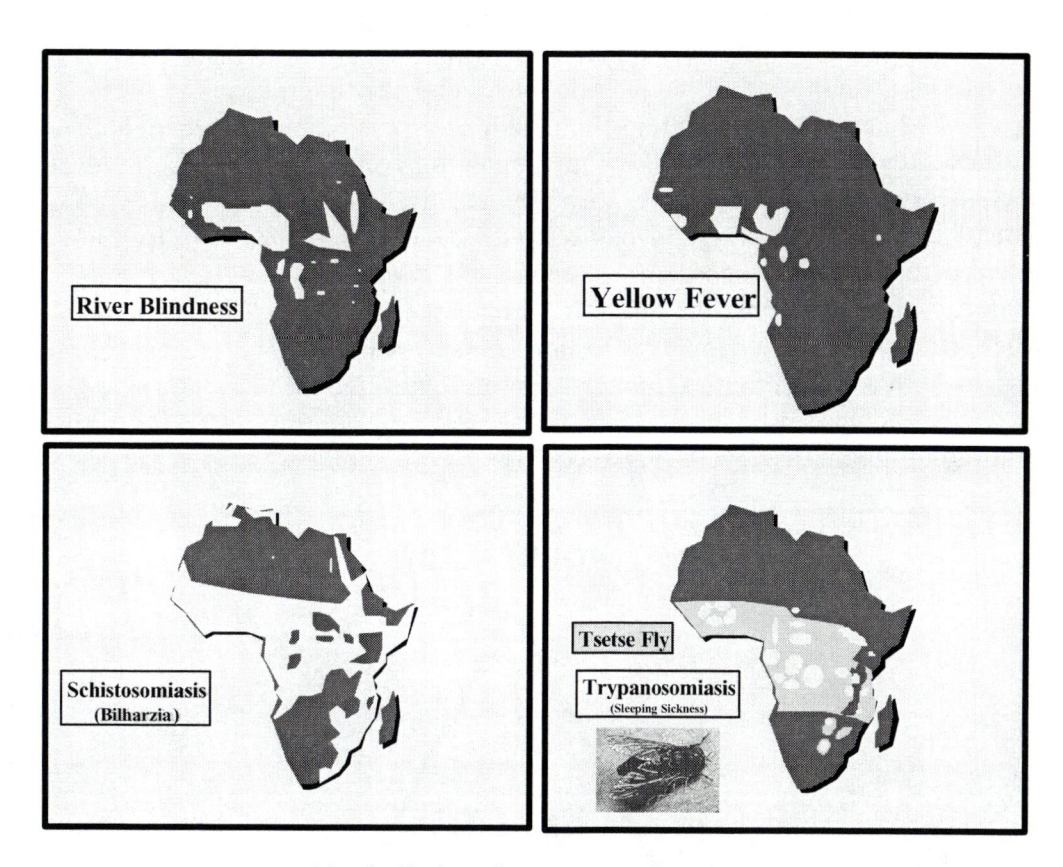

Map 3. Various disease patterns in Africa.

The Triangle of Human Ecology

One model that facilitates the understanding of the relationship between health and a particular place is the triangle of human ecology, conceptualized by Meade et al (1988). The triangle is formed by three vertices- population, behavior, and habitat, and encloses the state of health (see Figure 1).

Within the context of the above model, habitat is that part of the environment within which people live. It includes houses, workplaces, settlement patterns, recreation areas, and transportation systems. The population considers humans as the potential hosts of various diseases. Factors affecting, and yet characterizing the population include nutritional status, genetic resistance, immunological status, age structure, and psychological and social concerns. Behavior includes the observable aspects of the population and springs from cultural norms. It also impacts on who comes into contact with disease hazards and whether or not the population elects other alternatives.

The key to understanding the health of a population in a particular place is based on an appreciation of how the population, through its behavior, develops and interacts with its habitat. Virtually all diseases have an ideal habitat or natural nidus. Many exist in nature, independent of human activity, and are referred to as silent zones, unbeknownst to people until the latter come into contact with and contract the disease. The human factor, however, explains who, when, where, and why people come into contact with various diseases. Through various beliefs and activities, humans may become exposed to a disease, enhance or reduce the spread of it, or establish a variety of cultural buffers as means of controlling the disease cycle (Palka, 1995).

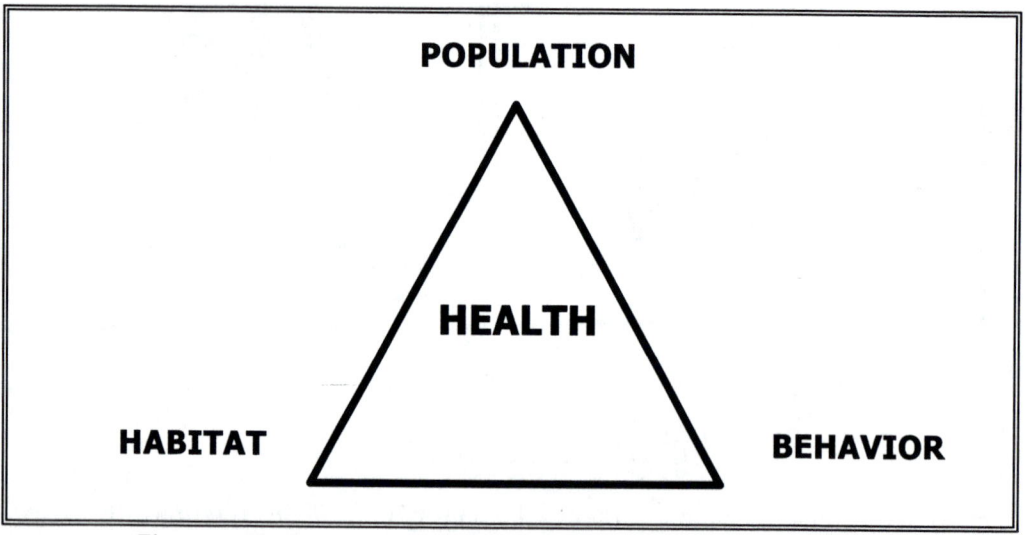

Figure 1. The Triangle of Human Ecology (After Meade et al., 1988).

Nutrition and Health

The triangle of human ecology emphasizes that the state of human health is linked to the interrelationships between people, their behavior, and the environment. "Good health" depends on much more than the successful avoidance of disease. A two-way interaction occurs between nutrition and health. A nutritional diet can enhance one's health, while deficiency diseases like malnutrition and undernutrition can have the opposite effect. Malnutrition results from deficiencies in protein, vitamins or minerals within a person's diet. Caloric intake may be adequate, but the lack of a well-balanced nutritional diet has a negative impact on the body's ability to function. Perhaps more importantly, malnutrition renders a person more susceptible to other diseases or illness, and renders the person incapable of rallying from new insults. Whereas malnutrition is related more to a lack of protein rather than inadequate food, undernutrition is directly related to a lack of calories or the quantity of food intake. Like malnutrition, undernutrition renders a person more susceptible to other maladies, and thus lowers the state of health.

Both malnutrition and undernutrition have a distinct geography. The underlying patterns are related to both environmental and cultural factors. Environmental factors, including rainfall, temperatures, soil quality, sunlight, and relief, may limit crop varieties or harvests. Agricultural productivity can be further limited by periods of drought or other natural hazards. The latter can have a devastating affect on millions of people in underdeveloped countries, who are totally dependent on subsistence agriculture.

Cultural preferences also play a part in both malnutrition and undernutrition. Given the population pressure in India, cattle could provide a much-needed source of protein, if not forbidden by Hinduism. Similarly, in various parts of the Middle East and northern Africa, hogs would provide the protein necessary to enhance the diet of many who are poverty stricken, yet Muslims will not consume pork. Even in the United States, where there are far less deficiency diseases, people go hungry rather than consume any of the dog meat from the thousands of dogs that are killed daily in animal pounds across the country. Dogs could provide a much-needed source of protein to those in desperate need; however, a cultural bias prevents Americans from even considering such an option.

On a global scale, maps of calorie supply and protein consumption reveal a distinct pattern in *Goode's World Atlas* (2000). While we could explain part of the pattern in terms of environmental factors and cultural preferences, we might also consider the impact of poverty on nutrition and health. Moreover, we might consider the availability of health care in the same locations, and then try to correlate all of the above factors with the map of life expectancy or to infant mortality rates in various countries.

With a better understanding of some of the basic concepts of medical geography, an appreciation of disease ecology, and an awareness of the impacts of nutritional status environmental conditions on health, let us now turn to an historical vignette to emphasize the relevance of medical geography to military operations. Bear in mind that the Buna-Gona campaign is only a representative example. We can surely look to virtually any war and discover specific examples of where disease and/or adverse environmental conditions rendered soldiers and units combat ineffective.

The Buna-Gona Campaign

Introduction

Papua-New Guinea (see Map 4) is the world's second largest island and is located at 8° south latitude. By late 1942 this island became a lynchpin in the Allied strategy for stopping and driving back the Japanese in the southwest Pacific (Weigley, 1973). Papua (as it was known in 1942) is a land of soaring mountain ranges, impenetrable tropical jungle, low-lying coastal swamps, boggy grasslands, swollen rivers and clouds of mosquitoes. The island's northern coast is a fever country, a landscape of torrential rains, steaming, insufferable heat, and a host of endemic tropical diseases (Martin, 1967). It was here that soldiers of the American 32d Infantry Division and 163d Regimental Combat Team, along with their Australian allies, won a pivotal battle of the Pacific War (Milner, 1989). The Americans paid a high cost for this victory. The fighting to take Buna cost the Americans some 10,879 casualties. Of these, 7,920 were casualties caused by tropical illness, poor food and water and appalling sanitary conditions. The jungle, a smothering tropical climate, illness and the Japanese were the enemy on New Guinea (Center of Military History, 1990). As one veteran recalled:

> "The men at the front in New Guinea were perhaps among the most wretched-looking soldiers ever to wear the American uniform. They were gaunt and thin, with deep black circles under their sunken eyes. They were covered with tropical sores. They were clothed in tattered, stained jackets and pants. Often the soles had been sucked off their shoes by the tenacious, stinking mud. Many of them fought for days with fevers and didn't know it . . . malaria, dengue fever, dysentery, and, in a few cases, typhus hit man after man. There was hardly a soldier, among the thousands who went into the jungle, who didn't come down with some sort of fever at least once. (Kahn, 1943)

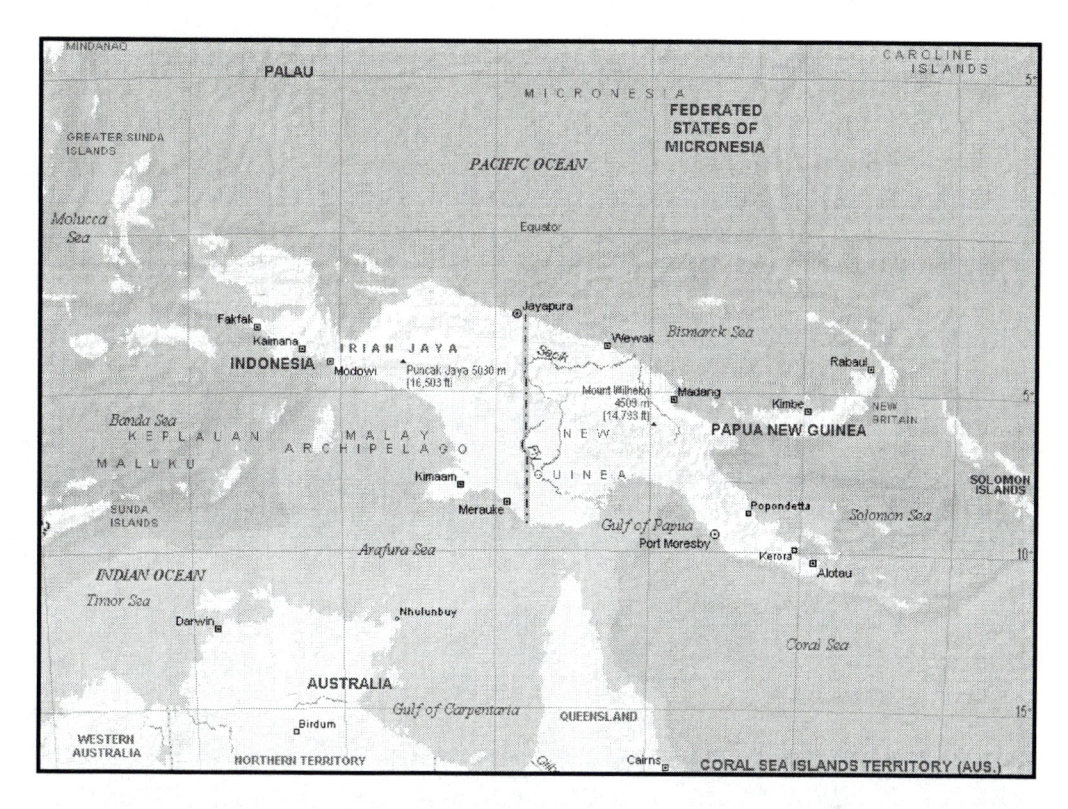

Map 4. The Island of Papua-New Guinea.
Source: Microsoft® Encarta® Interactive Atlas 2000.

Tropical disease and the stifling climate took a fearful toll of Allied soldiers. Poor nutrition and unhealthy water supplies compounded the effect of these problems. Not only did the soldiers endure malaria, dengue fever and a host of other tropical illnesses; but, they also suffered from food poisoning, dysentery, depression and lethargy (Center of Military History, 1990). The 32d Infantry Division was composed of soldiers from the American mid-west. These soldiers were not acclimated and were physically unprepared for the conditions they would face in Papua. Furthermore, their equipment and diet was ill-suited for the climate, leadership was sometimes lacking and the medical staff was unprepared to deal with the medical problems they faced. Within two weeks after entering the jungle, the rate of sickness began to quickly climb and it was not unusual for more than half the soldiers in an infantry battalion to be in the hospital with a tropical malady, festering skin ulcers, dysentery, dehydration, food poisoning or upper respiratory infection. The casualty figures indicate that for every two battle casualties, five soldiers were out of the front line because of a health related problem (Milner, 1989). It is remarkable that these soldiers were able to finally drive the Japanese out of Papua considering the harsh conditions and rates of non-battle casualties, which approached 80% in some units (Center of Military History, 1990).

Photograph 1. American soldiers in the dense tropical jungle.
Source: U.S. Army photograph.

The Geography and Climate of Papua

A brief description of Papua's geography and climate is useful to explain the conduct of the battle and why disease played such a major role during the operation. The campaign took place on the elongated tail of the island Papua-New Guinea (see Map 4) between Port Moresby and Buna, with the heaviest fighting taking place along the narrow strip of land along the northeastern coast of Papua extending on both sides of Buna (see Map 5). The coastal plain was a mix of jungle, impassable swamp, coconut plantations, and open fields of shoulder-high kunai grass (Center of Military History, 1990). In most places the vegetation was very dense and it made for an ideal breeding ground for vectors such as mosquitoes, chiggers, fleas, ticks, rats and biting flies (Milner, 1989).

The coastal plain around Buna is low-lying and seldom exceeds five feet above mean sea level. Hence, the water table was very high and the prodigious rainfall simply inundated the surface. Further, the lack of relief meant that rivers were wide, sluggish and easily flooded. Waterlogged land is pervasive and these swamps extend to the coast in many places. The highest ground is actually found along the coast and was occupied by the Japanese. Hence, Allied soldiers were forced to live and fight in fetid swamps or on a few patches of dry ground where the water table was only six inches below the surface. Consequently, Allied soldiers lived in a state of perpetual wetness. Their uniforms and boots rotted, skin ulcerated, food spoiled and weapons rusted. Although there was plenty of water, it was undrinkable (Kahn, 1943). These conditions are highly conducive to the spread of a host of tropical diseases and health related problems.

The region had no roads or other infrastructure except along a narrow coastal margin. Dense jungle and swamps made mobility on foot very difficult and virtually impossible by vehicle (Martin, 1967). Hence, the distribution food, water and medical supplies was difficult. It was typical for units to go several days without food, further wearing down resistance to disease and infection. Extensive engineering effort was needed to carve roads out of this dense jungle (Milner, 1989). Therefore, the distribution of fresh water, food and medical supplies was accomplished by native porters, but this method was ineffective at best (Kahn, 1943). Heavy rains would typically inundate most of the jungle trails slowing the forward movement of supplies to a trickle. Moreover, the dense vegetation made air drops an uncertain proposition -- only 50% of the drops reached Allied soldiers (Milner, 1989).

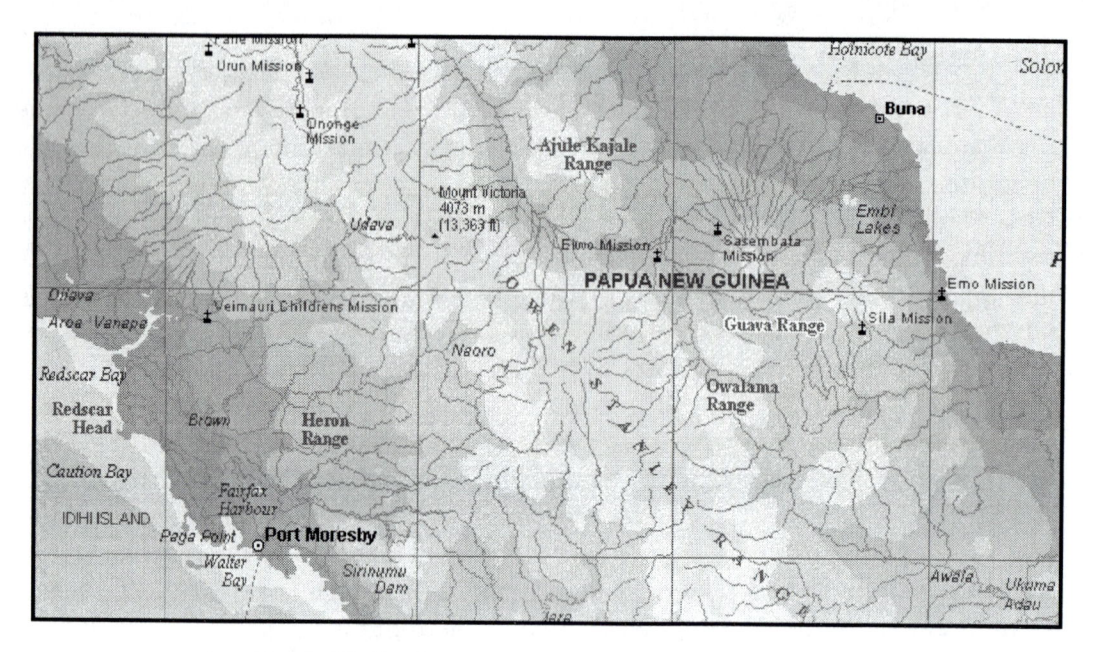

Map 5. The battle area during the Buna-Gona Campaign.
Source: Microsoft® Encarta® Interactive Atlas 2000.

Together with difficulties related to the region's terrain and vegetation, Allied soldiers were compelled to cope with the physical problems associated with Papua's tropical climate. Buna has a tropical rainforest climate (see Figure 2) with warm average monthly temperatures. Daily temperatures average near 85°F throughout the year with daily maximums that hover near 100°F. The temperature and humidity conditions at Buna were such that soldiers experienced increasing physical discomfort (sensible heat) for each 1° rise in temperature. The climate problem was compounded because most of the fighting took place between November and January (summer in the Southern Hemisphere); a period when precipitation, temperature and humidity are highest (see Figure 2). For example, during the month of December the daily temperature range extends between 72 and 89°F, humidity averages 82%, and the area typically receives nearly 16 inches of rain on average (Center of Military History, 1990). The effect of these conditions is evident in an entry of the field journal of Company F, 2nd Battalion, 128th Infantry Regiment for 9 December 1942, *"What is left of the company is a pretty sick bunch of boys. It rained again last night, men all wet, and sleeping in the mud and water."* (Milner, 1989).

CLIMOGRAPH: BUNA, NEW GUINEA

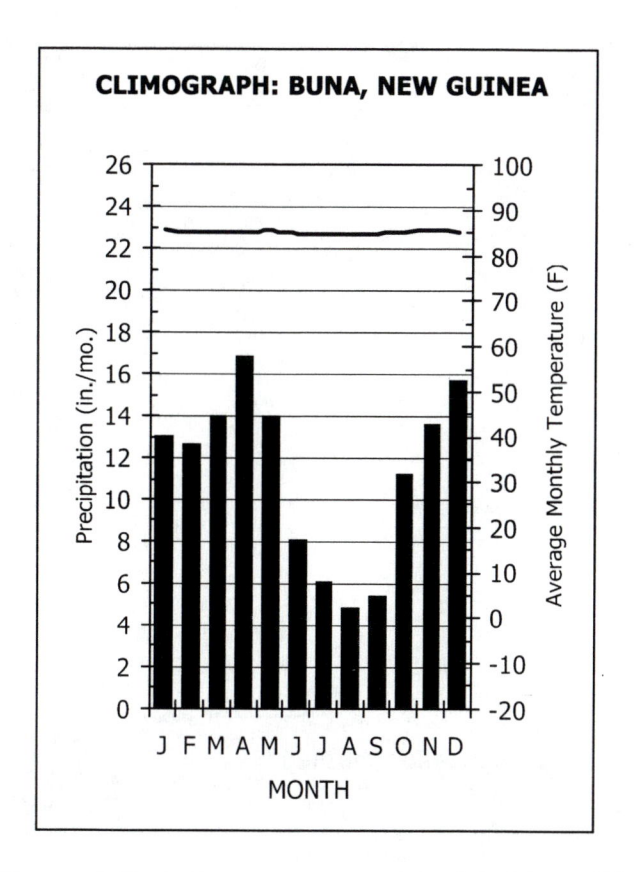

Figure 2. Climograph illustrating temperature and precipitation conditions at Buna. Source: National Climate Data Center, NOAA.

Campaign Summary: The Fight for Papua

On 22 January 1943, elements of the U.S. 32d Infantry Division captured Buna. This was the culmination of a grueling, 5-month counteroffensive to drive the Japanese back across the Owen Stanley Range, and in so doing blunting their planned invasion of Australia (Milner, 1989). The combined Australian-American force fought its way through rugged mountain rainforest and then across the dank, coastal swamps and jungles of Papua. This campaign was fought concurrently with that on Guadalcanal. Both were intended to stop Japanese eastward expansion in the South Pacific along the axis of the Solomon Islands and secure our lines of communications with Australia (see Map 6).

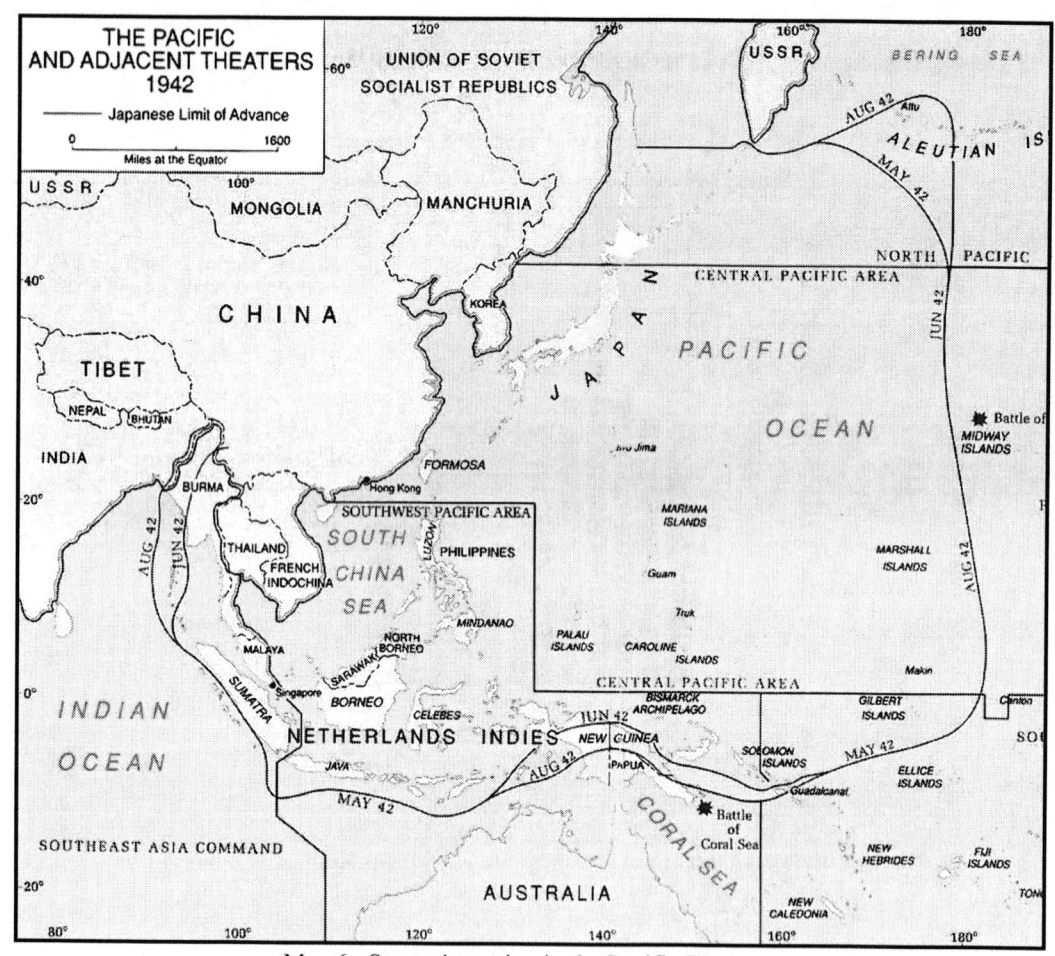

Map 6. Strategic setting in the Pacific Theater.
Source: Army Center of Military History (1995).

Strategic Overview: Prelude to the Buna-Gona Campaign

By late spring 1942 Allied prospects in the Pacific were bleak. Notwithstanding the "Doolittle" bomber raid on 18 April, the Allies were confronted with a litany of military disasters: 250,000 soldiers surrendered at Singapore on 15 February; the British evacuated Rangoon on 7 March; the Dutch surrendered Java on 8 March; on 9 April Bataan surrendered; and by 29 April the Japanese controlled Burma. The Japanese Empire extended over a vast expanse of the Pacific and Southeast Asia (see Map 6). In May 1942 a confident and virtually unstoppable Japanese military machine launched its long-planned sea and land operation against Australia to administer the *coup-d'gras* to Allied presence in the western Pacific. The first step in their plan was the seizure of Port Moresby, along the southern coast of Papua-New Guinea (see Map 5). Seizing Port Moresby would provide the Japanese with a staging area within striking distance of Australia, as well as bomber and fighter bases to cover their crossing of the Coral Sea.

On 1 May an invasion fleet that included four fleet carriers steamed from the Japanese naval base at Truk Atoll with the mission of landing Imperial Marines at Port Moresby. The fleet made passage through the New Britain Straits and into the Coral Sea where it was stopped by a small American carrier group. The so-called *Battle of the Coral Sea*, fought on 7 and 8 May 1942, was important for a number of reasons. First, it marked the first defeat of the relentless Japanese drive across the Pacific since Pearl Harbor. Second, it was the very first sea battle fought entirely between aircraft carriers. Third, it turned back the planned Japanese sea-borne invasion of Port Moresby and perhaps saved Australia (Keegan, 1989).

Initial Fighting on Papua-New Guinea

Although stunned by their defeat in the Coral Sea, the Japanese did not take long to continue their offensive against Port Moresby. They landed an invasion force near Buna on 21 July with the mission of taking Port Moresby from the landward side by crossing the formidable Owen Stanley Range (see Map 5). General Hori led this Japanese force across the island on a backbreaking march through the mountainous jungles, losing half of his force to the elements and disease along the way (Milner, 1989). When the Japanese reached Iwibiwa Ridge near Port Moresby in early September, they were a mere shadow of the force that departed Gona the previous summer -- mainly resulting from the effects of the oppressive climate, sickness and the physical exertion of crossing the Owen Stanley Range. On 18 September, General Hori ordered a retreat to the north coast. The Allies subjected the Japanese to intense bombing and a land pursuit by the Australian 7th and U.S. 32d Infantry Divisions (Center of Military History, 1995).

By 10 November, the Japanese were back at their start point near Buna Mission (see Map 7). However, the tables were turned -- now it was the Allied force that was exhausted and at the end of a tenuous supply line, fighting a skilled enemy that occupied favorable terrain near its base. The Allies, began their offensive near Port Moresby and were forced to cross the Owen Stanley Range in pursuit of the Japanese using narrow jungle trails over the mountains. In doing so they were compelled to fight not only the Japanese, but also with oppressive temperatures, heavy rainfall and difficult terrain, as well as a host of tropical diseases for which they were ill prepared (Keegan, 1989). Although ultimately victorious, the 32nd Infantry Divisions would pay a heavy price to drive the Japanese out of southeastern Papua-New Guinea. By the end of the battle, this 13,645-man force suffered 10,879 casualties; however disease would account for 7,920 (Center of Military History, 1990). Similarly, our Australian allies incurred disease-related casualties at a rate of six soldiers for every three killed by direct enemy action (Winters, 1998). A complete listing of American casualties is given in Table 1.

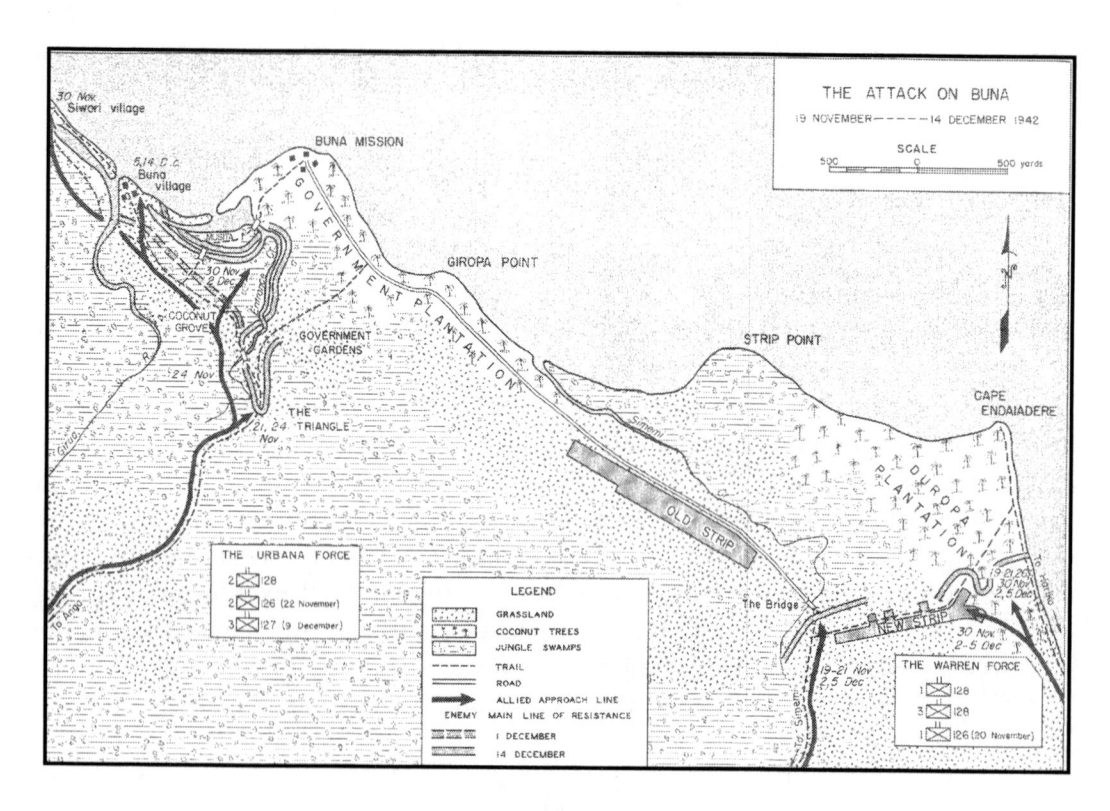

Map 7. The battlefield around Buna.
Source: Army Center of Military History (1990).

Table 1. I Corps Casualties during the Buna-Gona Campaign (Sep. '42 - Jan. '43).

Regimental Combat Teams	Beginning Strength	Killed in Action	Other Deaths	Wounded in Action	Sick in Action	Total Casualties
32d Infantry Division						
126th Infantry	3,791	266	39	816	*2,285*	3,406
127th Infantry	2,734	182	32	561	*2,813*	3,588
128th Infantry	3,300	138	29	557	*2,238*	2,962
41st Infantry Division						
163d Infantry	3,820	85	16	238	*584*	923
TOTAL	13,645	671	116	2,172	**7,920**	10,879

Source: Center of Military History (1990).

Disease, Health and the Buna-Gona Campaign

The casualty figures in Table 1 are quite revealing. It is clear that the United States I Corps was facing an unmitigated disaster during the fighting for Buna in terms of non-battle casualty rates and unit effectiveness. There were periods when commanders were uncertain if they could continue the fight because most of the battalions and regiments were actually in the hospital, or too weak to fight effectively (Martin, 1967). The reasons for these exceptionally high non-battle casualty rates are explainable and are in fact understandable, considering the state of our knowledge of medical geography and medical technology in 1942. This void was aggravated when combined with the conditions (i.e., terrain, climate and weather) that Allied soldiers endured in this campaign. The hot and exceedingly wet climate is a near-perfect breeding ground for a host of infectious agents and vectors -- some of which westerners had frankly never encountered up to this point (Milner, 1989). Papua experiences in excess of 135 inches of rain per year, and Allied soldiers fought and wallowed in fetid swamps and jungles along the north coast during the height of the rainy season. It was not uncommon to experience several inches of rain per day. This much rainfall combined with poorly drained soils and a very high water table turned fighting positions and bivouac areas into cesspools. As time went on, Allied soldiers became increasingly susceptible to disease and infection as the crushing heat and pervasive moisture drained their energy and weakened their immune systems. Further, in these conditions, minor injuries quickly festered and required extensive medical treatment for even the healthiest individuals (Kahn, 1943).

These debilitating conditions were further exacerbated by logistical and leadership failures. Once they began fighting the Japanese near Buna, the Allies were at the end of a very long logistical tether in a region with virtually no infrastructure (see Map 7). Although General MacArthur ordered an airlift, the pervasive cloud formations along the crest of the Owen Stanley Range, persistent rain, and dense tropical forest limited its effectiveness (Milner, 1989). Food and medical supplies were in increasingly short supply. Water purification equipment became almost as important as tanks and artillery ammunition in Allied supply priorities. Ironically, for a region with so much water, clean drinking water was a precious commodity. Local tribes used rivers as open sewers and fouled nearly every stream (Kahn, 1943). Other surface water supplies were equally unsafe because the water contained protozoa that caused dysentery that was so insidious, it could reduce a 500-man battalion to platoon strength in a matter of days. Water-borne disease accounted for almost half of the disease-related casualties in this campaign (Center of Military History, 1990). The combined effects of these conditions were profound. The average daily sick call rate in a World War II American division was 3.8% of the division's strength (roughly 15,000). The average daily sick call rate for the 32d Division during the Buna-Gona Campaign was 24%. The I Corps commander, MG Eichelberger, was shaken by these non-combat casualty rates. In some cases units were reduced to 50% strength and had not even fired a shot in anger. He observed that the 128th Infantry Regiment was "... riddled with malaria, dengue fever [and] tropical dysentery..." (Milner, 1989).

Leadership, Water and Disease

In this setting, discipline and leadership were critical, yet a host of water-borne and water-washed diseases were manifest because of lapses in this area. Strict water and hygiene discipline was absolutely necessary in this environment, but the realities of warfare too were stacked against the soldiers of the 32d Division. Combat tends to take a heavy toll of junior leaders, and as conditions deteriorated and increasing numbers of junior leaders were killed in action, discipline eventually broke down. Even ordinary things like water consumption resulted in large numbers of combat ineffective soldiers. Soldier hygiene and field hygiene standards quickly eroded as enemy action decimated the ranks of junior leaders and environmental conditions wore down soldier resolve and morale. An entry in the field journal of the 2nd Battalion, 128th Infantry Regiment for 12 December 1942 illustrates this point, *"The men haven't washed for a month, or had any dry clothing . . . "* (Milner, 1989).

Large numbers of soldiers stopped shaving and washing; and dirty uniforms quickly became filthy, but were not exchanged because clean ones were not available. Latrines were not prepared properly, or in many cases not at all, further contaminating the water supply. As the supply situation broke down, soldiers began drinking water out of local streams untreated. Even when chlorine tables made their way to the front, water proof storage bags were a rarity and the tablets disintegrated before they could be used (Kahn, 1943). Clearly, dysentery was the leading affliction in American units (Martin, 1967). Fresh, purified water was in short supply as were chlorine tablets. Furthermore, water heating equipment could not make it forward into the combat zone and there was no way for soldiers to properly clean cooking and eating utensils. In this type of hot and humid environment, these sorts of lapses were a recipe for the spread of debilitating water-washed and water-borne sickness among large numbers of combat soldiers. The acute diarrhea associated with dysentery became so bad that soldiers simply slit the backs of their trousers because they could no longer control their bowels (Milner, 1989). Eventually, most of them dehydrated and collapsed requiring weeks of hospitalization.

The Medical System and Technology Fails

The campaign in Papua was the first of its kind for Americans in World War II. In reality, Allied soldiers came into contact with endemic diseases of which they had no prior knowledge. The region revealed a host of endemic diseases such as malaria, dengue fever, scrub typhus, bacillary, and amebic dysentery. These maladies were complemented by conditions such as jungle rot, dhobic itch, ringworm and athletes foot (Milner, 1989). Soldiers quickly became more susceptible to these illnesses as their immune systems wore down with exhaustion and poor nutrition. In most cases the medical system was simply unprepared or at best overwhelmed. For example, the soldiers of the 32d Division were inoculated for European Typhus prior to their deployment or

Papua. This was problematic because Scrub Typhus existed in this region of the world and the initial inoculation had no effect on this disease.

The hot, humid climate was a haven for vectors, and mosquitoes, ticks, chiggers and flies inhabited the tropical forests and swamps in prodigious numbers. Nonetheless, American units did not have adequate insect repellents and mosquito netting was in short supply; hence it was difficult to break the cycle of disease at the lowest, simplest level (Kahn, 1943). Further, there was a general shortage of Quinine and Atabrine in the Pacific, and when supplies could be brought forward, they quickly broke down because waterproofing methods were ineffective. In Papua, soldiers were issued these tablets in small paper pouches, which offered no protection in the near saturated conditions. The existing malaria remedies merely treated symptoms and the malaria would recur with increasing intensity requiring many soldiers to be evacuated completely. Finally, the hospitals that could move forward had inadequate sterilization equipment and infection was rampant (Center of Military History, 1995).

Nutrition

Fighting in a tropical climate is exhausting business. Soldier health and unit efficiency can only be maintained if they are provided with the proper food in sufficient quantities. This was a major shortcoming in Papua. The supply system, as already discussed, was inadequate at best. It was not uncommon for soldiers to go without food for two and three days at a time (Martin, 1967). Furthermore, the diet was not suited for the conditions, and the method of storage was inadequate. The primary component of the Allied diet was an Australian canned beef product called "Bully Beef" by the soldiers. This notorious fare was packed in an oil that broke down in the high temperatures of Buna's jungles, tasted like fish oil, and caused soldiers to wretch (Kahn, 1943). Moreover, the storage tins quickly rusted along the seams because of the extreme humidity allowing the beef to spoil. Hence, the starving soldiers had no choice but to eat spoiled, wretched food that brought on a number of gastro-intestinal problems, dysentery and food poisoning. Those soldiers that survived those problems were still vulnerable to other illnesses as their general resistance eroded as a result of the poor nutrition (Milner, 1989).

Equipment

In 1942, American soldiers did not have specialized uniforms and equipment designed for tropical environments. The standard infantry uniform was made of wool, which is entirely unsuitable for Papua's climate. Wool retains moisture and is an excellent insulator; both undesirable properties in a jungle. A decision was made to have each soldier's uniforms dyed in a mottled green pattern to achieve some level of individual camouflage. This was accomplished by a local Australian contractor (Milner, 1989). This seemingly logical idea caused a great deal of difficulty for the soldiers of the 32d Division. First, the dye ran once it was exposed to the humid conditions and heavy rain in the Papuan jungle. This caused serious skin rashes, which quickly became infected. Second, the dye stopped-up the pores in the uniform material and they became unbearable in the extreme tropical heat (Milner, 1989). Heat exhaustion and severe dehydration were quick to follow.

As already mentioned, waterproofing methods were inadequate. Soldiers could not be issued more than one or two days worth of vitamins and other medications because the pills would simply disintegrate in the humid conditions (Martin, 1967). The most basic items such as soap, socks, underwear, oral hygiene material, shaving equipment and mosquito nets were in continual short supply. As the supply system deteriorated, clean uniforms were considered luxuries that received very low priory in light of the heavy demands for food and ammunition (Milner, 1989). Water purification equipment was in short supply and was too heavy to be brought forward. The net result was low morale and poor hygiene that permeated all ranks in the front lines.

The High Command Responds to the Medical Disaster

By early December 1942, the influence of tropical conditions on soldier health and diseases such as dysentery, dengue fever and malaria combined with poor water supplies and bad nutrition severely eroded the combat effectiveness of the 32nd Infantry Division. In most cases 250-man companies were reduced to platoon strength. The commander of the 2nd Battalion, 128th Infantry Regiment noted in his unit history that, *". . . all men display unmistakable signs of exhaustion and sickness."* (Milner, 1989). By the third week of December 1942 the division was nearly combat ineffective. It was at that point the Corps Commander, LTG Eichelberger, stepped in and established uncompromising hygiene and uniform polices to improve the health, discipline and morale of the beleaguered infantrymen. He enforced shaving and washing requirements, and improved supply to ensure that there was ample clean water, quality hot food and clean clothing for his soldiers. He redoubled efforts to airlift supplies into the combat area and demanded the improvement of trails to permit vehicle movement to bring forward hot food. These measures had an almost immediate effect on the overall strength of the 32nd Infantry Division. By early January, the division had regained sufficient strength and vitality to at last destroy the remaining Japanese forces around Buna Mission (see Map 7). Although LTG Eichelberger was able to alleviate the incapacitating effects of poor water supplies, dreadful hygiene standards and nutrition problems to a degree, he had no easy remedy for the pervasive number of infectious agents in the region. American soldiers would continue to incur diseases such as malaria, typhus and dengue fever until technology and our knowledge of medical geography improved.

Conclusion

It is often noted that the war in Papua was a war of mud, mountains, malaria and mosquitoes. The cost of taking Buna was staggering (see Table 1) although the tally of killed and wounded to enemy action was relatively small. The war on Papua-New Guinea would drag on for another year until the island was cleared in September 1943. However, by that time equipment, medicine and our knowledge of disease ecology improved to the extent that the near medical disaster experienced in the early fighting near Buna was not repeated.

Combat operations are challenging and the military operating environment (terrain, climate and weather, cultural landscape) further stresses the demand on human endurance and unit effectiveness. In this scenario disease and sickness can destroy a unit's effectiveness even before shots are fired in anger. Force protection is an important aspect of all operations -- clearly we must strive to minimize non-battle casualties in all cases. Although the Buna-Gona Campaign appears extreme on the surface, it is actually a typical case. For example, in the World War II China-Burma-India Theater 90% of all casualties were the result of disease and sickness. Likewise in the Southwest Pacific Theater 83% of all casualties were disease related. In spite of our vast improvements in medical technology and geographic awareness some 67% of all casualties in Vietnam were related to the depredations of disease and sickness (Collins, 1998). As this study of the campaign to take Buna-Gona all too clearly shows, medical geography is fundamentally relevant to military operations in any context.

Bibliography

Audy, J.R. 1965. Types of Human Influence on Natural Foci of Disease. In B. Rosicky & K. Heyberger (Eds.), *Theoretical Questions of Natural Foci of Disease*. Prague: Czechoslovak Academy of Science, 245-253.

Audy, J.R. 1971. Movement and Diagnosis of Health. In P. Shepard & D. McKinley (eds.). *Essays on the Planet as a Home*. Boston: Houghton Mifflen, 140-162.

Benenson, Abrams, ed. 1995. *Control of Communicable Diseases Manual.* FM 8-33. Washington, D.C.: American Public Health Association.

Center of Military History. 1990. *Papuan Campaign*. Washington, D.C.: U.S. Government Printing Office.

Center of Military History. 1995. *Papua*. Washington, D.C.: U.S. Government Printing Office.

Collins, John, M. 1998. *Military Geography for Professionals and the Public*. Washington, D.C.: National Defense University Press.

Gould, Peter. 1989. Geographic Dimensions of the AIDS Epidemic. *The Professional Geographer* 41 (1): 71-78.

Griess, Thomas E. (ed.), 1984. *The Second World War, Asia and the Pacific. The West Point Military History Series*. Wayne, New Jersey: The Avery Publishing Group.

Hunter, John M. and Young, Johnathan C. 1971. Diffusion of Influenza in England and Wales. *Annals of the Association of American Geographers* 61 (4): 637-653.

Kahn, E.J. 1943. *G.I. Jungle*. New York, New York: Simon and Schuster.

Keegan, John. 1989. *The Second World War*. New York, New York: Penguin Books.

Martin, Ralph, G. 1967. *The GI War*. Boston, Massachusetts: Little, Brown and Company.

May, J.M. 1950. Medical Geography: Its Methods and Objectives. *Geographical Review* 40: 9-41.

May, J.M. 1958. *The Ecology of Human Disease*. New York: MD Publications.

May, J.M. 1961. *Studies in Disease Ecology*. New York: Hafner.

Meade, Melinda S., Florin, John W., and Gesler, Wilbert M. 1988. *Medical Geography*. New York: The Guilford Press.

Meade, Melinda, S. and Earickson, Robert. 2000. *Medical Geography*. New York: The Guilford Press.

Milner, Samuel. 1989. *Victory in Papua. Center of Military History, United States Army in World War II, The War in the Pacific*, Washington, D.C.: U.S. Government Printing Office.

Palka, Eugene J. 1994. North Carolina: Natural Nidus for Rocky Mountain Spotted Fever. *The North Carolina Geographer*. Volume 3: 1-16.

Pyle, G.F. 1969. The Diffusion of Cholera in the United States in the Nineteenth Century. *Geographical Analysis* 1: 59-75.

Stock, Robert F. 1976. *Cholera in Africa*. London: International African Institute.

Thomas, Richard. 1992. *Geomedical Systems: Intervention and Control*. London and New York: Routledge.

Weigley, Russell, F. 1973. *The American Way of War*. Bloomington, Indiana: Indiana University Press.

Winters, Harold, K. 1998. *Battling the Elements*. Baltimore, Maryland: Johns Hopkins Press.

The MOOTW Context

Introduction

The fall of the Berlin wall and the demise of the former USSR have prompted significant changes in the National Security Strategy of the United States (Shalikashvili, 1995). The dramatic shift in strategic orientation has warranted substantial changes in the size, force structure, and disposition of the US Military. Moreover, although the numbers of forward deployed units have decreased considerably, the country's involvement in military operations other than war (MOOTW) has increased at an unprecedented rate. Between 1989 and 1997, the Military participated in 45 operations other than war, more than triple the total number conducted during the entire Cold War era from 1947-1989 (Binnendijk, 1998).

The end of the Cold War has provided an opportune time to re-examine the scope of military geography (Anderson, 1993; Palka, 1995). For Russell (1954) and Jackman (1962), military geography included the whole range of geographic research as it applied to military problems. Both, however, conceptualized military problems within a wartime context. Contemporary military problems are significantly different from those that were envisioned by either Russell or Jackman. To be sure, security remains the fundamental concern of the U.S. Military, but since 1991, there has been a considerable expansion of activities in the humanitarian sphere (Anderson, 1994; Palka, 1995) and environmental security realm (Butts, 1993, 1994).

The end of the Cold War and the Military's subsequent involvement with MOOTW have served to fuel the fire and to generate wider ranging possibilities for military geographic studies (Goure, 1995; Gutmanis, 1995; Palka 1995). Geographic applications to support warfighting, although invaluable, have not always been popular. Indeed, such applications encountered significant resistance and may have actually led to the demise of the subfield during controversial wars. Humanitarian and peacekeeping operations, however, have been much less disputable within academia and appear to be extremely attractive to a larger audience and new generation of geographers.

The MOOTW realm represents "uncharted waters" within an expanded military geography. The realm has commanded the Military's full-time attention during the post Cold War era and appears destined to demand the "lion's share" of emphasis within military geography in the future. Expansion into new realms will *not* dilute the subdiscipline, *nor* will it challenge the age-old definition of military geography. It *will* maximize the subfield's potential and enable practitioners to address military "problems" across the entire spectrum of contemporary military employment scenarios from peacetime to war.

Military operations other than war (MOOTW) include: nation assistance; security assistance; humanitarian assistance and disaster relief; support to counter-drug operations; peacekeeping; arms control; combating terrorism; shows of force; noncombatant evacuation; and support to domestic civil authority (CALL, 1993a). Recent examples are too many to list, but have been well publicized. Trends suggest that the Military will continue to be called upon to respond to refugee problems, floods, earthquakes, hurricanes, and forest fires, domestically and abroad. Additionally, the Military continues to be actively involved in operations to improve infrastructure and health conditions throughout Latin America, Africa, Asia, and the Pacific (Shannon and Sullivan, 1993; Binnendijk, 1998). These types of missions are innately geographic and afford ideal opportunities for military geographers to apply their expertise to worthwhile causes and showcase military geography as a versatile problem-solving subdiscipline.

Introduction 163

The current emphasis on MOOTW is indicative of America's global commitment, yet, also reflective of the increased flexibility (since the end of the Cold War) to employ military forces in a wide range of humanitarian scenarios. Similarly, the commitment to environmental security has been expressed in our National Security Strategy since 1991. These relatively recent changes in the US Military's orientation demand a military geography that is broader in scope and capable of focusing on the full range of contemporary military problems in peacetime and in war. The core of traditional military geography is still of timeless value to the Military, and will continue to be an integral aspect of all plans and operations during armed conflicts. Yet, now more than ever before, geographers specializing in various subfields or regions can make substantial contributions to the Military's humanitarian relief efforts and environmental programs by recognizing the previously untapped potential of military geography.

In this segment of the book, we will survey a number of critical issues. In Chapter 7, Eugene Palka examines the pattern of global instability and uncertainty characteristic of the past decade, and identifies the associated increase in military operations other than war to provide support, lend stability and promote peace. In Chapter 8, Kurt Schroeder presents a military geography of the war and subsequent peacekeeping operation in Bosnia. As illustrated in Chapter 7, MOOTW missions are typically derived from humanitarian assistance in a post-natural disaster environment. Further, there is a definitive geographic character to hazard analysis in a spatial and temporal sense. Ewan Anderson focuses on the geography of hazards in Chapter 9, and explains how the military apparatus manages disaster prediction and analysis. In the MOOTW context, the armed forces are postured in a force projection mode. Hence, they are required to travel sometimes long distances, moving large amounts of soldiers and heavy, outsized equipment. Consequently, force projection is a difficult problem that is intrinsically linked to geography. In Chapter 10, Mark Corson delineates the methodology and hardware associated with making force projection possible in a MOOTW environment.

The geography of conflict has always included the subject of insurgencies and counter-insurgency operations. These issues have taken on a greater focus in the post-Cold War era and clearly have a military geography context. Notwithstanding Western involvement in Vietnam, Malaya, and a host of African countries during the decades of sixties, seventies and eighties, the focus on insurgency movements, and especially the ethnic violence sometimes associated with them, has received renewed emphasis. Andrew Lohman investigates the geography of insurgent movements and proposes innovative counter-insurgency strategies in Chapter 11. Finally, in Chapter 11, Mark Corson and Julian Minghi examine the political and military geography of Bosnia from a pre- and post-Dayton Accord perspective.

Bibliography

Anderson, E. 1993. The Scope of Military Geography. Editorial. *GeoJournal,* 31/2: 115-117.

Binnendijk, Hans, editor in chief. 1998. *Strategic Assessment.* Washington, DC: Institute for National Strategic Studies, National Defense University.

Butts, Kent Hughs, editor. 1993. Environmental Security: What's DOD's Role? Special Report of the Strategic Studies Institute. Carlisle, PA: US Army War College.

Butts, Kent Hughs. 1994. Environmental Security: A DOD Partnership for Peace. Special Report of the Strategic Studies Institute. Carlisle, PA: US Army War College.

Center for Army Lessons Learned (CALL). 1993. Somalia. Newsletter NO. 93-1, January Fort 1993. Leavenworth, KS: U.S. Army Combined Arms Command.

Goure, Daniel. 1995. Non-Lethal Force and Peace Operations. *GeoJournal* 37/2: 267-275.

Gutmanis, Ivars. 1995. United States International Policies and Military Strategies in the Era of Defunct Aggressor. *GeoJournal* 37/2: 257-266.

Jackman, Albert. 1962. The Nature of Military Geography. *The Professional Geographer,* 14: 7-12.

Palka, Eugene J. 1995. The US Army in Operations other than War: A Time to Revive Military Geography. *GeoJournal*, 37/2: 201-208.

Russell, J.A. 1954. 'Military Geography.' In *American Geography: Inventory and Prospect*, by James, P.E., and Jones, C.F., eds. Syracuse, NY: Syracuse University Press.

Shalikashvili, John M. 1995. 'National Military Strategy of the United States of America.' Washington, DC: US Government Printing Office.

Shannon, The Honorable John W., and Sullivan, General Gordon R. 1993. 'Strategic Force-Decisive Victory: A Statement on the Posture of the United States Army, Fiscal Year 1994.' Presented to the Committees and Subcommittees of the United States Senate and the House Of Representatives, First Session, 103rd Congress.

Military Operations Other Than War. U.S. soldiers on patrol in a village in Bosnia. Source: U.S. Department of Defense, Defense Link Photograph, 1998.

Military Operations Other Than War

A young girl is amused to find U.S. soldiers lined up against the walls of her house in Mitrovica, Kosovo. The soldiers and United Nations' police are conducting a search for weapons.
Source: U.S. Department of Defense, Defense Link Photograph, 2000.

7

A Decade of Instability and Uncertainty:
MISSION DIVERSITY IN THE MOOTW ENVIRONMENT

By

Eugene J. Palka

Introduction

The war in the Persian Gulf reaffirmed the need for geographic expertise among military personnel in planning, training, and conducting operations. The military also continues to rely on professional geographers within the Department of Defense's research and development community to appropriately design uniforms, vehicles, arms, equipment, and supplies to enhance and sustain the force. Military "problems," however, are certainly not restricted to wartime preparation or actual wartime operations. In fact, mission diversity within the range of military operations other than war (MOOTW) demands an assortment of geographic information, tools, and techniques. Moreover, the military spends far more time operating within the MOOTW context than it does in war.

The fundamental purpose of this chapter is to examine the multifaceted mission of the current U.S. Army and highlight operations where geographers can make contributions that not only enhance mission accomplishment, but support noble and humanitarian causes. Inasmuch as military geography needs to keep pace with the changing role of the military, I have advocated (Palka, 1995; 2000) broadening the scope of the subdiscipline in order to address the new range of problems posed by contemporary military employment scenarios.

The scope of the chapter spans the past decade. As such, it is necessary to acknowledge that Army doctrine is in the midst of transitioning from the concept of MOOTW to stability and support operations (SASO). The latter term can be interpreted as being more descriptive because it emphasizes what is entailed, as opposed to what is not involved in the mission. Nevertheless, I will continue to employ the concept of MOOTW because of its widespread presence in the literature during the study period. Moreover, to geographers, there is virtually no distinction between the concepts, other than the name and associated acronym.

The chapter begins with a brief review of the "traditional" form of military geography that has long been applied in a wartime context. I will subsequently explain the impact which the end of the Cold War has had on the U.S. military, then proceed to examine the plethora of current Army missions which fall into the category of military operations other than war (MOOTW). I conclude by reemphasizing that we must broaden the "traditional," yet narrowly construed, military geography to take into account the Army's expanded role and current preoccupation with MOOTW. Undoubtedly, professional and academic geographers have much to contribute to mission success in these humanitarian and socially responsible endeavors.

Traditional Military Geography

Geography is literally the study of the earth, but is more specifically recognized as a broad science that focuses on the spatial distribution of phenomena and is concerned with their location, origin, evolution, and diffusion. Geographers seek to understand and explain the variable character of the surface of the earth and the diversity of human activity and interaction within the biosphere, using a wide variety of methods and techniques, many of which are unique to the discipline. Within this domain, military geography seeks to apply geographic principles and tools to military situations or problems. The subfield links geography and military science, and is a type of applied geography, employing the approaches, methods, techniques, and concepts of the discipline to military affairs, places, and regions.

"Traditional" military geography is primarily a European innovation, with the French, Germans, and British pioneering most work during the nineteenth and early twentieth centuries. The subfield, however, has had unquestionable utility during America's armed conflicts of the twentieth century, and so it has attracted a following of academic and professional geographers. Contemporary military leaders readily acknowledge that understanding and interacting with diverse cultures all over the world are fundamental requirements for the military professional. Besides being cognizant of the cultural geography of allies and potential adversaries, the military has long recognized the need for its officers at all echelons to have a basic understanding of physical geography and an assortment of geographic tools, such as maps, aerial photographs, LANDSAT images, and more recently, geographic information systems, and the global positioning system. Ironically, most of these tools that have come to be recognized as belonging exclusively to the geographer developed from their infancy within military environments.

As a career military officer who has been trained as a geographer, I can personally appreciate and attest to the utility of geographic skills and knowledge during the training, deployment, planning, and execution phases of military operations. Unfortunately, despite the requirements to operate within the constraints and restraints imposed by the geography of any area of operation, military professionals are not specifically trained to be experts on matters geographic. Some officers are required to take a couple of basic geography courses as part of their undergraduate curriculums at the military academies, but the latter account for only a small percentage of the professional military officer corps. Most training in geography is conducted on short notice, or comes in the form of "on the job training" in units. Under these circumstances, a broad, multifaceted discipline that holds a wealth of applicable information for the military officer is usually reduced in scope to the basics of terrain analysis and meteorology. The above approach is hardly a recipe for success in an environment where the stakes for misunderstanding or poorly applying military geographic principles are unquestionably high.

Despite the military officers' shortcomings in geographic education, wartime problems generally provide an unquestionable focus, receive the necessary resources, attract additional expertise, and instill a sense of urgency that helps to alleviate the lack of training. These conditions, however, do not exist in the typical MOOTW scenario. Another contrast between the wartime and MOOTW contexts relates to the involvement by professional and academic geographers. Both of the latter have a tradition of "rising to the occasion" and supporting the country's wartime efforts. The vast collection of "Area Handbooks" contained in the government documents section of any university library in America is a testament to the significant role played by geographers. In the MOOTW arena, however, there is no long-standing tradition of geographic inquiry, nor has there been much recent interest despite the significant trends that have developed since the end of the Cold War.

End of the Cold War: Beginning Of A New Era?

Since the fall of the Berlin Wall and the reorganization of the former USSR, the U.S. military has undergone considerable change. I strongly concur with Anderson (1993) that the end of the Cold War provides an opportune moment to re-examine the scope of military geography. Simply put, the consummation of a "bi-polar" world has prompted a drastic change in the force structure, size, and disposition of the U.S. military. Moreover, we have witnessed a significant decline in the numbers of "forward deployed" units, yet, an unprecedented increase in the country's willingness to employ its military in "operations other than war." Jackman's (1962) definition of military geography as "the application of geographical principles and knowledge to the solution of military problems," is *still* an appropriate way to define the subfield, but, the nature of the "military problems" are significantly different from those which existed at the time when he developed the definition (see Photograph 1). Moreover, the context is quite varied. As Anderson (1994a) notes, "security remains the fundamental concern of the military services, but, particularly since 1989, there has been a considerable expansion of activities in what could be broadly described as the humanitarian sphere."

Building upon Anderson's theses and concurring with Ramsbotham (1994), I have argued elsewhere (Palka, 1995; 2000) that the scope of military geography needs to be expanded to take into account the military's expanding role in "operations other than war. Current military undertakings within the context of MOOTW are noble causes that provide fertile ground for geographic inquiry. Geographers are well equipped to lend invaluable expertise and contribute to the Army's mission success in the MOOTW arena. To support my contentions and offer a number of possibilities, I will briefly consider the U.S. Army's participation in MOOTW since the end of the Gulf War.

Photograph 1. U.S. Army medical team examines victims at a mass gravesite in Kosovo.
Source: U.S. Department of Defense, Defense Link Photograph, 2000.

Mission Diversity In A Post-Cold War Context

The fundamental role of America's Army is to fight and win the nations' wars (Shinseki, 1999). A common misconception is that the role of the military, and the Army in particular, has changed drastically over the past ten years. The traditional, if not fundamental role remains; however, the Army *has* undertaken an increased number and variety of operations other than war. In fact, the Army's activity level during the first five years of the past decade has been higher than at any time during the Cold War. The trend has continued, as evidenced by an operations tempo that peaked at about 300 percent higher than at any time during the Cold War (Reimer, 1998). Consequently, the scope of traditional military geography needs to be broadened to keep pace with the Army's new challenges. It is within the MOOTW context where the Army continues to serve the nation by performing a wide variety of domestic and international tasks in support of national interests at home or abroad (Shannon and Sullivan, 1993).

Military operations other than war include: nation assistance; security assistance; humanitarian assistance and disaster relief; support to counter-drug operations; peacekeeping; arms control; combating terrorism; shows of force; noncombatant evacuation; and support to domestic civil authority (CALL, Jan 1993). Since the end of the Gulf War, these specific types of operations have involved large numbers of Army personnel and equipment. Specific examples include: protecting Kurdish refugees; responding to floods in Chicago; restoring peace after riots in Los Angeles; providing humanitarian relief and disaster assistance to those citizens affected by devastating hurricanes in Florida, Louisiana, Hawaii, and Guam; providing aid to United Nations forces in Bosnia and Macedonia; assisting starving people in Somalia; fighting forest fires throughout the western U.S.; conducting counter-drug operations along the U.S./Mexican border and in South America at the request of several host countries; restoring order and providing humanitarian relief in Haiti; and processing and caring for refugees in Cuba. To appreciate the magnitude of the Army's involvement in these types of operations since the Gulf War, one needs only to consider the following dollar costs which the *Army Times* published on 20 March 1995 for a *small* sample of contingency missions undertaken (but "unbudgeted") in 1994 *alone*: Southwest Asia-1.04 million; Somalia-17.3 million; Rwanda-17.2 million; Bosnia-311.9 million; Haiti-594.6 million; Cuba-367.1 million. When one adds the inventory replenishment and supply reimbursement for these missions, the total cost exceeds two billion dollars! More recently, the price tag attached to the U.S. commitment in Bosnia exceeds three billion dollars annually (Steele, 1999).

Two words that best describe the current nature of U.S. Army operations are "diverse" and "humanitarian." Diverse is an appropriate label because it alludes to the wide range of locations and environmental conditions under which the mission must be performed. It also refers to the varying types of packages (number and types of units, personnel, vehicles, equipment, and supplies) required to accomplish the mission, as well as the appropriate type of command and control structure and deployment support. Moreover, the Army may be required to coordinate and operate alongside or subordinate to a wide variety of federal or state agencies. Additionally, the nature of the mission may range from disaster assistance to peacekeeping. The relatively new operating environment that confronted the Army of the 1990s and is anticipated for the foreseeable future poses new challenges and dilemmas, without any relief from traditional responsibilities. This complex transitional period is eloquently described and examined in separate articles by Ramsbotham, Camilleri, Beach, and Anderson (a and b) in the 1994 special edition of *GeoJournal*.

Most soldiers and Army leaders gain tremendous fulfillment from the successful accomplishment of missions that are classified as operations other than war, especially from those operations that can be describe as humanitarian (Eckstein, 1999; Steele, 1999). That is *not* to say that the latter type missions are without inherent problems (Anderson, 1994b; Whitman, 1994; Baumann, 1997; Kretchik, 1997). Frustration within units can easily arise from what may occasionally be perceived as a conflict in priorities. Moreover, increased stress is often associated with the spontaneous nature of some humanitarian missions, the lack of clearly defined objectives and/or the absence of a "desired endstate" at the outset of the operation, and the frequent lack of any baseline knowledge as units journey into "uncharted waters." Perhaps most troubling to the individual soldier and leader alike is the extremely high operations tempo and the continuous deployments away from home. Nevertheless, the overwhelming majority of soldiers and leaders hail operations within the humanitarian sphere (Anderson, 1994a) as personally and professionally rewarding (see Photograph 2).

In U.S. Army doctrine, the "humanitarian" aspect of many contemporary missions refers to the condition whereby the Army is employed in an unconventional nature to render assistance and provide emergency relief to victims of natural or man-made disasters, in response to either domestic, foreign government, or international agency requests for help (CALL, Jan 1993). Humanitarian assistance is designed to be limited in scope and duration and is intended to supplement the efforts of civilian authorities or agencies who have primary responsibility for rendering such assistance (HQDA, 1993; USA, 1994). In 1998, the Army National Guard alone participated in 308 emergency response missions throughout the country (Schultz, 1998). In 1999, Army Reservists, who account for the smallest percentage of the total force, responded to more than 100 disasters (Whitehead, 2000). The following examples are humanitarian missions undertaken since the Gulf War, and are representative, rather than exhaustive.

Photograph 2. An American Medic bandages a dog bite on a boy in the town of Vitina, Kosovo. Source: U.S. Department of Defense, Defense Link Photograph, 2000.

Natural Disasters

Most university geography departments offer a course on "natural hazards" or "natural disasters." The scopes of most include such physical phenomena as earthquakes, hurricanes, tornadoes, volcanic eruptions, flooding, and even forest fires. Although geographers cannot predict natural disasters with any degree of certainty, they can certainly recognize high-risk locations by understanding the underlying earth processes that produce natural disasters. Moreover, geographers can identify populations who are at risk based on a wide range of variables such as proximity to the potential hazard, health conditions, early warning systems, building codes and the quality of housing, medical evacuation capability and care, and a host of other conditions that would mitigate the affects of a natural disaster.

From a military perspective disaster relief can be conducted across the entire spectrum of conflict, from peacetime to war. Disaster relief includes efforts to investigate the affects of either man-made or natural disasters (HQDA, 1993). An oil spill, toxic chemical leak, or nuclear meltdown is an example of the former, while examples of the latter were mentioned previously.

Military units are normally involved after disaster strikes, although there are instances when early warning enables some degree of emergency preparation or evacuation. Thus, humanitarian assistance operations can encompass both reactive and proactive programs (USA, 1994). The following are examples of natural disasters that involved significant numbers of Active Duty, National Guard, or Reserve soldiers during the past decade.

Hurricanes

Tropical cyclones are referred to by various regional names throughout the world. The term typhoon is used in the North Pacific, and cyclone is common throughout the Indian Ocean. Hurricane is the term applied to a tropical cyclone that occurs throughout the Caribbean Sea, Gulf of Mexico, and North Atlantic.

Tropical cyclones display distinct spatial patterns around the world during specific times of the year (see map 1). Within the US Army's primary area of concern, the hurricane season extends from August through September, with storms occurring as early as May and as late as November. Knowing the general areas of vulnerability and the associated time frames should enable some military units to anticipate support missions such as evacuations prior to a storm or disaster assistance during the aftermath.

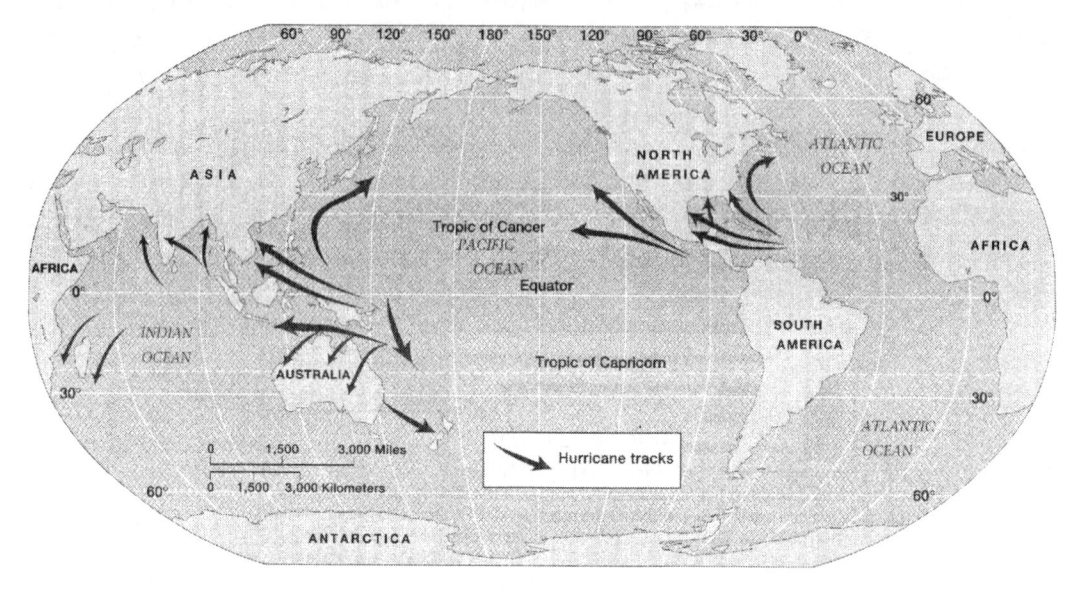

Map 1. Hurricane tracks.
Source: W.M. Gray, "Tropical Cyclone Genesis," *Atmospheric Science Paper 234*. Colorado State University, Fort Collins, CO 1975. In McKnight, T.L., 2000. *Physical Geography, A Landscape Appreciation*. New York: Prentice Hall.

In August 1992, Hurricane Andrew devastated parts of Florida and Louisiana. Thirty-eight people were killed, thousands were left homeless, and more than twenty billion dollars of damage was assessed (Yake and Hanson, 1995). Several of the Army's National Guard and Active Duty units were tasked to support the relief effort. The Army responded with engineer construction units, aviation assets, military police, and transportation, communication, and medical support. The number of soldiers assisting in the aftermath of Hurricane Andrew in southern Florida peaked at 21,988 Active Duty soldiers, 794 Reservists, and 5,703 Florida National Guardsmen (CALL, Oct 1993). Working in concert with civil authorities, the military provided emergency services and shelters, restored order, generated electrical power, cleared roadways, and cleaned up the aftermath. In addition to emergency medical care, Army personnel provided mental health screening and counseling, preventive medicine service, veterinary care, and dental treatment (CALL, Oct 1993). Army units also established "humanitarian support centers" at various locations to distribute supplies donated to victims of the disaster, and to provide food, water, clothing, beds and bedding, and temporary shelter (CALL, Oct 1993). The Army responded in a similar fashion only a few months later, when Hurricane Iniki devastated the Hawaiian Islands.

The past two years have entailed major responses to the devastation caused by Hurricanes Georges and Mitch in the Fall of 1998. More than 11,500 National Guards troops and approximately 7,000 Reservists were deployed to Honduras, Nicaragua, the Dominican Republic, and Puerto Rico to deal with the aftermath (Whitehead, 2000). Soldiers removed debris, conducted search and rescue operations, drilled wells, and repaired roads, bridges, houses, and schools (Eckstein, 1999; Whitehead, 2000).

Tornadoes

The Central Plains region of the United States is the most favorable location in the world for tornadoes to form (Forrester, 1981). Tornadoes are violent storms that can occur throughout the year and virtually anywhere in the United States (Forrester, 1981). Nevertheless, the largest percentage occurs from March through September and constitutes a distinct spatial pattern (see map 2).

Although tornado damage is generally on a smaller scale than hurricanes, 1999 was a particularly bad year in the U.S. when the country experienced its 10th worst outbreak of tornadoes on record. During that year, tornado outbreaks throughout Oklahoma, Kansas, Arkansas, and Ohio killed 43 people, injured more than 800, and resulted in more than one billion dollars in damage (Whitehead, 2000). In Oklahoma, there were more than 50 tornadoes. In that state alone, Army Reservists and National Guardsmen removed more than 500,000 cubic yards of debris from public property (Whitehead, 2000).

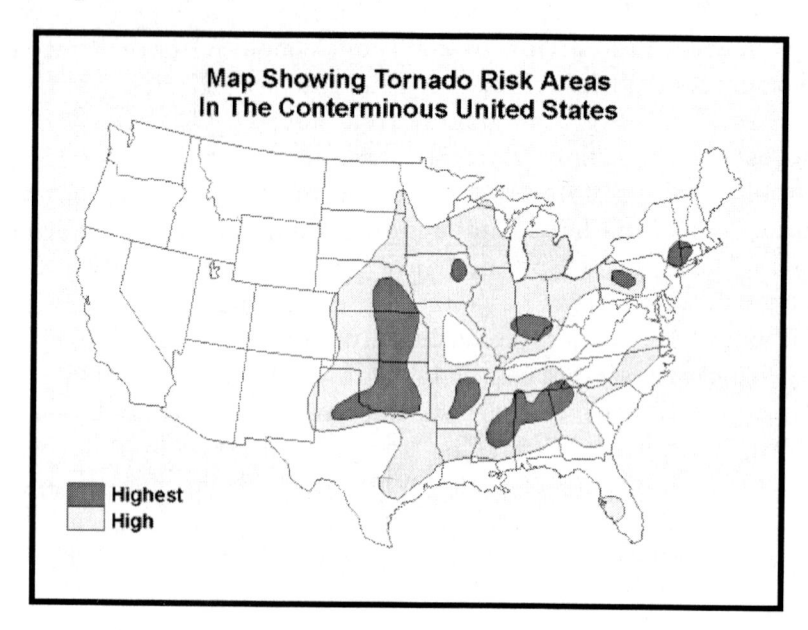

Map 2. Pattern of tornado risk areas within the continental United States.
Source: United States Geological Survey.

Flooding

In many cases, flooding occurs naturally as a consequence of hurricanes. In such cases, the two closely related phenomena are treated as a single event and military units that have responded to the first are forced to deal with the added challenge. In other instances excessive snowmelt, torrential rains, or even prolonged rain may trigger devastating floods. An example of the latter occurred in 1993, when the Mississippi River flooded thousands of acres throughout the American Midwest, in what was labeled as the "flood of the century." Less than a year later, the Army responded to a similar situation when flooding struck southern Georgia. In both cases, the flood relief included rescuing and evacuating flood victims, providing emergency shelter, food, water, and medical care, reestablishing communications services, and cleaning up the devastating effects (Photograph 3).

Earthquakes

When a massive earthquake rocked the Los Angeles area in January 1994, Active and Guard units were called to the area to provide disaster assistance and relief. Army units established and operated temporary shelters for displaced persons, provided emergency medical treatment, food, and water supplies, and assisted with the clean-up effort (Yake and Hanson, 1995). As with the other natural disasters, it is a relatively easy task for geographers to locate earthquake prone areas. Superimposing a map of recorded earthquakes onto a map of plate boundaries reveals an indisputable, but not surprising correlation (see map 3). Unfortunately, it is virtually impossible to anticipate the timing and precise location of a quake. Moreover, active tectonic activity occurs throughout a vast area within the Western Hemisphere.

Photograph 3. A Marine Corps helicopter participating in evacuation and food distribution operations during the floods in Mozambique, Spring of 2000.
Source: U.S. Department of Defense, Defense Link Photograph, 2000.

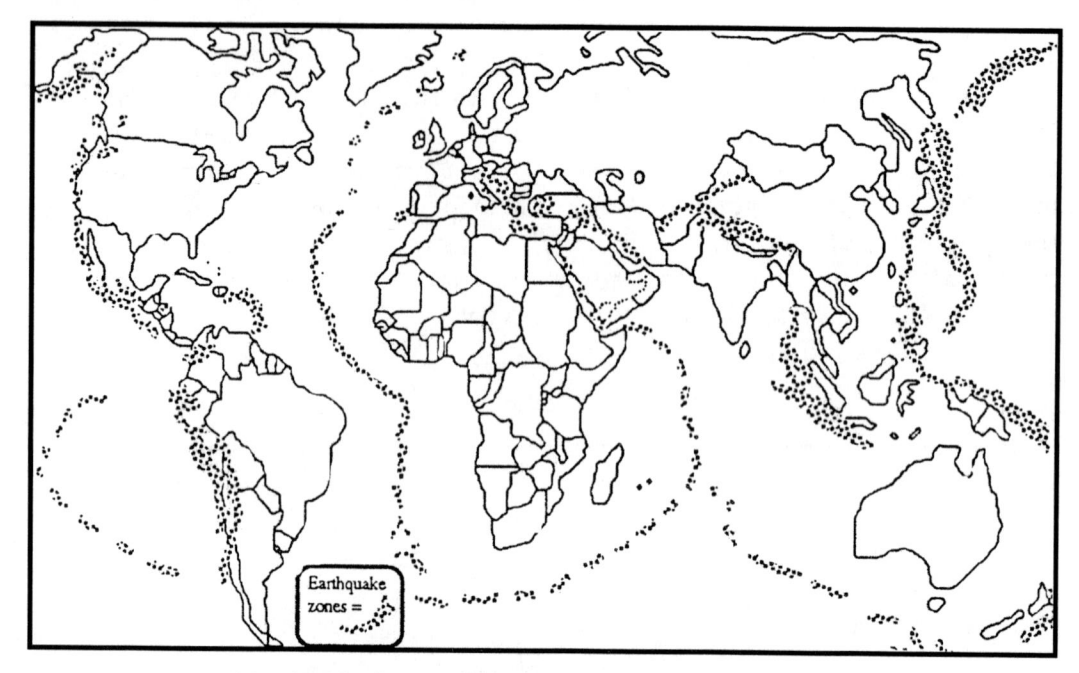

Map 3. Distribution of earthquakes along plate boundaries.
Source: Data from National Geophysical Data Center/NOAA.

Forest Fires

Fighting fires is a form of disaster assistance that has become an annual requirement for Army units stationed in the West and Alaska. Soldiers from Fort Hood, Texas, Fort Carson, Colorado, and Fort Lewis, Washington routinely support the National Interagency Fire Center. Since 1988, Army units have been involved in fire-fighting programs throughout fire-prone areas of the western part of the country. In 1994 alone, more than 3,500 soldiers worked in concert with federal agencies to help combat some 70,000 wildfires that burned more than four million acres (Yake and Hanson, 1995). During the summer of 1996, 1,160 soldiers battled fires that consumed more than 3 million acres (*Army Times*, 2000a). Less predictable was the recent forest fire that swept through Florida in 1999, aggravated by a prolonged drought. Nevertheless, National Guard, Reserve, and Active Duty troops combined efforts to battle the blaze. Most recently, during the summer of 2000 more than 500 soldiers joined U.S. Forest Service regulars in an effort to control a massive forest fire in the Salmon River Mountains of Idaho. The task force from Fort Hood received only a one-day warning prior to deploying with engineer equipment and Blackhawk helicopters from the flatlands of Texas to the mountains of Idaho (Army Times, 2000b).

Although the root cause of any forest fire can stem randomly from either human activity or natural phenomena (such as a lightning strike), fire prone areas are easily mapped based on climate and vegetation patterns. Understanding the conditions that result in a rather predictable pattern of forest fires enables a degree of preparation by Army units that are routinely tasked to provide assistance.

Photograph 4. A Honduran man departs a U.S. food distribution site with a sack of rice after Hurricane Mitch destroyed much of that country's harvest in the Fall of 1999.
Source: U.S. Department of Defense, Defense Link Photograph, 1999.

Other Forms of MOOTW

Nation Assistance

Aside from the wide range of humanitarian missions that occur in response to disasters, the Army is routinely involved in other forms of MOOTW in support of national security strategy. Missions occur at home and abroad and range from short-term crises to well planned, long-term endeavors.

Nation assistance strives to promote internal development and growth of sustainable institutions within a host country (USA, 1993). In simplest terms, it involves helping a country to help itself. Nation assistance entails political, economic, informational, and military cooperation between the U.S. and the host country. During the early 1990s, operations to improve infrastructure and health conditions were undertaken in Central and South America, Africa, Asia, and the Pacific (Shannon and Sullivan, 1993). Highly publicized humanitarian missions were more recently conducted in Rwanda, Cuba, Haiti, Panama, Somalia, and Honduras. Nation assistance continues throughout Latin America, Africa, and Asia, and based on our country's strategy of engagement, will likely continue in the foreseeable future (The White House, 1998).

Refugee Evacuation and Support

Refugee evacuation and support can occur as a prelude to or a consequence of war or internal conflict. In Rwanda, the Army deployed more than 2,000 soldiers (including medical personnel, communications specialists, water purification teams, transporters, aviation assets, infantrymen, quartermaster units, and engineers) to provide humanitarian relief to victims of civil war (Vogel, 1994). Units provided food, water, shelter, and emergency medical care to refugees, and furnished transportation, communications, security, and logistical support to international agencies working in country (Photograph 5). Additionally, units cleared sites and established medical treatment facilities and water purification units, repaired roads and bridges, and have dug hundreds of graves to bury the victims of Rwanda's tragic civil war (Vogel, 1994).

More than 2,000 military personnel established operations at Guantanamo, Cuba, providing support to nearly 16,500 Haitian refugees

(Lawson, 1994). U.S. Army personnel established shelters, and provided food, water, medical care, and protection to the refugees. The Army has also established refuge camps in Panama in support of Haitian refugees seeking to gain immigrant status to the U.S. or any of several Central American countries.

During operations in Bosnia, Macedonia, and Kosovo, the refugee problem was enormous. Support to Kosovar refugees included transporting more than 4,000 refugees to Fort Dix, New Jersey for medical support, processing, and resettlement in the United States (Whitehead, 2000).

Photograph 5. Rwandan refugees enroute to a refugee camp.
Source: U.S. Department of Defense, Defense Link Photograph, 1996.

Peace Operations

The term peace operation is an all-encompassing one that includes peace making or building, peacekeeping, and peace enforcement. Peace making or building involves diplomatic activities that may or may not be supported by military action. Peacekeeping activities are military operations that are undertaken by a neutral party with the consent of all belligerents, and are intended to facilitate implementation of truce or long term political agreement (USA, 1994). In contrast, peace enforcement involves restoring order under broadly defined conditions by the international community in order to support diplomatic efforts to achieve a long-term political settlement (USA, 1994).

Throughout the 1990s, the Army has been involved in peace operations in Bosnia, Kosovo (Photograph 6), Kuwait, Korea, and the Sinai Desert. The U.S. Military has been intimately involved in Bosnia since December of 1995, attempting to enforce, and then maintain the peace among warring factions. Army units entered Bosnia initially as part of Operation Joint Endeavor and served as an implementation force for the NATO peace enforcement plan. Operation Joint Guard followed, and Operation Joint Forge began in June of 1998. The number of American troops peaked at around 23,000 and has been gradually reduced to about 6,000 (Steele, 1999a). The overall objective of bringing hostilities between warring factions to a halt and establishing and maintaining a safe and secure environment has been accomplished. The fact that the deployment is approaching 5 years in duration (4 years beyond initial estimates) is indicative of the complicated nature of the operational environment and surely reflects the difficulty, if not futility, in trying to resolve conflict that is rooted in centuries of hatred between cultures.

The U.S. participated in Operation Allied Force in March of 1999, using air strikes to target Serb forces and positions in Kosovo. In June of 1999, U.S. forces and other international units moved into Kosovo to restore order following the Serbian withdrawal. Operation Joint Guardian has been ongoing since then. Similar to the role in Bosnia, the objective in Kosovo is to provide a safe and secure environment for citizens under the NATO mandate. The situation in Kosovo, however, is a bit more complicated than the scenario in Bosnia, since there are no clearly defined boundaries between adversaries (Steele, 1999b).

Civil Disturbance

The problem of civil unrest surfaced within the U.S. on numerous occasions during the civil rights movement and particularly during the Vietnam era. More recently, on 29 April 1992, Los Angeles was swept by racial violence that erupted after the acquittal of four police officers on trial for beating motorist Rodney King. More than fifty people died, hundreds were injured, an estimated 1,000 businesses were destroyed, and approximately 6,000 fires ravaged the city, resulting in damage in excess of one billion dollars (Yake and Hanson, 1995). In response to the widespread civil unrest, a Presidential Executive Order was issued on 1 May, establishing Joint Task Force Los Angeles, federalizing units of the California National Guard, and authorizing Active Duty military forces to assist in the restoration of law and order (CALL, Nov 1993). Army National Guard and Active Duty units subsequently restored order, fought fires, provided emergency medical care and communications, and assisted in cleaning up the aftermath.

Photograph 6. U.S. soldiers and U.N. police (light helmets) conduct a house-to-house search for weapons in the Serbian quarter of the town of Mitrovica, Kosovo.
Source: U.S. Department of Defense, Defense Link Photograph, 2000.

MOOTW Trends

As others have noted, the types of missions described above are not new to the Army. What is noteworthy, however, is that they are being conducted as a matter of routine; the dollar cost is staggering; and units frequently are required to write the pertinent doctrine during the course of the mission (CALL, Jan; Sep; Oct; Nov 1993). Nevertheless, all of these types of missions are innately geographical and afford an ideal opportunity for military geographers to apply their expertise to worthy causes. In all of the examples presented herein, regional and systematic geographies can be combined and supplemented with appropriate geographic tools and techniques to assist military decision-making throughout the training, planning, and execution phases of the operation.

Photograph 7. Army Military Policeman man a checkpoint during a civil disturbance exercise.
Source: U.S. Department of Defense.

Active involvement in the MOOTW arena will continue given the country's strategic goal of remaining engaged with other countries of the global community. The U.S. strategy of engagement will surely rely on the military to undertake a wide variety of missions as an extension of our foreign policy. Regardless of the terminology, MOOTW now, or SASO in the near future, the nature of the context is the same-complicated scenarios and diverse missions in both foreign and domestic environments.

Conclusion

Since the close of World War II, the United States has played an increasingly greater role in world affairs. The number of U.S. military bases overseas has increased dramatically in response to the country's global commitment. Current Army doctrine envisions units that are capable of deploying to special environments (arctic, desert, jungle, mountainous, or any combination thereof) throughout the world to protect the country's interests. For Army units, a scenario becomes increasingly complicated when required to operate in a "joint" environment, as part of an alliance, or deal with the multifaceted problem of low-intensity conflict. In light of the numerous possibilities described above, the need to understand world regional geography and to apply geographic principles in a military context is greater now than ever before.

Meanwhile, the fundamental mission of the U.S. Army has remained virtually unchanged over the past 225 years. The Army's doctrine, however, is always in the process of evolving in response to national objectives, new technology, the changing nature of the threat, and the world's ever-changing balance of power. Moreover, the Army must adapt to each administration's specific goals and areas of emphasis, as the latter guides the country within a rapidly changing social, political, and economic world. Throughout these turbulent times, the Army must retain its role and responsibility to respond to the country's needs across the spectrum of armed conflict or in military operations other than war, be they foreign or domestic.

The current emphasis on MOOTW is indicative of this country's global commitment, yet, also reflective of the increased flexibility (as a result of the end of the Cold War) to employ military forces in a wide range of humanitarian scenarios. Although the Army's fundamental purpose remains unchanged, there are indications that the Army will continue to be active in the MOOTW arena. The current trend demands a military geography that is broader in scope and specifically focuses on humanitarian assistance and disaster relief missions. The core of traditional military geography is still of timeless value to the military, and will continue to be an integral aspect of all plans and operations during armed conflicts. Yet, now more than ever before, geographers specializing in a wide range of subfields or regions can make substantial contributions to the military's humanitarian relief efforts by recognizing the previously untapped potential of military geography.

The examples of military operations other than war examined in this chapter constitute a relatively new operating environment for the Army, yet these types of problems have long been studied by geographers. Geographical studies on natural disasters and hazards, famine, refugee problems, social unrest and riots in urban areas, problems of health and infrastructure in developing countries, and regional instabilities stemming from cultural collisions, are too many to list. Applying the "already established" geographic information, concepts, and tools to these new types of military problems would considerably strengthen military geography as a versatile problem-solving subdiscipline.

Since the end of the Cold War, the Army's operational tempo has increased 300 percent (Reimer, 1998). Today, soldiers are stationed in 90 different countries all over the world, engaged in training, peacekeeping, and humanitarian assistance missions (Shinseki, 1999). Military geographers, as well as other systematic or regional specialists can enhance the success of MOOTW through professional writing, guest lectures in Army units, seminars with the senior leadership of the Army, and formal teaching at the military's institutions of higher learning, just to name a few initiatives. It is my hope that this chapter and those that follow help to revive military geography in the U.S. by rallying traditional practitioners, and by luring experts from across the discipline to contribute to the socially responsible and politically correct missions which the U.S. Army is currently undertaking under the rubric of military operations other than war.

Editor's notes: This chapter was adapted from: Palka, E.J. 1995. "The U.S. Army in Operations Other Than War: A Time to Revive Military Geography." *GeoJournal*, 37 (2): 201-208. Reprinted by permission.

Bibliography

Anderson, E. 1993. The Scope of Military Geography. Editorial. *GeoJournal* 31 (2): 115-117.

_____. 1994a. The Changing Role of the Military. Editorial. *GeoJournal* 34 (2): 131-132.

_____. 1994b. Disaster Management and the Military. *GeoJournal* 34 (2): 201-205.

Army Times, 2000a. "Military to Help Fight Fires." *Army Times*, August 7, 2000, p. 3.

Army Times, 2000b. "Fight Fire with Thunder." *Army Times*, August 14, 2000, pp. 8-9.

Baumann, Robert F. 1997. Operation Uphold Democracy: Power Under Control. *Military Review* LXXVII (4): 13-21.

Beach, J.M. (1994). Confronting the Uncomfortable: Western Militaries and Modern Conflict. *GeoJournal* 34 (2): 147-153.

Camilleri, J.A. 1994. Security: Old Dilemmas and New Challenges in the Post-Cold War Environment. *GeoJournal* 34 (2): 135-145.

Center for Army Lessons Learned (CALL). 1993. Somalia. Newsletter NO. 93-1, U.S. Army Combined Arms Command, Fort Leavenworth, KS, January 1993.

_____. 1993. Counterdrug Operations II. Newsletter NO. 93-5, U.S. Army Combined Arms Command, Fort Leavenworth, KS, September 1993.

_____. 1993. Operations Other than War, Volume II: Disaster Assistance. Newsletter NO. 93-6, U.S. Army Combined Arms Command, Fort Leavenworth, KS, October 1993.

_____. 1993. Operations Other than War, Volume III: Civil Disturbance. Newsletter NO. 93-7, U.S. Army Combined Arms Command, Fort Leavenworth, KS, November 1993.

Clausewitz, K.V. 1943. *On War*. (Initially published in 1832.) Translated by O.S. Matthijs Jolles. New York: Modern Library.

Cole, D.H. 1950. *Imperial Military Geography*, 10th edn. London: Sifton Praed & Company, LTD.

Davies, A. 1946. Geographical Factors in the Invasion and Battle of Normandy. *Geographical Review* 36 (4): 613-631.

Dorpalen, A. 1942. The World of General Haushofer; Geopolitics in Action. New York: Farrar and Rinehart.

Eckstein, Major Jeffry R. 1999. A Short and Successful Operation. *Army*. (September, 1999): 67-70.

Fishel, John T. 1997. Operation Uphold Democracy: Old Principles New Realities. *Military Review* LXXVII (4): 22-30.

Forrester, Frank H. 1981. *1001 Questions Answered about the Weather*. New York: Dover Publications, Inc.

Gawrych, George W. 1997. Roots of Bosnian Realities. *Military Review* LXXVII (4): 79-86.

Gray, W.M. 2000. "Tropical Cyclone Genesis," *Atmospheric Science Paper 234*. Colorado State University, Fort Collins, CO 1975. In McKnight, T.L., 2000. *Physical Geography, A Landscape Appreciation*. New York: Prentice Hall.

Headquarters, Department of the Army. 1993. *Operations*, Field Manual 100-5. Washington, D.C.: U.S. Government Printing Office.

Jackman, A. 1962. The Nature of Military Geography. *The Professional Geographer* 14: 7-12.

James, P.E.; Martin, G.J. 1979. *The Association of American Geographers: The First Seventy-Five Years, 1904-1979*. Washington: Association of American Geographers.

Johnson, D.W. 1921. Battlefields of the World War, Western and Southern Fronts: A Study in Military Geography. Research Series No. 3. New York: American Geographical Society.

Keegan, J. and Wheatcroft, A. 1986. *Zones of Conflict: An Atlas of Future Wars*. New York: Simon and Schuster.

Kretchik, LTC Walter E. 1997. Force Protection Disparities. *Military Review* LXXVII (4): 73-78.

Lavallee, T.S. 1836. *Geographie Physique, Historique et Militaire*. Paris.

Lawson, C. 1994. Safe haven in Cuba. *Army Times* (August 15, 1994).

Mackinder, H.J. 1902. *Britain and the British Seas.* New York: D. Appleton and Co.

Mahan, A.T. 1890. *The Influence of Sea Power upon History*, 1660-1783. Boston: Little, Brown and Co.

Martin, G.J. and James, P. E. 1993. *All Possible Worlds: A History of Geographical Ideas.* New York: John Wiley & Sons, Inc.

Munn, A.A. 1980. The Role of Geographers in the Department of Defense. *The Professional Geographer* 32 (3): 361-364.

O'Loughlin, J. and van der Wusten, H. 1986. Geography of war and peace: notes for a contribution to a revived political geography. *Progress in Human Geography* 10 (3): 434-510.

_____. 1993. Political geography of war and peace. In: Taylor, P.J. (ed.), *Political Geography of the Twentieth Century: A Global Analysis.* London: Belhaven Press.

O'Sullivan, P. and Miller, J.W., Jr. 1983. *The Geography of Warfare.* New York: St. Martin's Press.

Palka, LTC Eugene J. 1995. The U.S. Army in Operations Other than War: A Time to Revive Military Geography. *GeoJournal* 37 (2): 201-208.

Palka, Eugene J. 2000. Military Geography: Its Revival and Prospectus. In *Geography in America at the Dawn of the 21st Century.* Edited by Gary L. Gaile and Cort J. Willmott. Oxford: University Press.

Peltier, L.C. and Pearcy, G.E. 1966. *Military Geography.* Princeton, NJ: D. Van Nostrand Company, Inc.

Peterson, D.M. 1995. Time to pay the bill. Editorial. *Army Times* (March 20, 1995).

Ramsbotham, D. 1994. The Changing Role of the Military. *GeoJournal* 34 (2): 133-134.

Reimer, General Dennis J. 1998. The Army is People. *Army* (1998-1999 Green Book). October 1998: 17-26.

Rosen, S.J. 1977. *Military Geography and the Military Balance in the Arab-Israel Conflict.* Jerusalem: The Jerusalem Post Press.

Schultz, MG Roger C. 1998. The National Guard's Secret to Success. *Army* (1998-1999 Green Book). October 1998: 95-100.

Shannon, J.W. and Sullivan, G.R. 1993. Strategic Force - Decisive Victory: A Statement on the Posture of the United States Army, Fiscal Year 1994. Presented to the Committees and Subcommittees of the United States Senate and the House Of Representatives, First Session, 103rd Congress, Washington.

Shinseki, General Eric K. 1999. Beginning the Next 1000 Years. *Army* (1999-2000 Green Book). October 1999: 21-28.

Steele, Dennis. 1999a. The U.S. Army in Bosnia. *Army.* April 1999: 12-19.

_____. 1999b. Kosovo: A Special report. *Army.* September 1999: 18-66.

United States Army. 1994. *HA Multiservice Procedures for Humanitarian Assistance Operations*, Field Manual 100-23-1. Fort Monroe, VA: U.S. Army Training and Doctrine Command.

Vogel, S. 1994. U.S. troops in Rwanda busy making a difference. *Army Times* (August 15, 1994)

Whitehead, LTC Ray. 2000. The Soldiers Almanac. *Soldiers.* January 2000, vol 55, no 1.

Whitman, J. 1994. A Cautionary Note on Humanitarian Intervention. *GeoJournal* 34 (2): 167-175.

Yake, D. and Hanson, T. 1995. When Disaster Strikes. *Off Duty*, America Edition (February/March 1995).

Marines from the 2nd Division, Light Armored Reconnaissance, Delta Company erect tents which will become shelter for homeless earthquake victims in Turkey on Aug. 28, 1999.
Source: U.S. Department of Defense.

Mission Diversity in the MOOTW Environment

Montana Army and Air National Guard troops hack out a fire break in the dried grass during training at Fort Harrison near Helena, Mont. on Aug. 3, 2000.
Source: U.S. Department of Defense.

A Light Amphibious Reconnaissance Craft filled with U.S. sailors, Marines and Russian media departs the shores of a mock disaster area for a U.S. landing craft during Exercise Cooperation From the Sea '96 near Vladivostok, Russia, on Aug. 14, 1996.
Source: U.S. Department of Defense.

8

The Human Landscape and Warfare:
THE MILITARY GEOGRAPHY OF THE WAR IN BOSNIA

By

Kurt A. Schroeder

Introduction

The conflict in Bosnia had its course shaped largely by its human geography and that of the surrounding region. Although the military capabilities and operational possibilities of each side were limited in large measure by terrain, the human landscape played an equally decisive role in the war. Even though Bosnian Serb and Bosnian Government forces enjoyed the advantages of internal lines of communication, their ability to deploy and engage in large-scale offensive operations was been limited by the constraints imposed by terrain and the ethnic landscape.

The first section of this chapter will propose a geographic rationale for military operations in an ethnically mixed area and will examine the limitations imposed on possible operations by the three main parties. Next, the chapter will briefly survey the course of the war to illustrate how each side used the human landscape to achieve their operational goals. This analysis will focus on the early part of the war when Bosnian Serbs were able to mount major offensive operations, and the last phase of the war, which was dominated by combined Croat and Bosnian government offensives. Third, the chapter will examine a historical parallel to the conflict in Bosnia—that of the Spanish Civil War. Finally, the chapter will briefly consider the situation at the end of 1995, and the possible implications for peace.

The Geography of Bosnia – Operational Goals

One of the salient characteristics of a civil war in an ethnically mixed area is the need for each faction to employ its population for logistical support. Areas populated by ethnic groups friendly to a given side are critical to sustaining participation in the war. They provide manpower to continue the fight, food, intelligence, space in depth, which makes successful military operations possible, and the economic basis for production or purchase of advanced weapons. This situation can express itself in both offensive and defensive goals. Defensive goals are typically articulated as a desire to 'protect our people' or to 'protect our base of operations in the current struggle.' Offensive goals are expressed as a desire to gain base areas and to achieve strategic goals.

A logical geographic strategy at the beginning of a war in such an ethnically mixed area would include operations, which accomplish three types of objectives. These are, in order of probable urgency:

1. *Control Operations* bring under control areas that are inside the core area but are contested, or those contiguous to core territory and predominantly inhabited by the friendly ethnic group. This may include areas inside the core area, which are of mixed ethnicity. Being inside or contiguous to the core, they are relatively easy to control and increase the geographic base of support. At the same time, defensive operations should be undertaken to prevent enemy offensives from occupying the home territory.

2. *Connect Operations* are conducted to connect isolated areas under the ethnic group's control. In other words, link exclaves to the core and solidify the geographic base of support. Exclaves, which are closer and of significant size, are more likely to be selected for control operations. Those,

which are smaller and more distant, are less likely to be chosen. At the same time, the force must isolate enemy enclaves within friendly territory from similar connecting attacks from competing ethnic groups.

3. *Pacification Operations* are conducted to pacify and absorb enemy enclaves within the core territory, and enclaves created by control and connect operations. Pacification operations would be the last operations conducted based on ethnic distributions. After the successful conclusion of pacification operations, military operations shift to strategic objectives, such as control of major population centers or capital cities, usually deep inside enemy territory, which would decisively end the war on favorable terms.

Each of the three sides in the Bosnian conflict—Muslim, Bosnian Serb and Croat—had their operational needs dictated by the human landscape and by sources of outside aid. At the beginning of the war, given the extremely mixed ethnic distribution in Bosnia, control of much of the country was uncertain. The highly confused ethnic distribution at the beginning of the war is evident on Map 1, which shows the majority Serb, Croat, and Muslim areas in pre-war Bosnia (Department of State, 1993). Hence, the war would essentially be focused on operations to consolidate ethnic core areas and link disconnected ethnic enclaves—sometimes by purging an area (i.e., ethnic cleansing) of a competing ethnic group. The military potential of each ethnic faction was based in the major core areas (see Map 1).

The Bosnian Serbs dominate the eastern portion Bosnia, in the primarily rural area contiguous to Serbia and Montenegro, and in a large region of Serb population in west central Bosnia. This second area of Serb population extends over the Bosnian border into Croatia, and surrounds the sizeable Muslim enclave around Bihac (see Map 1). The areas of Serb population that extended into Croatia are known as the Krajina, although the word *krajina* (etymologically related to the Russian territorial designation *kray*) in Serbo-Croatian simply means frontier, and can be applied to a number of frontier areas, such as the area of Montenegro near Albania.

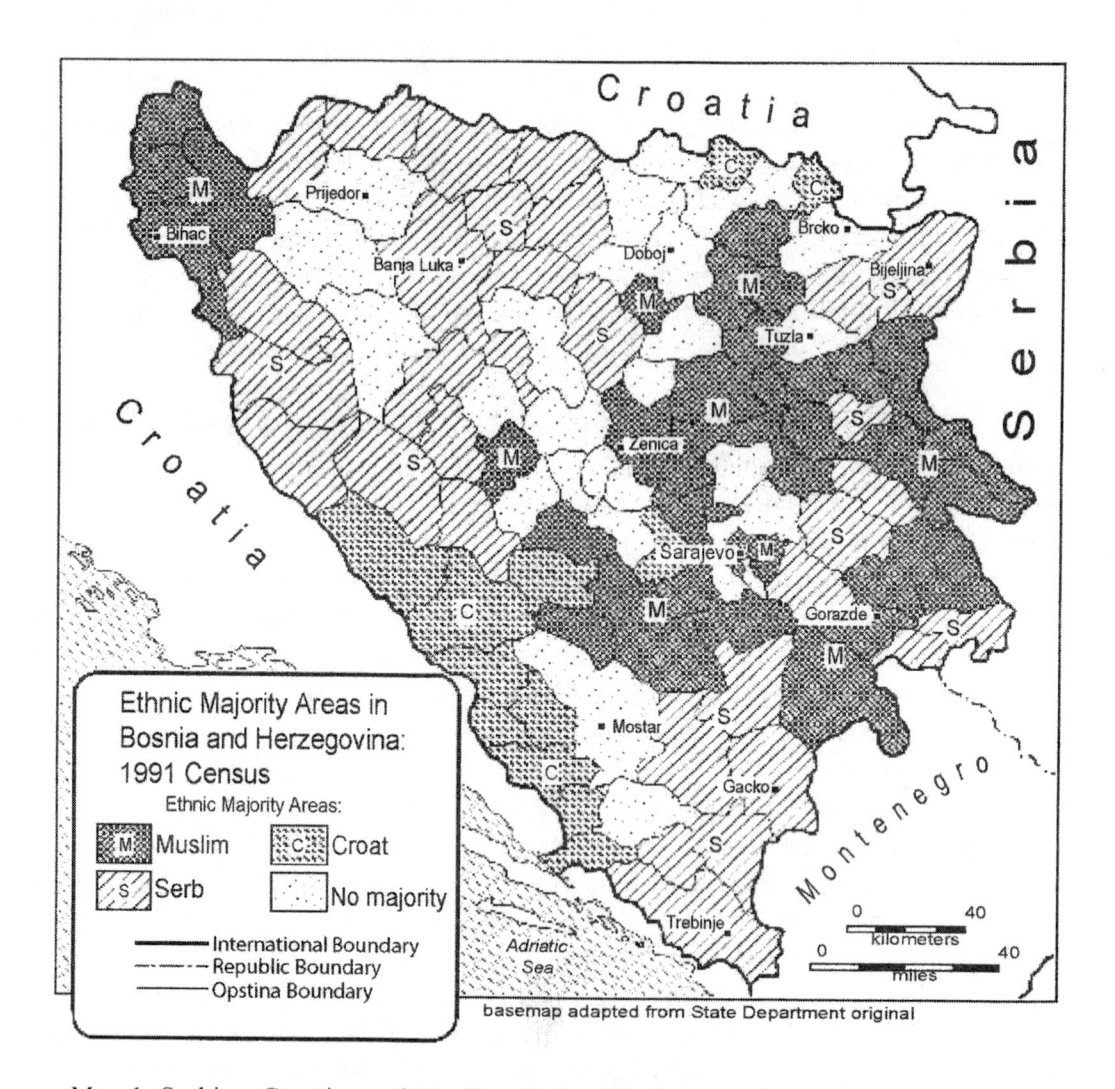

Map 1. Serbian, Croatian and Muslim ethnic majority areas in pre-war Bosnia. The ethnic pattern on the map illustrates major core areas
Source: U.S. Department of State, 1995.

Bosnian Serbs had one major geographic problem, which they were able to solve because of their military superiority. This problem was the security of the Brcko (or Posavina) corridor between their base of support in Serbia and their base of operations at Banja Luka (i.e., a classic connect operation). This corridor is evident in Map 1, between Bijeljina in far northeast Bosnia and Banja Luka in north central Bosnia. At the beginning of the war, this was an area of mixed ethnic population. With that corridor secure, the Serbs, operating from their base around Banja Luka, enjoyed what in traditional military terminology is known as internal lines of communications. Meaning that, because of a central position, they could shift forces rapidly from one sector to another in order to take advantage of opportunities or prevent enemy offensives from succeeding. Thus, once the corridor was secured, Serbs could undertake centrifugal offensives outward from Banja Luka, and could shift their forces rapidly from one threatened sector to another. Significantly, the Brcko corridor is the only area of geographic uncertainty left by the recent Dayton Peace Agreement (Department of State, 1995). The Serbs also had to keep open bridges across the upper Drina, which runs generally along the eastern edge of Bosnia. These bridges connect areas of Serbian ethnic predominance in eastern Bosnia to Serbia, which was their major external source of military aid. These bridges needed to be secured in order for the Serbs to sustain offensives in Eastern Bosnia, and they were able to accomplish this task easily.

A major problem for the Bosnian Serbs throughout the war was their numerical inferiority to the combined forces of Bosnian Croats and Bosnian Muslims. Only 31% of the pre-war population of Bosnia was Serbian (Central Intelligence Agency, 1991). Another major difficulty for the Serbs was the division of command between the forces of the so-called Republic of the Krajina in Croatia, Serb warlords in Bosnia, and the Serbian-dominated government of the new Yugoslavia. This disunity made it difficult for the Serbs to use their material and geographic advantage (i.e., interior lines) more fully.

Bosnian Government forces, also possessed internal lines of communication south of Sarajevo, but had the geographic liability of controlling several major exclaves. One such important exclave was situated around Bihac in far western Bosnia (see Map 1). The Serb-controlled territory of the Krajina surrounded this exclave. Other exclaves were the urban areas of eastern Bosnia, which became so notorious during the course of the war. The towns of Srebrenica and Gorazde, and halfway between these two, the village of Zepa were the scene of Serbian-initiated ethnic cleansing and other atrocities. These urban sites are surrounded by rural areas, which are predominantly Serbian. Furthermore, Sarajevo and its surrounding valleys are areas of mixed population, with large and frequent concentrations of Serbs in many of the suburbs. Sarajevo was essentially a semi-exclave. Thus, the Bosnian government, when it was not on the defensive, generally pursued *connect operations* to sustain and connect the archipelago-like set of exclaves they nominally controlled at the beginning of the conflict. These geographic and operational problems, coupled with their weakness in matériel, greatly hampered their effort. The Bosnian government was also the only side in the conflict that had no readily available outside base of support. Because of pre-war positioning of war stocks, they typically lacked heavy weapons, including tanks and artillery. Hence, Bosnian operational goals in the war were mainly defensive.

The Bosnian Croats enjoyed one major advantage in the conflict. Their outside base of support in Croatia surrounds Bosnia on three sides and contains better transportation links than are generally found inside most of Bosnia itself. In traditional military terminology, they held the inherent disadvantages of external lines of communications. The major disadvantage associated with external lines of communication was that the Croats were vulnerable to centrifugal attacks by Serbian units, since the band of Croatian territory around Bosnia is very thin and subject to possible penetration at several places, most notably west of Bihac at Karlovac, and south-southwest of Bihac at Maslenica. Early in the war, successful Serbian attacks at each location threatened to divide Croatian territory.

However, a major geographic advantage enjoyed by the Croats was the mobility afforded by the excellent external transportation network. The Croats could potentially shift troops and supplies around the outside of the area of conflict to meet enemy threats and take advantage of operational opportunities. Hence, the Croats could shift army units around the periphery of Bosnia within Croatia and concentrate at the decisive location. It took the Croats several years to build up their mobility and secure transport routes around Bosnia. This finally enabled them, in the later stages of the war, to mount successful centripetal offensives against the Serbs. The predominantly Croatian areas within Bosnia are near the Croatian border adjacent to

Dalmatia, along the southwestern border of Bosnia, and some of the river towns along the Sava River, which defines the northern boundary of Bosnia.

The Progress of The War And Operational Goals

The war in Bosnia resulted from the collapse of the second attempt at a Yugoslav nation. The first Yugoslavia, a post-World War I orphan of the Austro-Hungarian Empire, survived for about twenty years on the basis of an uneasy truce between dominant Serb and Croat ethnic groups. Other ethnic groups, such as Slovenians, Bosnian Muslims, Macedonians, and Bulgarians were in a decidedly inferior position. This first attempt at a united Yugoslav state was destroyed by a German invasion in Spring of 1941.

The second attempt at a Yugoslav state was built on the organization of the communist-led partisans of World War II, but slowly destabilized during the decade after the death of Marshall Tito in 1980, and began further disintegrating with the withdrawal of Slovenia from Yugoslavia in 1990. Complete summaries of the breakup of Yugoslavia can be found in Bennett (1995) and Cohen (1993). Donia and Fine (1994) and Malcolm (1994) are excellent examinations of the history of Bosnia. Other useful summaries of Bosnian and Yugoslav history are given in Kerner (1949), Clissold (1966), and Singleton (1976)

This section of the chapter will trace the spatial and temporal progress of the war in Bosnia using operational goals described in the previous section as a base of reference. This geographic analysis will be segmented into discrete periods, each with its own operational and geographic characteristics. Of course, any periodization of a conflict, especially one as complicated as the one in Bosnia, represents a considerable but useful simplification. This investigation is based on an analysis of place names and descriptions of military operations in Western media, primarily the *Times* of London, the *New York Times*, and *The Washington Post* for the period of January 1990 to September 1995.

As we examine the military geography of the war in Bosnia, one must bear in mind that 'warlords' controlled much of the conflict. Largely independent, local commanders controlled critical areas, such as Bihac and the Krajina. This inevitably makes short descriptions of this conflict somewhat misleading, as it will seem that the major participants controlled events.

I. February to April, 1992

This period was characterized by disorganized fighting and the beginnings of the first major offensives (see Map 2). Serbian forces were principally involved with *control operations* aimed at securing the contiguous areas in eastern Bosnia, with attacks along the Drina from Serbia on either side of Zvornik and through Visegrad. They also began *connect operations* aimed at securing the Brcko corridor to central Bosnia, with attacks into northeastern Bosnia at Bijeljina. During this period, the Croats consolidated the areas of Bosnia contiguous to Dalmatia through *control operations*, such as those west of Gornji Vakuf. Toward the end of this period, the Serbs renewed *control operations*, this time further south from the area near Montenegro towards the Croat-Muslim city of Mostar. By the end of April, fighting quieted down as factions prepared for the next round.

II. May to July, 1992

In May, the Yugoslav government in Belgrade yielded control of perhaps as many as 100,000 Serbian troops in Bosnia to local commanders, effectively creating a Bosnian-Serb Army. Intense fighting again broke out in June, with the Serb offensive to the Dalmatian coast of Croatia at Dubrovnik, southwest of Trebinje (see Map 3). This resulted in the destruction of much of the ancient city through repeated artillery bombardments. This could be classified as a *pacification operation*, since the Serbs were attempting to occupy an isolated portion of the Dalmatian coast. The Serbs also continued *connect operations* eastward and westward into the Brcko corridor, consolidating their control of this important area. At the same time, the Serbs consolidated their control of eastern Bosnia, beginning sieges of Zvornik (which was occupied quickly), Srebrenica, and Gorazde. With these *pacification operations* successfully completed, the Serbs had essentially completed their control of the of the eastern Bosnian region adjacent to Serbia. Simultaneously, Croatian *control operations* against the part of the Krajina south of Bihac were attempted to secure routes connecting the Dalmatian coast to the rest of Croatia.

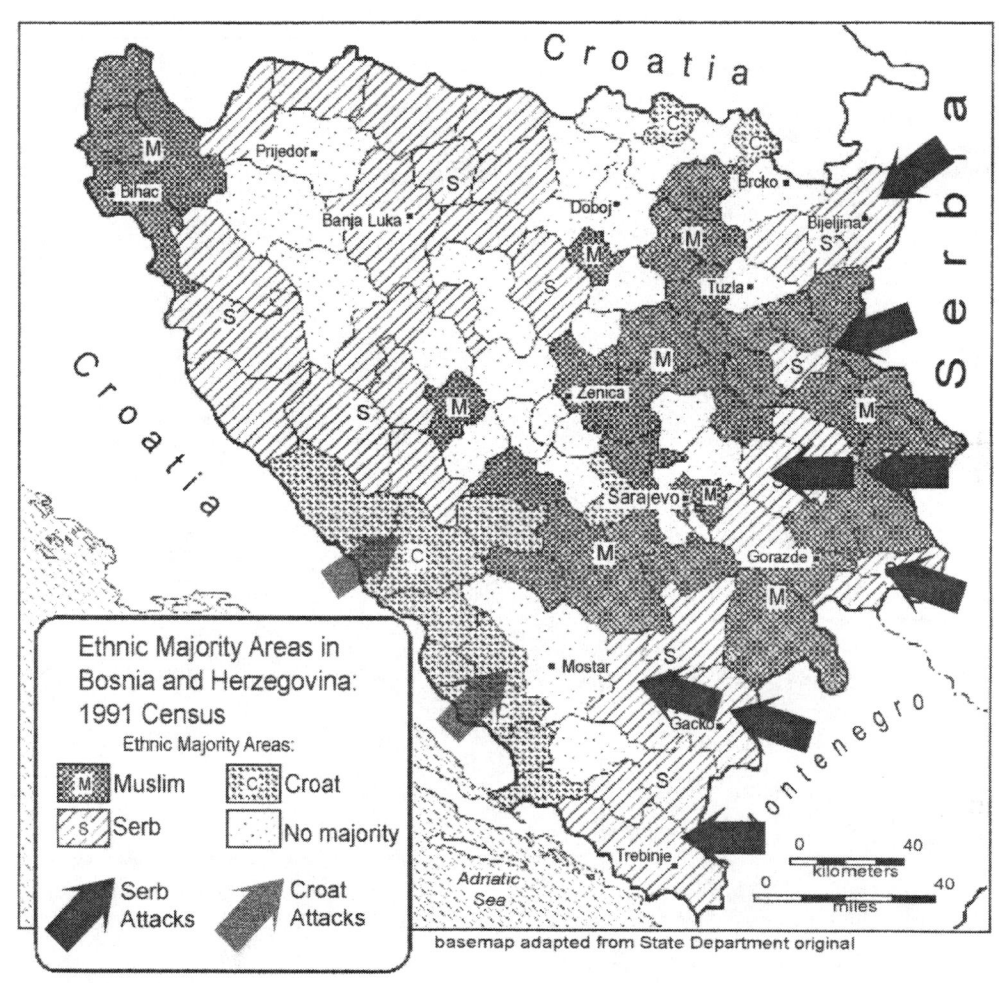

Map 2. Phase I of the Bosnian War - control and connection of ethnic enclaves.
Source: U.S. Department of State, 1995.

After a lull in late June, fighting resumed in July. Croatian forces staged an offensive from Croatia toward Mostar to secure that area against further Serb and possible Bosnian government attacks. The Serbs tightened their siege around Gorazde, and began clear peripheral areas around the Muslim enclave of Bihac.

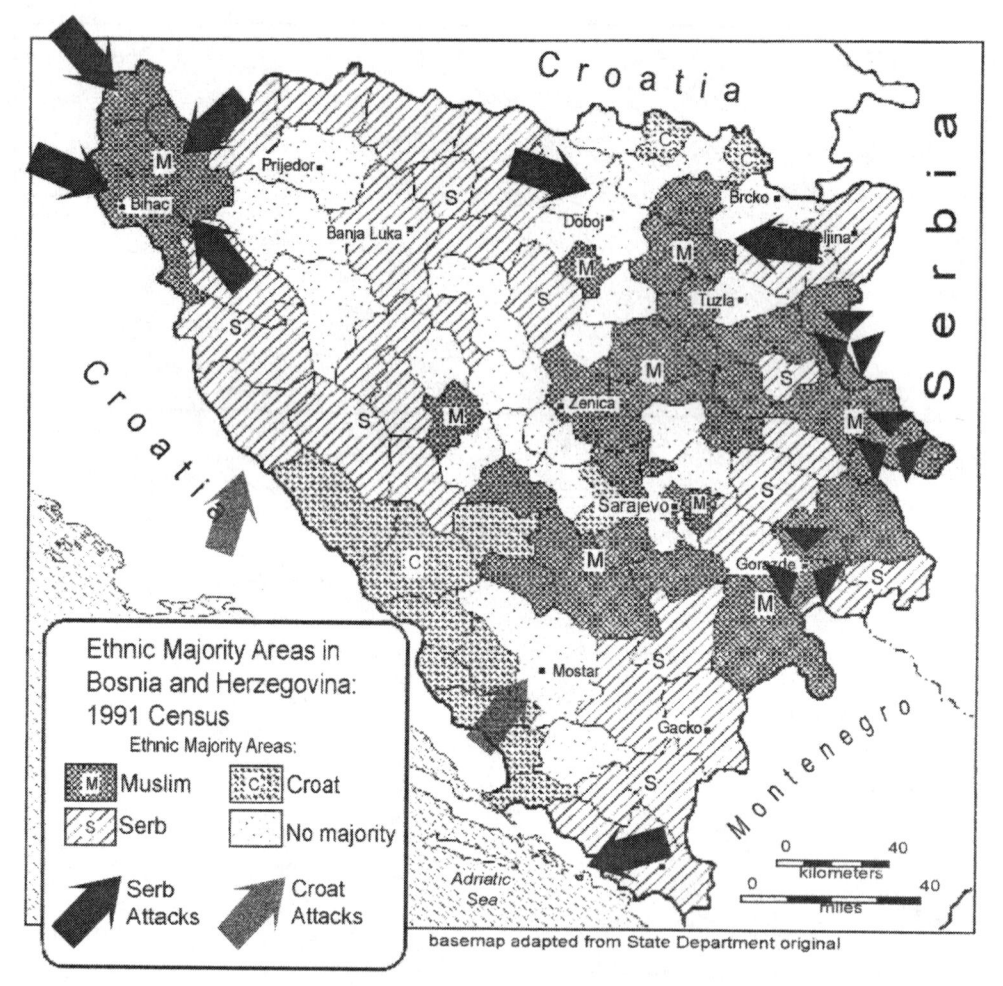

Map 3. Phase II of the Bosnian War.
Source: U.S. Department of State, 1995.

III. September 1992 to July 1993

The beginning of this period saw a major change in the war—the outbreak of overt Croatian attacks against the Bosnian government (see Map 4). Although the events leading to this change are still somewhat mysterious, it appears that the Serbs and Croats came to an understanding about the status of the Brcko corridor in October 1992. In early October, the Croats pulled back from Bosanski Brod, a town on the Sava River in northern Bosnia and the town fell to Serb forces two days later. A few days after that, Serbian attacks began against Gradacac and Brcko, the last Muslim-held towns in the corridor. These *connect operations* continued into December, and were resumed in the Spring of 1993. However, in early November, the Serbs abandoned their siege of Dubrovnik (i.e., a *pacification operation*), and the Croats re-occupied the Prevlaka Peninsula at the far southeastern extremity of Croatia.

The Prevlaka Peninsula is important because it controls the outlet to the Gulf of Kotor and the harbor of Kotor, which is the only major harbor in the region. Fighting during the winter of 1992-1993 was confused, with three-way battles for local objectives in central and southern Bosnia, and in the Brcko corridor. In January 1993, the Croats launched a major *control offensive* against the Serbs at Maslenica, in Croatia, south of Bihac, where the Serbs threatened to cut off Dalmatia from the rest of Croatia. In early 1993, the Serbs continued pacification *operations* in eastern Bosnia with attacks against what became designated the "U.N. Safe Areas" of Gorazde, Srebrenica, and the village of Zepa. The U.N. Safe Areas were small areas, centered on cities, towns, or villages, in which U.N. troops were stationed to provide temporary havens for refugees from less secure areas.

April 1993 saw the start of the most successful foreign intervention of the war—Operation Deny Flight. This American-led operation was designed to prevent any warring faction from using military air power. In effect, this was directed at the Serbs, who were the only force with significant air resources, and were operating out of their base at Banja Luka. Operation Deny Flight, which lasted until the Dayton Peace Agreement in December 1995, was generally successful, but with a few exceptions. By July of 1993, areas under the control of the three sides had reached approximately the limits outlined in Map 5 (Crampton, 1996).

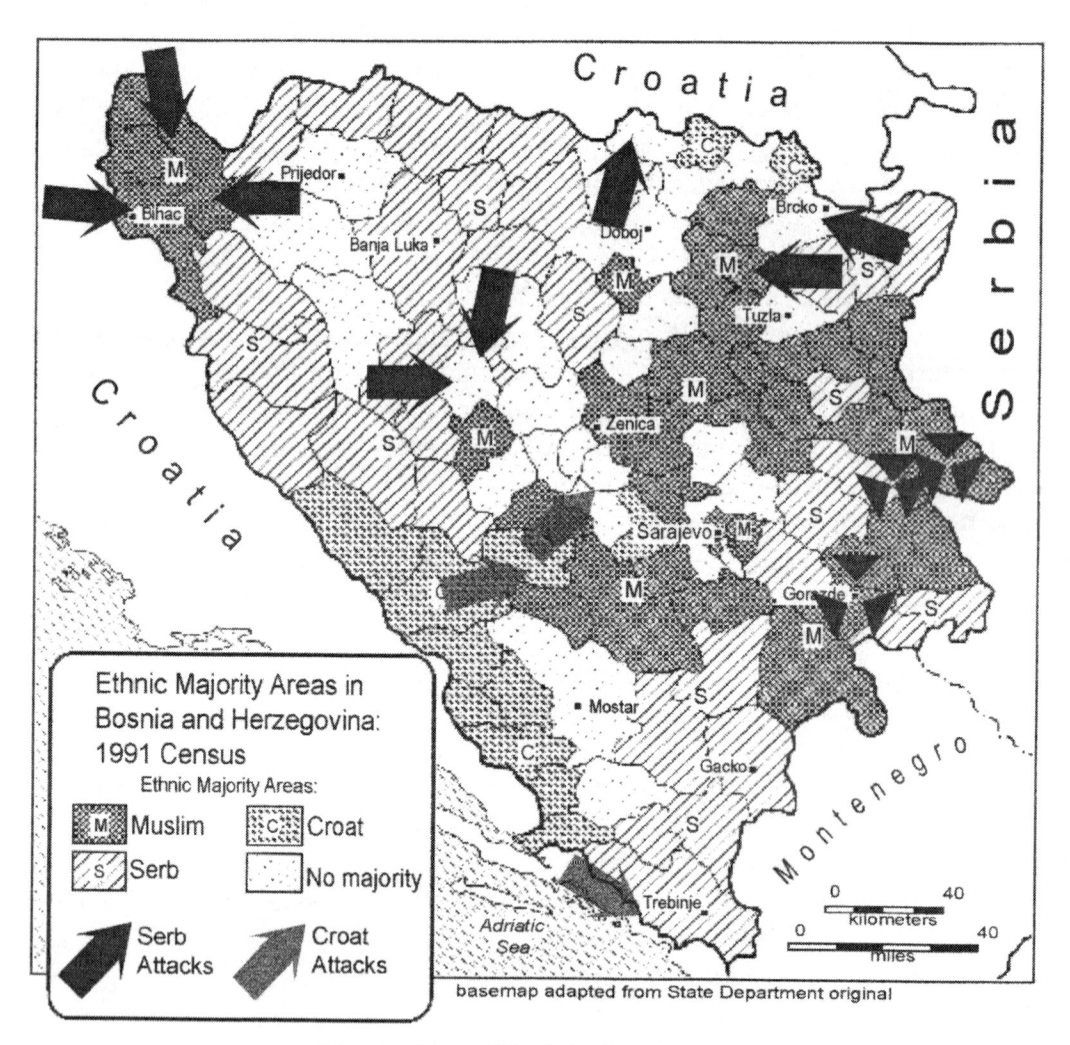

Map 4. Phase III of the Bosnian War.
Source: U.S. Department of State, 1995.

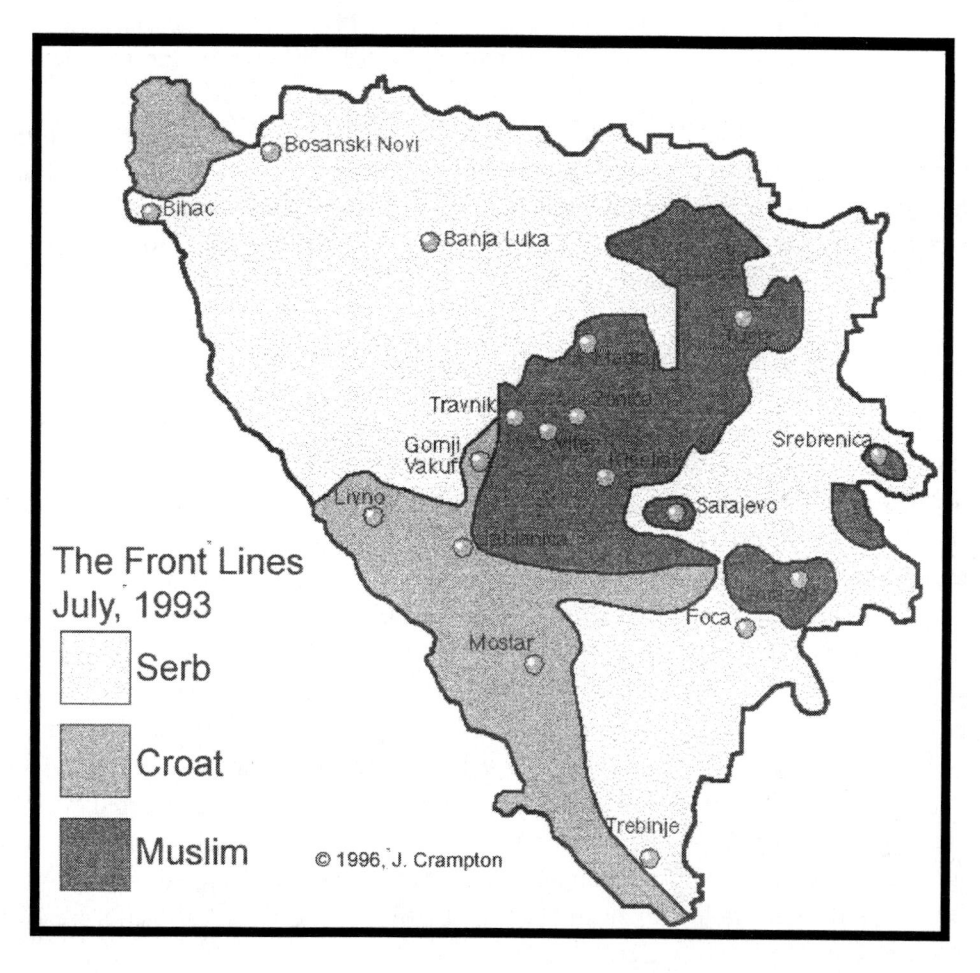

Map 5. Approximate areas of ethnic control in Bosnia in July 1993.
Source: U.S. Department of State, 1995 and Crampton, 1995

IV. July 1993 to April 1995

By July 1993, the apparent, if never acknowledged, truce between the Croats and Serbs was failing badly. The Serbs were also losing momentum because of their continued shortages of manpower. Their clear material superiority began to be counter-balanced by their numeric inferiority in the face of increasing Croat cooperation with government forces in Bosnia. There was mixed fighting along all fronts, especially in central and southern Bosnia, around the Bihac pocket, and in Sarajevo. A year of stalemate began after the end of Croatian-Bosnian government conflict in a U.S.-brokered truce in March of 1994. The truce marked the beginnings of the Croat-Bosnian Federation, which is one of the two Bosnian parties to the Dayton Peace Agreement. Nonetheless, Serbian sieges in eastern Bosnia, around Bihac, and Sarajevo continued unabated.

V. May 1995 to October 1995

By the spring of 1995, the Croats built their army to a point where they could successfully undertake major offensive operations against Serbian units (see Map 6). The Croats adroitly used their external lines of communication around the periphery of Bosnia to launch three major offensives. The last of these was in conjunction with Bosnian government forces. With the advent of the Croat-dominated Federation of Bosnia and Herzegovina, Bosnian government forces had in effect become the junior partner. This occurred because of the Bosnian Croats' ability to draw upon their base of support in Croatia. In effect, Croats inside and outside of Bosnia fought as one force.

The first Croat offensive in Bosnia was "Operation Flash" in early May 1995. In five days, the Croats retook Western Slavonia. The Serbs had controlled Western Slavonia since 1991. This *control operation* reopened a major highway and rail line connecting central Slavonia with Zagreb. In July, the Serbs managed to end two of the sieges of Muslim-held towns in Eastern Bosnia when the U.N. 'safe areas' of Srebrenica, and Zepa, fell to Serb forces.

The second Croat offensive was "Operation Storm" in August 1995. This was a Croat *control operation* against the entire Krajina. The Croats, in a decisive and surprising operation, were able to rapidly overrun the remaining Serb-controlled areas in Croatia with the exception of Eastern Slavonia. Operations "Flash" and "Storm" were also the first contemporary military operations with their own pages on the World Wide Web, a trend, which has been followed by the U.S. Department of Defense (Croatian Academic and Research Network, 1996; U.S. Department of Defense, 1996a; 1996b).

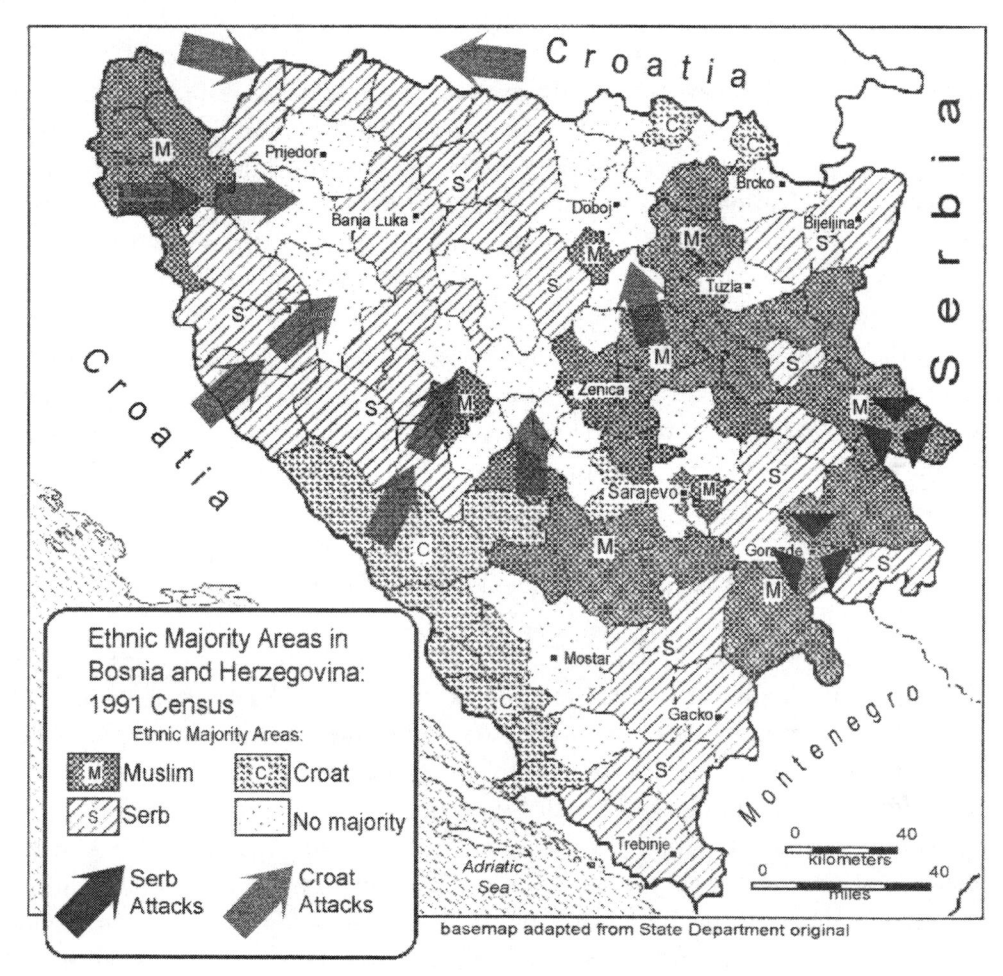

Map 6. Phase V of the Bosnian War.
Source: U.S. Department of State, 1995.

The final offensive of the war, which was conducted by Croat and Bosnian government forces in September 1995, was aimed at the southern part of Serbian-controlled areas in central Bosnia. An offensive here would liberate territory and had two other major geographic goals. First, the Croats and Bosnians wanted to secure the Bihac enclave, although connection of the enclave to Croatian territory had in effect been accomplished by the Croatian offensive in August. Second, they wanted to bring under Croat-Bosnian control the major east-west highway within Bosnia (the M5), which runs generally from Sarajevo through Jajce to Bihac. This was accomplished easily, and was probably aided by Serb disorganization. Although the ultimate aim of the operation may have been the central Bosnian Serb stronghold of Banja Luka, the operation never reached that city, either because of stiffening Serb resistance, or perhaps because the Croat-Bosnian Federation realized that a negotiated peace would have been difficult if they controlled the majority of the territory claimed by the Serbs.

At the end of this last offensive, the areas controlled by each side had reached the limits they would have at the time of the signing of the Dayton Peace Agreement (United States Department of State, 1995). Mutual exhaustion seems to have set in at this point, and it appears that all sides realized that further fighting would not be worth the cost.

An Historical Analogy

Analogies have been drawn in the American press between the war in Bosnia and previous wars involving the United States, usually the Vietnam and the Gulf Wars. However, the most valid historical analogy to the conflict in Bosnia is The Spanish Civil War of 1936-1939 (Mitchell, 1983; Beevor, 1983; Bolloten, 1991). That conflict has several points of similarity to the war in Bosnia.

Although Spain is significantly larger than Bosnia, the terrain is similar. Mountains, poor weather, and poor communications constrained military operations in the same manner as in Bosnia. Spain, like Bosnia, is an area with good communication and transportation around the perimeter, but, with the exception of the hub of Madrid, the interior communications are less adequate for major military operations. The overall level of military technology was similar as well; World War II era weaponry predominated both conflicts. The Warsaw Pact tanks and aircraft used by Serb forces and to some extent by the Croat were a notable exception. However, the conflict in Bosnia saw less air power than was apparent in the Spanish Civil War. The air power evident in Bosnia was provided by the 'interventionist powers' of the United States and Western Europe, just as Germany, Italy, and the Soviet Union provided air power for the Nationalists and Republicans in Spain.

There are political similarities as well. The internationally recognized government in each case was faced with a well-armed and powerful opponent. Intervention by outside powers on the side of the recognized government was generally ineffective in Spain, and was generally ineffective in Bosnia. Volunteer forces of various forms from outside the region came to the aid of the various sides in Bosnia, as they did in Spain. World opinion played a strong role in shaping the course of the Spanish Civil War, and continues to play a leading role in determining the actions of the major parties to the Bosnian conflict. Although the Spanish Civil War, unlike the predominantly ethnic conflict in Bosnia, was caused by ideological and social cleavages in Spain, that war did have ethnic elements, with the Basques, Galicians, and Catalans generally siding with Republican forces (Bolloten, 1979). The war in Bosnia did have elements of social and economic cleavage, with the Bosnian Moslems being predominantly urban, and the Serbs being predominantly rural.

The Spanish Civil War followed the pattern described in the second section of this chapter, with each side at first consolidating its base of power, then moving against enclaves controlled by the other side. The Nationalists (in Spain) were able to divide areas of Republican control via offensives at Castellon in July 1938, then proceeded to reduce the remaining areas of Republican control with centrifugal offensives from the center of Spain (Mitchell, 1983; Beevor, 1983). In Bosnia, the Serbs and Croats first consolidated their bases of power, and then moved to enlarge them by absorbing friendly exclaves and seizing enemy enclaves.

One major difference between the two conflicts is that, in Spain the Nationalists were able to achieve complete operational and strategic victory, and were able to determine the course of events for the next generation. However in Bosnia, no side was able to impose its will, and the current stalemate seems to hold the possibility of forming the basis for a more-or-less permanent end to the conflict. This holds with it the risk for renewed conflict, but also maintains the promise of a solution that may be able to meet the basic needs of the three parties to the conflict.

Conclusion

It is evident from a cultural map of Bosnia that the human landscape is a patchwork of ethnic enclaves, separated by imposing, mountainous terrain in many places. This landscape played a key role in the conduct of the war as each of the three warring factions conducted various operations (i.e., control, connect and pacification) to achieve their goals. Relative location had an equally influential role in the war as each of the warring factions exercised interior and exterior lines of communications and sought to secure links to outside sources of military aid from geographically contiguous states. Transportation links were a vital factor during the final, Croatian-led, offensives in which they were able to use external lines to slide forces around the periphery of Bosnia.

It seems clear that all sides in the Bosnian conflict were ready for an end to the struggle by 1995. The offensives described in this chapter were driven in large measure by geographic necessity. The Serbs began the war constrained by the requirement to open and maintain the Brcko corridor. This necessity was manifest during peace negotiations when the Serbs demanded the widening of that corridor to 20 km as one of their main issues. Clearly the Serbs were concerned about maintaining a direct land link to that enclave and they felt threatened by two Croat enclaves on the south bank of the Sava River, which might conceivably threaten the Brcko corridor... These issues have been deferred to arbitration, and were not tentatively settled by the Dayton Peace Agreement (United States Department of State, 1995).

The Brcko corridor also provides possible evidence that the peace agreement may actually be rooted in the attitudes and goals of the participants. After the successful Croatian-Muslim offensives of July-September 1995, the Federation could have pressed its advantage by continuing attacks on the Serbs in Bosnia. One obvious target should have been the Brcko Corridor. An attack across the corridor would have isolated the Serbs in western Bosnia from their sources of supplies, which were critical for the Serbs given their chronic shortage of manpower. An attack across the narrow, if well-defended, corridor could have been a prelude to the complete destruction of Serb military forces in central and western Bosnia. Nevertheless, such an offensive never materialized. This may have been because of the heavy defenses in this area, or may have been due to logistical difficulties for the Croats, who would have been operating at the end of their supply lines.

An alternative interpretation would be that the Croatian and Bosnian governments, having brought under their control approximately half of the territory of Bosnia, now controlled the approximate territory, which would be awarded them under any peace accord. An attempt to destroy Serb power in central and western Bosnia would have hardened Serb resolve, both inside and outside of Bosnia. This certainly would have made continued conflict almost inevitable, with a strong likelihood of the more active involvement of Serb-controlled Yugoslavia. In other words, such an offensive would almost certainly have meant a widened war, with Croatia and Yugoslavia openly entering the conflict. With the beginnings of a settlement drawn up in the Dayton Peace Agreement, and the five years of stability, which it has brought, hope may finally be dawning for the people of Bosnia.

Bibliography

Beevor, Antony. 1983. *The Spanish Civil War.* New York : Harper & Row.

Bennett, Christopher. 1995. *Yugoslavia's Bloody Collapse: Causes, Course and Consequences.* New York: New York University Press.

Bolloten, Burnett. 1979. *The Spanish Revolution: The Left and the Struggle for Power during the Civil War.* Chapel Hill, NC: University of North Carolina Press.

Bolloten, Burnett. 1991. *The Spanish Civil War: Revolution and Counterrevolution.* Chapel Hill, NC: The University of North Carolina Press

Central Intelligence Agency. 1991. *Bosnia and Herzegovina Summary Map.* Washington, DC: Government Printing Office.

Clissold, Stephen (ed.). 1966. *A Short History of Yugoslavia from Early Times to 1966.* Cambridge: Cambridge University Press.

Cohen, Lenard J. 1993. *Broken Bonds: The Disintegration of Yugoslavia.* Boulder, CO: Westview Press.

Donia, Robert J., and Fine, John V. A. 1994. *Bosnia and Hercegovina: A Tradition Betrayed.* New York: Columbia University Press.

Kerner, R. J. (ed.). 1949. *Yugoslavia.* Berkeley, CA: University of California Press.

Luttwak, Edward N. 1987. *Strategy: The Logic of War and Peace.* Cambridge, Massachusetts: Harvard University Press.

Malcolm, Noel. 1994. *Bosnia: A Short History.* New York: New York University Press.

Mitchell, David J. 1983. *The Spanish Civil War.* New York: F. Watts.

The New York Times. New York. January 1990 to present.

Singleton, Fred B. 1976. *Twentieth Century Yugoslavia.* New York: Columbia University Press.

The Times. London. January 1990 to present.

The Washington Post. January 1990 to present.

World Wide Web Sites:

Crampton, Jeremy. 1996. *The Bosnian Virtual Fieldtrip.* http://geog.gmu.edu/gess/jwc/bosnia/bosnia.html

Croatian Academic and Research Network. 1996. *News from Croatia.* http://www.carnet.hr/hr_news/index.html

Foelckers, Joerg. 1996. *JF's Bosnia Pages.* http:// mac-absynt-1.Informatik.Uni-Oldenburg. DE/ Pages/ War/ Bosnien.HTML

Irfanoglu, A., Kirac, A., and Cataltepe, Z. 1996. *Bosnia HomePage at Caltech.* http://www.cco.caltech.edu/~bosnia/bosnia.html

Tele/Telecom Finland. 1996. *Bosnia On-Line.* http://www.inet.fi/jugoslavia/info_e.html

Terviö, Marko. 1996. *Information About Croatia and Bosnia-Herzegovina.* http://www.helsinki.fi/~tervio/info.html

United States, Department of Defense. 1996a. *America's Army in Bosnia.* http://www.dtic.dla.mil/bosnia/army/

United States, Department of Defense. 1996b. *BosniaLINK.* http:// www.dtic.dla.mil/bosnia/index.html

United States, Department of State. 1995. *Dayton Peace Agreement--Official Texts.* http://dosfan.lib.uic.edu/bosagree.html

Members of the Dutch, French, German, and U.S. military watch as an Italian honor guard hoists the new Stabilization Force flag during the Stabilization Force (SFOR) activation ceremony in Sarajevo, Bosnia and Herzegovina, on Dec. 20, 1996. The ceremony recognizes the transition from the Implementation Force (IFOR) and the change of mission from the implementation of the Dayton Peace Accords to SFOR and the stabilization of the changes brought about by that treaty.
Source: DoD photograph by Staff Sgt. Michael Featherston, U.S. Air Force.

9

The Geography of Hazard Analysis:
DISASTER MANAGEMENT AND THE MILITARY

By

Ewan W. Anderson

Introduction

Some disasters can be prevented, but others can be merely limited in their effect. From time immemorial the military has been involved with both categories of disaster.

> *"It is doubtful if any of the armed forces of the world would deny that they have the capability for making a constructive, sometimes vital, contribution to the conduct of disaster relief operations."* (Harbottle, 1991)

The aims of this chapter are: 1) to examine the phenomena commonly characterized as disasters, natural hazards or environmental hazards; 2) to identify and assess vulnerabilities and preparedness; and 3) using these as a background, to identify the changing actual and potential role of the military.

A hazard can be defined as:

> *potential risk to humans and their welfare* (Smith, 1992).

The hazard poses the threat and the risk is an assessment of the probability that the hazard will actually occur. The realisation of the hazard can be a disaster. On 11 December 1987, the United Nations General Assembly designated the 1990s as the "International Decade for Natural Disaster Reduction." The aim was to reduce the disruption and loss of life caused by earthquakes, wind storms (e.g., cyclones, hurricanes, tornadoes, typhoons), tsunamis, floods, landslides, volcanic eruptions and wildfires (Degg, 1992). This was a timely intervention in that the previous 50 years appear to have witnessed an increasing rise in the occurrence of natural hazards, which have resulted in disasters. This may mean that the frequency or magnitude—or both—of natural hazards have been rising, or it could indicate that reporting has simply been more efficient.

> *"No major geological or climatological changes over the last 50 years adequately explain the rise* [in natural hazards]." (O'Keefe et al, 1976)

The most likely explanation for the increase in hazard incidence would therefore seem to be the effect of human influences. Hazards may frequently be triggered by natural phenomena, but then compounded by human actions so that they are classified as complex disasters. This brings into question the epithet "natural" and certainly any classification, which purports to distinguish between natural and man-made or technological disasters. Therefore, the preferred term, covering all possibilities, is "environmental" hazard and the study of such events is clearly multidisciplinary, since the focus is at the interface between human and natural or physical systems.

These points can be illustrated by the tragic and well-documented loss of 500,000 lives in East Pakistan (now Bangladesh) in November 1970. The tragedy was initiated by a cyclone, but the more appropriate term was that used by Paul Richards (O'Keefe et al, 1976): "a social disaster". That such a disaster occurred on such a scale could be largely attributed to two human factors. Firstly, farmers had been encouraged by the government to settle on reclaimed coastal land that was not protected from storm surges. Secondly, the

Meteorological Bureau did not issue a warning as soon as they had detected the advancing cyclone, three days before the event. In fact, they only transmitted the warning to the radio station after it had closed at 11:00 p.m. on the night that the cyclone struck (Bryant, 1991).

"To blame "nature" for unavoidably and unexpectedly killing 500,000 people is to fail to view "hazards" as an integral part of the spectrum of man-environmental relations." (Hewitt, 1983)

Therefore, any increase in the occurrence of disasters is most obviously explained by the growing vulnerability of the world's population. This is evident and results from human influence.

"The growing numbers of people in the underdeveloped world are becoming increasingly vulnerable to extreme geophysical events, due partly to their tendency to live in severely overcrowded conditions in areas characterized by poor building design, inadequate infrastructure and often non-existent communication systems, and partly due to so called "development" schemes having negative feedbacks on the environment such as famine and soil erosion." (O'Keefe, 1976)

Therefore, given the interplay between man and the physical systems, clearly demonstrated in this discussion, environmental hazards are best studied in an ecological framework. As Jones (1991) argues, the natural environment is not a neutral background for human life, but it is an active component in a complex interrelationship with human societies, which varies in both time and space. Attributes of the natural environment, valued by society, are defined as resources, those which inflict disruption, damage, destruction or death are referred to as hazards or disasters.

"It is essential to recognize that both "resources" and "hazards" are human assessments or cultural appraisals and are, as such, intimately interrelated. Few phenomena can be deemed wholly good or wholly bad, the vast majority are combinations of the two ... Whether phenomena are deemed "beneficial" resources or "damaging" hazards depends on the balance between their perceived "costs" and their perceived "benefits".

This, in turn is largely dependent on the state of cultural development of the affected society." (Jones, 1991)

Key factors in assessing environmental hazards are the number of events and the magnitude of each. Environmental hazards, the effects of which cause great costs, are known as disasters and are of relatively low frequency and high magnitude. Such disasters are at one end of the continuum that includes numerous occurrences of medium size and large numbers of small events. For example, earthquakes: each year 50,000 to 100,000 shocks are detected; approximately 1,000 are large enough to be felt by humans; possibly 100 are sufficiently large to entail costs; but only 10 would be of sufficient size to cause catastrophic losses.

Classification

To develop theory on environmental hazards and thereby to assess vulnerability and develop relief procedures, the various sources of potential disaster need to be classified. Bryant (1991) makes a basic distinction between climatic and geological hazards. The former comprise: wind, drought, floods and fire; and the latter: earthquakes, volcanoes and instability. Smith (1992) distinguishes atmospheric (single element or compound), hydrologic, geologic, biologic and technologic. This classification is illustrated by a spectrum of hazards from geophysical events to human activity: earthquake, tsunami, cyclone, volcanic eruption, tornado, avalanche, flood, drought, bush fire, transport accidents, industrial explosions, water pollution, radioactive fallout, civil riot, food additives, smoking and mountaineering. The spectrum also moves from the involuntary to the voluntary and from the intense to the diffuse. Both are, in their way, comprehensive, but neither demonstrates clearly the interaction between man and the physical environment. This is shown more clearly in the taxonomy adopted by Jones (1991), which can be summarized as:

a) <u>natural</u>:

    i) *meteorological*: fog, snow, hail, lightening, tornado, tropical cyclone, temperature extremes, and drought. Many of these may occur as compound hazards (e.g., thunderstorms, blizzards or hurricanes, etc.)

ii) *geomorpholoical/hydrological*: mass movements, flooding, erosion, silting sand movement, soil degradation, wave action, sea ice, glacier advance
iii) *geophysical*: earthquakes, volcanic eruptions, tsunamis
iv) *biological*: severe epidemics, plant or animal invasions, and fires.

b) <u>quasi-natural</u>: smog, acidification, desertification, flooding, salinization, avalanching, soil erosion, landsliding, global warming, sea-level change, and ozone depletion.

c) <u>anthropogenic</u>: pollution, photochemical smog, social hazards, transport hazards, production hazards, terrorism/sabotage, war.

This list can be further modified by the following key variables: 1) intensity; 2) duration; 3) response predictability; and 4) responses. Many hazards are characterized as being violent, high energy events. This is not, however, universally true. For example, drought, desertification, heatwaves and disease all inflict damage in an insidious non-violent fashion.

Death tolls have been over-emphasised as a measure of hazard impact significance, largely because they are immediately measurable. However, only before God are people equal. The effect of massive loss of life in rural Bangladesh has only a relatively minor impact upon the national economy. Losses of a similar scale in an urban environment would be infinitely more damaging. Furthermore, the risks of loss of life are not equal among hazards and costs may be far greater in property than life. In addition, in the developed world, casualty figures can be greatly reduced through warning systems and emergency procedures.

A distinction must be made between the short, sharp, violent events and those which are gradual. The chaos caused by a major earthquake is likely to be far more obvious than that resulting from an epidemic. Desertification is clearly a relatively slow process and as such, is relatively difficult to measure. However, once measured, it is easier to predict.

Hazards have causes and patterns of behavior, which can be studied. For example, recurrence intervals or return periods can be used to indicate average intervals between events of a particular magnitude. Thus, a hundred year flood is the value likely to be equalled or exceeded, once in a hundred years. These events are therefore not freaks of nature, but extremes. A major problem is that many of the recurrence intervals are far longer than the time-scales normally employed in planning and certainly in political decision-making. Furthermore, if the interval is particularly long, memories are erased and society becomes once more vulnerable.

Some defense can be offered by technology in terms of both structural and non-structural responses. Structural responses would include the provision of dams or the strengthening of buildings against earthquakes, while non-structural responses would be illustrated by the use of satellites for forecasting or telecommunications for issuing warnings. However, technology may cause sufficient disturbance to the environment that hazards are actually engendered. For example, dams are said to provide, among many advantages, flood control. In fact, if the lake behind the dam is sufficiently high to allow maximum generation of power, there is no spare storage for any flood. Therefore, multi-purpose dams are designed to allow flood water to pass without damaging the dam.

Vulnerability

Since disasters, as defined and classified, can be in part the result of physical forces and in part that of human activities, it is obvious that vulnerability will vary from place to place. The same physical hazard, such as an earthquake of a particular intensity, is likely to have a different effect in different locations. Further, two earthquakes of very different intensities may have the same effect, depending upon the state of local preparedness. The International Federation of Red Cross and Red Crescent Societies (IFRCRCS) has prepared a guide to vulnerability (8 March 1993) in which the concept is discussed and the production of an assessment is outlined. Vulnerability is about the potential for something to occur and therefore involves probabilities and is not a fixed condition.

In identifying threats, the IFRCRCS has produced a classification based on the practicalities and therefore, rather different from the scientific taxonomies already discussed. It is more broadly based, but separates the physical from the human:

a) events based in nature (e.g., earthquakes, floods, etc.);
b) events based in violence (e.g., war, intimidation, harassment, conflict, sexual assault, prejudice, etc.); and

c) events based in economy/society (e.g., environmental degradation, structural adjustment, AIDs and other health hazards, technological discovery, trade shifts, changes of government, etc.).

Because of these threats, the vulnerability of particular societies can be examined. Three basic characteristics are considered to make some groups more vulnerable than others: 1) proximity/exposure; 2) poverty; and 3) exclusion/marginalization.

Poverty of itself does not necessarily indicate vulnerability, which depends upon the capacity for action. Therefore, the idea of exclusion or marginalization is introduced. Clearly, people who are both poor and marginalized are particularly vulnerable. The capacity to counter or overcome threats depends upon the physical or material state of the particular group, its social and organizational level and the prevailing attitudinal/ motivational/psychological stance. Using this framework, the IFRCRCS has developed a vulnerability assessment in which for any country or area, the threats, and the vulnerable groups and their response capacities can be identified. The U.S. Office of Foreign Disaster Assistance (OFDA), based in the State Department, in monitoring disasters and providing a variety of types of assistance, promotes:

"Self-reliance so that countries are able to anticipate, mitigate, and manage disasters." (OFDA 1998)

In 1998, aggregating natural disasters and complex emergencies, OFDA identified 64 countries as disaster-affected: 22 in Africa, 11 in Asia, 14 in Eurasia and 17 in Latin America, including the Caribbean. According to OFDA's records the number of complex emergencies has increased dramatically over the past decade. The change is indicated by the number of refugees from and internally displaced people (IDPs) in crisis affected countries. Examples are given below:

Table 1: Refugees and Internally Displaced Peoples (IDPs)[1]

Country	1991	1999
Afghanistan	6,500	4,600
Sudan	4,900	4,400
Palestine	2,600	3,900
Former Yugoslavia	300	2,500
Angola	1,300	2,200
Burma	800	2,100
The Democratic Republic of the Congo (DRC)	80	1,500
Iraq	2,500	1,400
Burundi	200	1,200
Sierra Leone	10	950
Somalia	900	800
Republic of Congo (Brazzaville)	10	800
Russian Federation	10	750
Indonesia/Timor	400	750
Turkey	40	730
Sri Lanka	1,200	720
Azerbaijan	300	700
Rwanda	250	650
Eritrea	200	600
Lebanon	800	350
Liberia	1,300	300
Vietnam	120	300
El Salvador	450	280

[1] Total of refugees fleeing from the country, plus the number of IDPs remaining within the borders. Scale in 1,000s.

In the 1990s, the concept of complex emergency in many ways replaced natural disasters as the paramount concern of the international humanitarian community. An event is normally considered a complex emergency when emergency conditions are prolonged by strife. Although having features in common with international wars, complex emergencies also include ill-defined low intensity conflicts, civil wars, genocide, ethnic cleansing and situations where effectively there is no government control. However complex emergencies are defined, a common characteristic is that the human response itself is complex and difficult. Since they are likely to affect significant proportions of the civilian population and to result in the destruction of social structures, they pose extraordinary challenges. Whereas natural disasters may require a direct operational response to food or shelter, complex emergencies necessitate additional attention to security and political and social issues. Obviously, the effect of armed conflict in a complex emergency situation magnifies the impact upon civilians. Examples of combat strategies that targeted civilian areas occurred in Mozambique, Angola, El Salvador and Guatemala. The other term sometimes encountered is that of the "failed state". Complex emergencies in failed states have occurred in Somalia, Bosnia, Liberia and Sierra Leone. However, natural disasters can still be of major significance. In the annual report of OFDA (1998) it is stated that:

"In the span of one year, natural disasters attributed to El Nino affected more than 129 million people world-wide: an estimated 22,000 people lost their lives, approximately 378,000 were infected with water and vector-borne disease, and some 4.6 million people were displaced from their homes."

Preparedness

The IFRCRCS defines disaster preparedness as:

". . . a readiness to predict, prevent, respond to and cope with the effects of a crisis. Preparedness may be maintained by the potential victims themselves as well as external support systems at any level, from local to international."

While local preparation and planning are vital, it is important to identify the key level of preparedness management. In small and medium sized countries this is likely to be at a national level, but in large countries at the state, regional, provincial or district level. The suggested requirements at each level are set out in a document by the IFRCRCS, which also indicates the potential role of the Federation itself. With its vast experience, the Federation will clearly continue to be the focus for improvements in global preparedness.

Disaster relief appears to be making increasing use of the military, but this is a controversial issue as recognised by the IFRCRCS in their paper *The Use of Foreign Military Resources in Natural Disaster Relief*. In a presentation at the International Peace Academy (1986), Field Marshall Lord Bramall summarised the military role:

> *"The armed forces have special qualifications (for disaster relief). They possess strategic and local mobility – essential in inaccessible areas – and a wide range of specialised equipment: helicopters, aircraft, earth-moving machinery, respirators, medical supplies, power and lighting equipment and under-water capability. Most important, they are self contained with their own rations and transport. They operate a command and communications system invaluable in liasing with the host government and other forces and agencies in the area; and they are specialists in reconnaissance. If required, they could provide a command and coordination network to run an operation."*

The essential resources identified are therefore: a) transportation; b) engineering; c) communications and intelligence; and d) medical skills. These capabilities may be deployed at various scales of military involvement, identified by the IFRCRCS as: a) simple involvement in natural disasters; b) formal partnership in large scale disasters; and c) cross-border operations.

According to the degree of involvement, the question of leadership of the operation arises. The arguments against the use of the military focus on the military aspect. The question is, can the Red Cross/Red Crescent maintain its humanitarian, neutral and independent role if there are standing arrangements with the military? Disasters may result directly from military action and it is clearly somewhat incongruous for humanitarian agencies to work closely with totalitarian military governments. For some, to rely on the military as part of the standard framework of disaster prevention is incompatible with the traditions and objectives of the Red Cross/Red Crescent.

Therefore, the question raised is how best to utilise the resources and infrastructure of the military for peaceful disaster relief. It is already clear from their role in supporting the Kurds in Iraq (1991) that the military system can interface with non-governmental organizations. The lesson to be learned from that exercise is summarized by Harbottle (1991):

> *". . . there is scope for a wider appraisal and appreciation of the military's potential for humanitarian relief operations, and not just as an emergency commitment of a limited term. While coping with the emergency is essentially a civilian undertaking, the quick response capability of the armed forces should be a built-in element of any world disaster relief organization; provided the rapid deployment needed while maintaining in situ for as long as it takes to achieve its purpose and only with-drawing when it can hand over to a competent and well founded civilian operation. In this wider and wholly humanitarian role the military would certainly not constitute a barrier to the furtherance of world peace, but would be more likely to enhance it."*

The involvement of the UN in the Kurdish disaster in 1991 and in Somalia in 1992 has demonstrated both the potential and the weaknesses of involving the military in a role where they are working alongside a variety of civilian charities and relief organizations without a clear command strategy. It is not surprising that the strength of intention has not always been equalled by the quality of useful aid delivered to the people. The fact that a variety of UN agencies based in New York have been operating alongside the Red Cross and Red Crescent based in Geneva and a multiplicity of international agencies and national charities is an impressive testimony to humanitarian commitment. However, the lack of a co-ordinating agency to oversee these multi-faceted responses is evidence of the incompleteness of international planning in the area. The response to famine, flood and other natural disasters is often generous, but frequently disjointed. A co-ordinating agency is vital if humanitarian commitment is to be adequately directed through greater efficiency of the disaster relief effort.

Maharashtra

During research into the feasibility of establishing a global database and co-ordinating agency, the author was privileged to receive personal briefings on the Maharashtra earthquake disaster from the General Officer Commanding Maharashtra and Gujarat Area, the Additional Director General of Police, the General Secretary of the Maharashtra State Branch, Indian Red Cross Society and the Secretary and Special Commissioner for Earthquake Relief and Rehabilitation for the State. These briefings indicated clearly that, while there may be problems internationally, at the sub-national level, there could be highly effective collaboration. There is no Disaster Management Agency for India nationally, or for any of the constituent states and therefore the procedures had to be developed during the emergency. The only organization with the logistical capacity to mount such an operation was the military, which therefore assumed the leading role. Furthermore, partly by design and partly by default, little effective international assistance was involved.

In retrospect, the complete operation has been considered both in India and elsewhere, a notable success. The response to the Maharashtra earthquake can therefore be seen as a good example of co-operation between the government, the military, the police and non-governmental organizations (NGOs) and as a model for the use of the military in disasters. It was particularly fortunate that the General Officer Commanding was such an experienced and perceptive officer as Major General R Mohan, VSM.

The village at the epicentre was Killari, located some 400km from Poona and 250km from Hyderabad, which suffered a 40-second earthquake, registering 6.4 on the Richter Scale. Immediately the news was received, troops and medical teams were put on standby and helicopter reconnaissance commenced, revealing that 11 villages had been severely damaged and that there was large-scale devastation. The troops were alerted at 0400 on 30 September 1993 and within 12 hours, General Mohan was establishing his headquarters at the scene of the disaster. In all, 15,000 troops, engineers and doctors were involved. The immediate tasks included the rescue of people from under the rubble, the provision of medical attention for those seriously injured and the recovery and disposal of bodies. Next, attention was given to the infrastructure. Village tracks were cleared, vehicles were removed from approach roads and a start was made on the restoration of water and electricity supplies. At the same time, reception centres, relief camps, soup kitchens, medical aid posts and distribution points for food and relief stores were established. Most importantly, a radio network was set up to keep the Chief Minister informed and to allow integration with the police. Finally, mobile reaction and relief teams were detailed to visit each of the 43 villages affected by the earthquake. All these activities were monitored and reviewed at the Task Force headquarters in Killari.

At the end of this initial phase, 25,000 people had been housed in 43 tented villages, 50 medical teams had been operational and three 45-bed hospitals had been established. The framework for the operations had been laid down by the military and within three to four days, as other assistance arrived, close cooperation was achieved with the police and the Indian Red Cross. Within two weeks, as rehabilitation measures were being started, the military was able to withdraw.

The disaster illustrates how, in the appropriate environment, the military can mount an effective response and can rapidly ensure conditions within which the other services can operate. During the initial stages, command and control is provided by the military, but, as soon as the immediate emergency has passed and rehabilitation has begun, it is essential that the civilian authorities assume responsibility. Based on the lessons learned at Killari, it is intended to establish a Disaster Management Centre for Maharashtra.

Conclusion

Disasters, whether defined as natural, man-made, technological or complex, present an increasing challenge. The global disaster system, including the agencies of the United Nations, the NGOs, governments and practitioners, at all levels from international to local, has yet to be effectively integrated. As the potential military contribution to emergency response is increasingly realized, it is essential that a military component is induced in the system. However, before this can be achieved, further research is required into the establishment of an acceptable and effective relationship between the military and the key civil authorities: the UN agencies and the NGOs.

Editor's note: this chapter has been adapted from: Anderson, E.W., 1994. *Disaster Management and the Military.* GeoJournal, 34 (2): 201-205. Reprinted by permission.

☙ ❧

Bibliography

Bryant, E.A. 1991. *Natural Hazards.* Cambridge: Cambridge University Press.

Degg, M. 1992. "Natural Disasters: Recent Trends and Future Prospects." *Geography, 336:* 198-209.

Harbottle, M.N. 1991. *What is Proper Soldiering?* The Centre for International Peacebuilding, Chipping XXX.

Hewitt, K. 1983. "The Idea of a Calamity in a Technocratic Age." In: Hewitt, K. (ed.), *Interpretations of Calamity*, Boston Massachusetts: Allen & Unwin.

Jones, D.K.C. 1991. "Environmental Hazards." In: Bennett, R., Estall, R., (eds.), *Global Change and Challenge*. London, Great Britain: Routledge.

OFDA. 1998. "Annual Report FY 1998." *U.S. Department of State,* Washington, District of Columbia: United States Government Printing Office.

O'Keefe, P. Westgate, K., Wisner, B., 1976. "Taking the Naturalness out of Natural Disasters." *Nature, 260:* 566-7.

Smith, K. 1992. *Environmental Hazards.* London, Great Britain: Routledge.

10

Strategic Mobility In the 21st Century:
Projecting National Power in a MOOTW Environment

By

Mark W. Corson

Introduction

Somalia, Iraq, Haiti, Bosnia—this litany of places where American armed forces have recently been involved is the best indicator that the Cold War is over, and we are indeed experiencing a new world order. This new world order, however, is anything but orderly. The international scene today is characterized by a resurgence of ethnic nationalist conflict, regional powers flexing their muscles, the potential for proliferation of weapons of mass destruction, and regional instability caused by refugees, environmental catastrophes, political persecution and economic under-development. The challenges of this Post-Cold War Era involve the interest of the United States and its western allies such that the application of military power is sometimes necessary. However, the geography of American military power has changed in the Post-Cold War Era as well. No longer are massive American military forces deployed overseas in close proximity to trouble spots. Today these forces are based in the continental United States and must be transported across vast distances to hot spots. Projecting American military power is the focus of strategic mobility.

Strategic mobility, or the ability to rapidly move large military forces from the continental United States to foxholes in distant theaters, is a critical component of America's military strategy. Simply stated, the best military force in the world is useless if it cannot get to the fight. As the President's *National Security Strategy for a New Century* put it:

> '*Strategic mobility is a key element of our strategy. It is critical for allowing the United States to be first on the scene with assistance in many domestic or international crises, and is a key to successful American leadership and engagement. Deployment and sustainment of U.S. and multinational forces requires maintaining and ensuring access to sufficient fleets of aircraft, ships, vehicles and trains, as well as bases, ports, pre-positioned equipment and other infrastructure.*'
>
> (White House, 2000)

Strategic mobility is a fundamental geographic issue in that it involves the geographic themes of place, space, and movement. Yet, the American defense establishment has historically had a problem in getting mid-level leaders to appreciate the importance of preparing units to get to the battle. The United States Transportation Command (USATRANSCOM) is the agency responsible for strategic mobility issues. In 1994, they conducted a study in which they traced the flow of information and actions in the deployment process from the President to the deploying unit. This study identified a problem in that middle-level leaders were focused on planning for operations in the contingency area to the exclusion of other concerns—such as getting there. In geographic terms, these leaders did not understand the necessary scale of their focus. This lack of a macroscale view manifests itself in that bwer level leaders fail to emphasize deployment planning, training, and preparation. The results are errors and delays that slow the deployment process, increase the time it takes to put forces in the contingency area, and may ultimately risk allied lives (Fletcher, 1996).

This chapter is an effort to educate current and future military and civilian leaders on the nature and importance of this critical, but often overlooked military geographic theme. We begin with a background discussion of how our national strategy and force deployments changed with the end of the Cold War. The chapter then introduces a military geographic transportation model that provides a conceptual framework in understanding the flow of material from the fort to the foxhole and the systems used to make that happen. The chapter concludes with a case study of the recent deployment of North Atlantic Treaty Organization (NATO) forces to Kosovo.

Threats and Responses

The Cold War National Military Strategy focused on containing communism by maintaining forward-deployed forces. The Army's mission was to defend Central Europe. As late as 1990, the U.S. maintained over 300,000 troops in Europe, as well as pre-positioned stocks of unit equipment. With the end of the Cold War, the situation has changed. The National Military Strategy is now a regionally focused defense with the Army having the mission of being a continental (CONUS) based power projection force, capable of deploying anywhere and conducting operations ranging from conventional warfare to humanitarian relief operations (Department of the Army, 1994). The change in the location of forces and the requirement to overcome long distances poses a fundamentally geographic challenge and strategic mobility is the linchpin.

Operation Desert Storm/Shield demonstrated just how critical strategic mobility operations are to vital U.S. interests. While the Allied Coalition enjoyed considerable military success, the operation would have been impossible without a Herculean logistics effort. This deployment effort had many problems, took months to complete, and illustrated significant deficiencies in American sealift capability, deployment doctrine and planning (Department of Defense, 1992).

The contemporary National Military Strategy foresees a regionally based force with a capacity to handle two major regional contingencies (MRC) at the same time. While the MRCs could be anywhere, the primary focus is on the potential threat from North Korea, and threats by either Iraq or Iran to oil producing states in Southwest Asia. The current strategy envisions stopping an enemy advance in both theaters, holding the enemy in one theater while launching a decisive counter-offensive in the other. Finally, forces are shifted to the second theater to launch a war-winning counter-blow. Clearly this strategy involves substantial movement of large forces around the globe.

Nevertheless, major regional wars are the most likely threat as recent history has shown. Other conceivable operations include limited regional contingencies such as Grenada or Panama, or even more likely, Military Operations Other than War (MOOTW). MOOTW missions include: humanitarian aid, peacekeeping/enforcement, disaster relief operations, support of allies through exercises and shows of force and, support of allied deployments.

Given this new strategy, Congress mandated that the Department of Defense conduct a mobility requirements study to identify assets needed to accomplish the strategic mobility mission. This study identified the need for improved sealift and maritime pre-positioning capability, the continuation of the C-17 program, and improvements in the CONUS transportation infrastructure. The Army's response was to develop the Army Strategic Mobility Program or ASMP (Department of the Army, 1994).

The ASMP is the Army's plan to implement the recommendations of the Mobility Requirements Study. The Army's position is that to support the National Military Strategy it must be able to deploy five and one third divisions with sustainment assets anywhere in the world within 75 days (Wilson and Capote, 1995). This is the equivalent of transporting a city of over 100,000 people with all of their vehicles and supplies. The timeline for this deployment is shown in Table 1.

Table 1. Army Deployment Timeline. Source: Department of the Army, 1994.

Timeframe	Event
C + 4	Lead Brigade (airborne insertion) Closes
C + 12	Lead Division (airborne or light) Closes
C + 30	2 Heavy Divisions (via sealift) Close
C + 75	2 Heavy Divisions (via sealift) Close
C + 75	Corps Support Command Closes

Notes:
1. Army Pre-Positioned Afloat heavy brigade arrives C+7 to C+21.
2. C Day is the date the deployment order is issued.

From Fort to Foxhole

Transporting over 100,000 soldiers with all of their vehicles and equipment to the other side of the world in 75 days or less is a tall order. And remember, that is just the Army—the Air Force, Marines, Navy, and Coast Guard must go as well. The entire joint force must be sustained with food, fuel, ammunition, medical, and other support. Consequently, how do we get all of these people with their vehicles and equipment to the place where they are needed (i.e., "from fort to foxhole")? Transportation geography provides some models for understanding the movement process. The simplest model involves an origin, a destination and intervening distance that must be overcome with a transportation system that moves the cargo (Taaffe et al, 1996). A modification of this system applied to the strategic mobility function yields a model that takes the cargo from fort to foxhole (see Figure 1). U.S. military doctrine recognizes three stages of strategic movement: from fort to port, from port to port, and from port to foxhole (Department of the Army, 1995b).

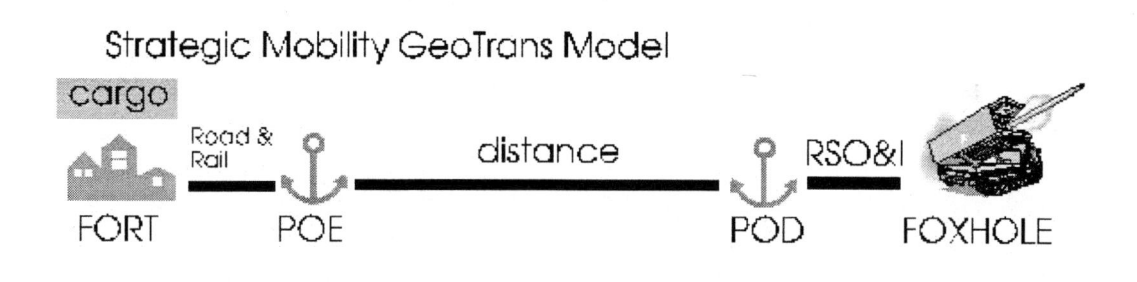

Figure 1. Strategic Mobility GeoTrans Model.
Source: Department of the Army, 1995b.

While this model may look simple, each segment of the movement requires a complex set of actions and resources, which must be fully integrated across the length of the model. The fort or depot in CONUS is the origin of the cargo. The cargo moves via rail, truck, or convoy to the Port of Embarkation (POE), which can be either an airport or seaport. The cargo changes modes to aircraft or ship, and makes it way to the Port of Debarkation (POD) where it begins the process of Receipt, Staging, Onward Movement, and Integration (RSO&I). This RSO&I process is the critical link between the strategic deployment and tactical maneuver levels of military operations. It is currently an evolving doctrine and has received much attention because this part of the process proved to be one of the weakest links during the Gulf War. The utility of this model stems from its simplicity and macro scale schematic of how the entire strategic deployment process operates.

Strategic mobility is the responsibility of the United States Transportation Command (USTRANSCOM) headquartered at Scott Air Force Base, Illinois.

USTRANSCOM is one of the nine U.S. Unified Commands and is the single manager of America's global defense transportation system. USTRANSCOM is composed of three component commands. The Air Force's Air Mobility Command (AMC) provides strategic airlift, aerial refueling, and aeromedical evacuation. The Navy's Military Sealift Command (MSC) provides ships for strategic sealift. The Army's Military Traffic Management Command (MTMC) provides the overland lift component and is the primary traffic manager for USTRANSCOM (USTRANSCOM, 2000).

From Installation to Port of Embarkation

Well before equipment and personnel move, a sophisticated planning process is at work identifying and prioritizing cargo and planning the move. Transportation planners at the installation or fort use computer systems such as the Transportation Coordinators Automated Command and Control Information System (TCACCIS) to develop automated unit equipment lists that identify the size, weight, and dimensions of each piece of equipment to be shipped.

With a deployment order, units stage their equipment on the installation in preparation for movement to the Port of Embarkation or POE. The Military Traffic Management Command arranges for the equipment to be moved to the POE by commercial rail, commercial or military trucks, or even by barge on inland waterways. If the distance from the installation to the port is not too great, wheeled vehicles will often move in unit convoys on interstate highways, and helicopters will fly directly to the port.

Transportation specialists use handheld scanners to scan the barcode labels of the cargo as it leaves the installation. This information is then fed back into the computer systems that feed USTRANCOM's Global Transportation Network computer system. The cargo is tracked in this way throughout its journey thus allowing commanders to maintain In-Transit-Visibility or ITV and accountability of their equipment and cargo. The technology and methods are similar to those pioneered by overnight package delivery services that enable customers to track the progress of a shipment and anticipate when it will arrive.

A major problem during the Gulf War was the bottleneck at the installation created by inadequate rail loading facilities and a lack of super-heavy railcars for transporting oversize equipment. The Army Strategic Mobility Program provided millions of dollars to upgrade installations identified as "Power Projection Platforms" with improved rail loading facilities and other infrastructure. By way of example, the rail loading facility at Fort Riley, Kansas can load more than eight trains at a time. The ASMP also funded the purchase of 1069 heavy-lift rail cars necessary to transport the 70-ton M1A2 Abrams tanks, and over 12,000 containers used to transport equipment, supplies, and ammunition. (Department of the Army, 1994)

The Port of Embarkation

It is a complex process to load efficiently thousands of pieces of equipment and containers—some of which contain hazardous material—on multiple ships in a port. The United States has 17 designated strategic seaports. The Military Traffic Management Command (MTMC) is responsible for running these ports and uses a combination of military labor and civilian longshoremen to load the ships. During a crisis, U.S. Army Reserve Transportation Terminal units, who operate many of the "contingency" ports not normally used by the military, reinforce full-time military and civilian MTMC personnel.

Cargo is staged by equipment type, inspected for hazardous materials and proper documentation (labeling). Helicopters are shrink wrapped in plastic to prevent corrosion and other damage. The equipment is then loaded on ships using the Improved Computerized Deployment System or ICODES, which is essentially a geographic information system that has the layout and dimensions for each deck and hold of strategic sealift ships. The system also contains digital templates of all military vehicles and cargo that might be transported. The information collected at the installation is electronically forwarded to ICODES which then creates an optimum stow plan for each hold of each ship. This capability dramatically reduces the time it takes to load a ship. As the cargo is secured on the ship, transportation specialists once again scan the barcode labels and feed the information into the Worldwide Port System (WPS). The WPS system tracks the specific location of each piece of cargo and reports this information to GTN thus continuing to provide ITV and maintain accountability. The WPS also creates manifest lists and other documents necessary for the ship to sail. This information from the port of embarkation is electronically transmitted to the port of debarkation (Department of the Army, 1995b).

Aerial Port of Embarkation operations are done in similar fashion but take place at airports and involve moving people as well as critical equipment. While the process is similar, taking care of a large number of people creates logistics requirements for food and shelter. Additionally, having exact information on exactly who is on what airplane (creating a manifest) is critical to maintain personnel accountability. Aircraft transport the great majority of the military's most critical cargo—its people.

From POE to POD

The longest and most difficult part of the journey is across the globe between the port of embarkation (POE) and port of debarkation (POD). The strategic mobility triad of strategic airlift, strategic sealift, and prepositioning accomplishes this segment. The following sections examine maritime and ashore preposititoning, strategic sealift, and strategic airlift with a focus on our aging airlift fleet and the introduction of the C-17 *Globemaster III*.

One way to reduce the complexity, costs, and time of strategic deployments is to simply maintain the combat forces in proximity to the threat. Forward basing involves the stationing of troops outside CONUS and was the preferred strategy during the Cold War. This approach overcomes the problem of distance by maintaining the force at or near the foxhole. There are, however, significant limitations in terms of cost and political issues in maintaining large forces abroad. Forward based U.S. forces today have been reduced to approximately 100,000 troops in Europe, 42,000 troops in Korea, and 47,000 troops in Japan.

An alternative strategy to forward basing is to pre-position stocks of equipment and supplies in proximity to a potential threat. Personnel can then be rapidly airlifted into the theater to linkup with their equipment and get to the foxhole in a matter of days, rather than weeks. Prepositioning is not a new idea. During the Cold War, the U.S. maintained large sets of pre-positioned overseas material configured to unit sets, or POMCUS stocks that were to be used by CONUS-based forces reinforcing the defense of Western Europe. In the 1980s, the U.S. Marine Corps pioneered maritime prepositioning. The U.S. now maintains large stocks of equipment and supplies both afloat and ashore for all services. The principal advantage of maritime pre-positioning is that it eliminates the POE, thus rapidly speeding up the process (see Table 2). Additionally, the afloat prepositioned assets can deploy to any port (Department of the Army, 1995a)

Table 2. Distance/time factors from port to port.

Distance Nautical Miles/Sail Days (At 17 Knots/Hour)

	DESTINATION	
ORIGIN	Korea	Southwest Asia
Diego Garcia (Indian Ocean)	4800/12	2700/7
Guam/Saipan	1560/4	7100/18
Charleston, South Carolina	14857/36	9100/23

The Navy's Military Sealift Command operates the Maritime Prepositioning Force consisting of 14 ships organized into three Maritime Prepositioning Ship Squadrons (MPS). MPS 1 has five ships and is usually located in the Atlantic Ocean or Mediterranean Sea. MPS 2 has five ships and is located at Diego Garcia in the Indian Ocean. MPS 3 has four ships and is located in the Guam/Saipan area. Each MPS Squadron carries enough equipment and supplies to sustain a 17,000-Marine Air Ground Task Force for up to 30 days. The MPS Squadrons optimally discharge their cargo in fixed ports, but if a fixed port is not available they can discharge their cargo using lighters. This over the shore capability is important in underdeveloped parts of the world. Note that MPS Squadrons are not designed as combat forces to affect a forced entry on a hostile shore. However, once Marine amphibious forces secure a port or a beachhead, they can be rapidly reinforced to build substantial combat power.

The pre-positioned equipment and supplies of the Army Prepositioned Stocks (APS) Program provide starter stocks in sufficient quantities to support committed forces until lines of communication are open (see Figure 2). The APS Program is a variation of forward basing, and only requires that troops be flown in to linkup with equipment. Currently APS has eight heavy brigade sets of equipment plus sustainment equipment. APS 1 includes sustainment stocks in CONUS. The Combat Equipment Group Europe (based mostly in the Netherlands) maintains three brigade sets and large quantities of sustainment material that constitute APS 2. Combat Equipment Group Northeast Asia maintains one brigade set in South Korea as well as other stocks in Japan and Hawaii (APS 4). Combat Equipment Group Southwest Asia maintains one brigade set at Doha, Qatar, and one brigade set in Kuwait (APS 5). While this method speeds up deployment time and is less expensive both monetarily and politically, it only supports a limited force for a limited time (Third U.S. Army 1995, MSC Webpage 2000, OSC Webpage 2000).

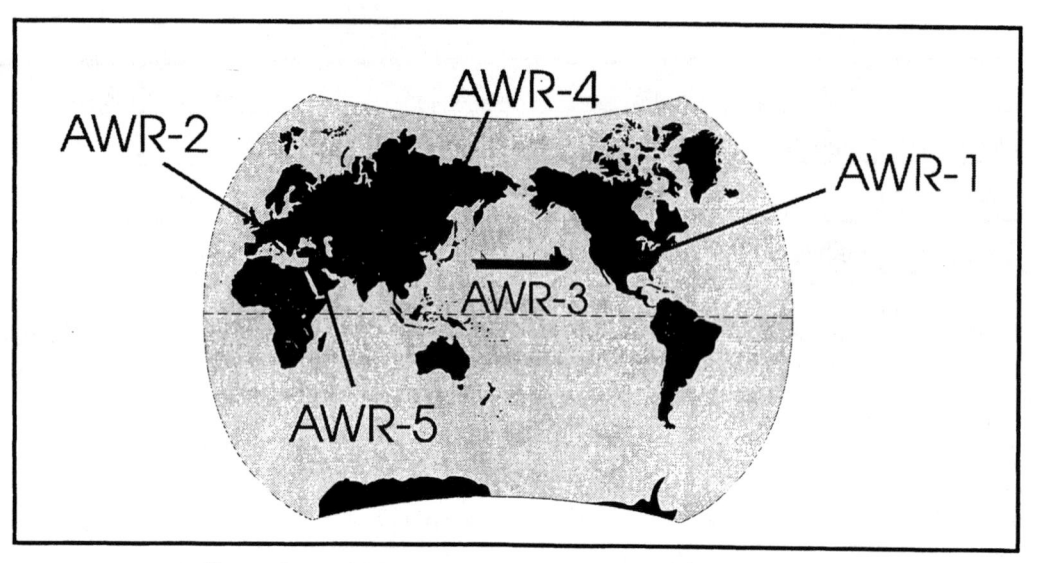

Figure 2. Map of the Army War Reserve storage locations.
Source: Third U.S. Army, 1995.

APS 3 is the afloat portion of the APS Program and is known as the Combat Prepositioning Force. Afloat Prepositioned Squadron 4 consisting of eight of the latest construction Large, Medium Speed, Roll-On, Roll-Off ships or LMSRs has two brigade sets of combat equipment and is stationed in the Arabian Gulf. Three LASH or Lighter Onboard Ships stationed at Diego Garcia carry the majority of the Army's prepositioned ammunition. The LASH ships carry their cargo in floating containers or lighters, which the ship can discharge directly into the water. Tugboats then push the lighters to their discharge location thus potentially freeing up valuable deep draft berthing space for the large equipment carrying ships. The Combat Prepositioning Force also has the Army's Port Opening Package with some very interesting ships. The *MV American Comorant* is a float-on, float-off of FLO-FLO ship that carries large army watercraft such as tugboats and a large floating crane. The ship partially submerges to discharge its cargo (see Figure 3). The *MV Strong Virginian* is a heavy lift ship with an 800-ton boom, and the *SS Gopher State* is an auxiliary crane ship with large cranes used to unload other ships. The port-opening package also includes materials handling equipment such as forklifts, cranes, and "yard-dogs" (trucks that tow semi-trailers) essential for vessel operations. These vessels as well as two combination container/RORO ships are usually based at Diego Garcia (MSC Webpage, 2000).

Figure 3. The FLO/FLO ship *MV American Cormorant* is not sinking. The ship partially submerges to discharge its cargo of tugboats and other Army watercraft used to open a seaport. Source: Author's collection.

The Logistics Prepositioning Force carries equipment and fuel for the Air Force, Navy, and Defense Logistics Agency. The Air Force maintains it major ammunition stocks on three container ships. The Navy maintains two ships with ordnance for its ships and planes, and equipment for a 500-bed field hospital to support the Marines. The Defense Logistics Agency has an underway replenishment tanker, and two tankers equipped with the Offshore Petroleum Discharge System. These three tankers support the force afloat and ashore with 650,000 barrels of fuel. Finally there are two Aviation Logistics Ships to support USMC helicopters; they are based on the East and West Coast and are maintained in a reduced operating status requiring five days to get underway (MSC Webpage, 2000).

Sealift is the backbone of the strategic mobility process. American sealift capability is designed to meet three requirements: maritime prepositioning, surge sealift capability for rapid power projection, and sustainment sealift to support extended combat operations (USTRANSCOM, 1997). Strategic sealift relies on government owned ships, American flagged commercial ships, and foreign flagged ships. Each element poses challenges for planners. During peacetime, most cargo is transported on commercial liners, as this is the most cost-effective method. However, during a contingency, oversized cargo must be moved into dangerous areas. Most commercial ships are container ships, designed to handle standard containers. Most military equipment, however, is oversize and is best transported in roll-on/roll-off ships of which there are few in commercial service. A second problem is the vitality and size of the U.S. Merchant Marine. The Department of Defense relies on it to transport supplies and ammunition. Yet, since W.W. II, the U.S. Merchant Marine has had a difficult time competing in the international market. Foreign flagships are one alternative, but there are questions of availability and reliability. The solution was to create a Ready Reserve Force of militarily useful cargo ships, partnered with industry to ensure the viability and availability of the US Merchant Marine, and otherwise rely on foreign flag vessels to make up any shortfalls.

The Ready Reserve Force (RRF) is a fleet of government owned ships held in reserve and maintained by the U.S. Maritime Administration (MARAD) for the purpose of deploying U.S. and allied forces (see Figure 3). The RRF consists of approximately 94 RO/RO, breakbulk, and other types of ships kept in a reduced operating status and based at ports around the country. During a contingency U.S. civilian contractors operate the ships and are required to sail within four days.

The RRF has been upgraded from 19 to 36 roll-on-roll-off (RORO) ships with plans for the addition of 19 Large, Medium Speed, RORO (LMSR) ships by 2001. The LMSRs (see Figure 4) are enormous ships (nearly the size of an aircraft carrier) and typically carry about 1500 pieces of military equipment and can travel at over 24 knots. These ships permit military vehicles to be rapidly loaded or unloaded (MSC WebPages, 2000). The Military Sealift Command also operates 8 Fast Sealift Ships (FSS) capable of traveling at over 30 knots. The FSS's can transport a tank division to Southwest Asia in two weeks rather than the four weeks required for a conventional ship (USTRANSCOM 1997). The RRF supplies surge sealift for wheeled and tracked vehicles and other bulky equipment, but the majority of sustainment sealift must come from commercial sources.

Figure 3. The *MV Cape Diamond* loading the 24th Infantry Division (Mechanized).
Source: Author's collection.

Supplementing the RRF are U.S. flagged civilian charter ships. These ships are expected to carry the bulk of ammunition and supplies to sustain the force but since they consist largely of container and break bulk ships they are less useful for carrying military equipment such as tanks. These ships also tend to be much slower, typically traveling at about 14 knots (Caprano, 1996). USTRANSCOM relies on civilian carriers for most peacetime and nearly all wartime sustainment sealift. Yet, as mentioned previously, the U.S. Merchant Marine has fared badly in international competition due to high U.S. labor costs and the trend for American ship owners to register their vessels under "flags of convenience" such as Liberia and Panama. The U.S. Government's response has been to provide incentives to U.S. flag carriers to ensure they are available in both peace and war. The Voluntary Intermodal Sealift Agreement (VISA) is a business incentive program that guarantees Department of Defense peacetime shipping business to carriers who agree to make their transportation assets available to support U.S. military operations in time of war. The VISA ensures access to U.S. commercial transportation resources (including trains, trucks, containers, port facilities, ships, etc) to transport military cargo from the installation to the port of debarkation while also giving American commercial shippers the government's business.

Figure 4. The USNS Bob Hope is one of the first new construction LMSR strategic sealift vessels of the Military Sealift Command.
Source: Author's collection.

Strategic airlift is the third leg of the strategic mobility triad. Airlift is very expensive in that the equipment costs 95% of the strategic mobility budget. However, the speed of delivery makes airlift an essential element in transporting high priority cargo such as personnel, critical equipment, and spare parts. The Air Mobility Command's fleet of C-141 and C-5 aircraft is aging, with many planes due to retire between 2005 and 2010. The average age of AMC's fleet is

30.6 years for transports and almost 39 years for tankers. The C5 *Galaxy* fleet provides 82 percent of outsized lift, but due to the advanced age of most planes, it is proving to be less reliable. The introduction of the C-17 *Globemaster III* will alleviate some of these problems (USTRANSCOM, 1997).

The replacement aircraft is the C-17 *Globemaster III* (see Figure 5) of which 120 are planned for production. The C-17 boasts many revolutionary features, most notable of which are its ability to transport outsize cargo such as tanks and land on austere airfields only 3000 feet long. These capabilities preclude the need to offload the aircraft at a large airfield away from the operating area (Kudalka, 1996; USAF 437th Airlift Wing, 1995). The C-17s are also being outfitted with the contingency precision approach capability which will enable them to serve as pathfinder aircraft in marginal weather conditions for AMC aircraft flying into austere airfields (USTRANSCOM, 1997).

A final element of the airlift program is the Civil Reserve Air Fleet (CRAF), which allows the military to charter commercial aircraft. The CRAF program is another business incentive program similar to the VISA program. USTRANSCOM trades peacetime business to U.S. airlines that contractually pledge aircraft, crews, and infrastructure in times of national emergency of force deployments. Thirty-five carriers have pledged 693 aircraft to the CRAF program. The majority (521) are for international flights, while 66 are for domestic service, and 33 support aero medical evacuation missions. The CRAF can be mobilized in three stages: Minor Regional Crisis, Major Regional Crisis, and National Mobilization. The National Mobilization level has never been used and would have a severe impact on civilian air travel (USTRANSCOM, 1997)

POD To Foxhole

Historically, the focus of strategic mobility was on the segment between the ports of embarkation to the port of debarkation. Failing to look at the entire process, however, led to bottlenecks and inefficiencies that dramatically slowed the movement of forces from the ports of debarkation to the foxhole. In the Gulf War it took 200 days to link soldiers with their equipment and get them to the front—ready to fight. Even as the ground war started there were still units waiting to get their equipment. The requirement now is to have a heavy corps deployed and ready to fight in 75 days. The solution to this problem is the adoption of the doctrine of RSO&I (Department of the Army, 1999).

Figure 5. A C-17 *Globemaster iii* of the 437th Airlift Wing at Charleston AFB.
Source: Author's collection.

The basic RSO&I concept is to take soldiers arriving at aerial ports of debarkation and quickly link them up with their equipment arriving at seaports of debarkation. Theses units are then fueled, armed, and moved forward to tactical assembly areas where they can join their higher headquarters and begin combat operations. This process however, is not nearly as simple as it sounds.

Equipment must be marshaled and transported to staging areas. Personnel must be provided with life support such as food, billeting, and medical care. Staging is the process of linking up personnel and equipment into units. Staging often occurs at a Theater Staging Base (TSB) that provides life support for personnel, and maintenance and other logistical support to get unit equipment ready for combat. Units remain in the TSB until they become self-sustaining and are ready to move to the battle area. Onward movement involves moving complete units from the TSB up to Tactical Assembly Areas in the battle area. Wheeled vehicles usually move via unit convoy over the road system. Tracked vehicles and heavy equipment move via rail, heavy equipment transporter trucks, or possibly by barge over inland waterways. Integration involves the transfer of authority over the units to the combat commander who will employ them in battle. Integration is not complete until the units are situated in the tactical assembly area in a combat ready posture.

Geography plays a critical role in the RSO&I process. Knowledge of the physical and human landscape is important in selecting air and seaports that are optimally located and have the attributes necessary to support the reception of the ships and planes that will transport the cargo. Knowledge of the cultural and economic geography of a region is critical in planning the staging and onward movement of military forces so they can ultimately be integrated into the combat force.

Selecting sea and aerial ports of debarkation requires a detailed assessment of the location and attributes of possible sites. The location of PODs must be assessed in relation to the distance and route from the POEs to the POD. Locations of APODs and SPODs must also be assessed in relation to each other, as the personnel arriving at the APOD will have to link up with their equipment arriving at the SPOD. The attributes of PODs also require careful assessment as even the best-located port cannot be used if lacks the necessary facilities to support our ships or planes.

Seaports are classified as World Class Ports such as Dammam in Saudi Arabia or Pusan in Korea, unimproved ports such as those in Somalia or Haiti, and bare beach or no port. Some of the attributes that must be considered in port selection are: number of berths and their characteristics, cargo throughput capacity, transportation links, and availability of contract labor or host nation support. Many of the ships that transport our military equipment are too large for the berths at smaller ports. A Large, Medium Speed, Roll-on, Roll-off (LMSR) ship has a draft of 35 feet, making many berths too shallow to accommodate these ships. A second consideration is the capacity of the port to handle cargo loading and unloading operations—known as "throughput." Throughput capacity is based on the number of berths in a port, the amount of space to stage cargo, and material handling equipment such as cranes. The best seaports also have good transportation links. These include rail-loading facilities and access to the rail system, good links to the highway system, and possibly facilities for loading barges and links to the inland waterway system. In many cases some material from the port will move via intra-theater airlift. Hence, the best ports will be in close proximity to an airfield that serves as a Sea-Air Interface Site (SAIS). It takes a lot of people to run a seaport and while the Army has units of stevedores, the number of soldiers involved can be reduced if contract labor or host nation support is available to unload the ships. Thus the availability of contract stevedores from the International Longshoremen's

Association or host nation labor is yet another attribute to consider when selecting a port to serve as a seaport of debarkation.

Obviously having an operational world-class port available will dramatically speed a deployment. However, unimproved ports can be improved by having channels dredged, berths repaired or improved, material handling equipment repaired or supplemented, staging areas cleared, and transportation infrastructure repaired or built. A worst-case scenario involves having no ports at all. In this case a Joint Logistics Over the Shore or JLOTS operation must be conducted. JLOTS uses Army watercraft and other lighterage to transfer cargo from the ship, over the beach, to staging areas ashore. A variation on this approach is to extend a floating causeway from the ship to the shore and drive vehicles right off the ship over the causeway to the beach. JLOTS operations are very slow and vulnerable to enemy attack, and can only be accomplished in calm seas. The optimum seaport of debarkation is almost always a world-class port because it will have the location, infrastructure, and host nation labor to facilitate rapid offload.

Assessing the suitability of airfields as aerial ports involves a similar analytical process, but with a different set of criteria. Just as with a seaport, the location of the airfield in relation to the battle area is a critical concern. However, the most critical attribute specific to airports is the number, length, and type of runways. Other critical attributes include amount of tarmac space, hanger and maintenance facilities, refueling and fuel storage facilities, space for personnel and cargo support and staging, and availability of contract labor and host nation support.

Prior to the introduction of the C-17 *Globemaster III*, AMC strategic airlift aircraft were limited to large airfields with long runways of 10,000 feet or more. Cargo was unloaded at these large airfields, then shipped forward either overland or on tactical aircraft. Trans-shipping cargo from the large plans to the smaller planes was a time consuming affair that required personnel and ground equipment. The introduction of the C-17 changed the nature of this problem in transportation geography. The C-17 can carry large payloads to include "outsize" cargo, such as a main battle tank, that was formerly limited to the C-5, while being able to land on small, austere airfields with 3000-foot runways. Thus unlike the C-130, C-141, and C-5 which perform parts of the airlift mission, the C-17 can accomplish the full spectrum of airlift faster and with fewer resources. While there are 847 airfields available worldwide to the C-5, there are over 10,000 airfields available to the C-17 (USAF 437th Airlift Wing, 1995).

A second consideration is tarmac space, which affects throughput. Tarmac space is a parking area for planes to load and unload, and throughput refers to the amount of cargo transported from origin to destination in a given amount of time. The "Maximum on Ground" (MOG) number is the number of aircraft that can be parked on an airfield at one time. A C-5 is 248 feet long with a wingspan of 223 feet thus requiring a great deal of space to park and unload. Many smaller airports that can handle the C-5's runway requirements (especially in the developing world), but do not have the tarmac space to park and load/unload larger aircraft. Consequently, the aircraft unloads on the runway closing down the airport until the cargo is removed. The C-17, however, has a wingspan of only 171 feet and can unload on taxiways only 50 feet wide. The C-17 also has thrust reversers on its engines that enable it to load/unload with engines running. Additionally, the C-17 can make a 180-degree turn in only half its own length (80 feet) making it very agile on the ground. The overall result is that C-17 throughput at short, austere airfields is over twice that of other airlifters while requiring fewer personnel and resources to operate (Kudalka, 1996; USAF, 1996).

Optimally aerial ports will have infrastructure to support airlift operations. Hangers and maintenance facilities allow aircraft to be parked and serviced between flights. Fixed aircraft refueling facilities allow for rapid refueling and turnaround of planes, while fuel storage facilities preclude the necessity for numerous fuel trucks. Given that thousands of troops will process through the aerial port there must be space and facilities to provide life support. Finally, host nation labor to support air operations and especially cargo handling and troop life support activities reduces the requirement for U.S. logistics personnel and increases throughput. Transportation planners must take into account all of these geographic (location and attribute) factors in planning a strategic movement of U.S. and allied forces.

The nature of a theater's infrastructure and the availability of Host Nation Support are two important issues that are often overlooked. A geographic analysis of a potential strategic deployment must obviously examine seaports and airfields, but it must also examine the nature and extent of interior infrastructure such as roads, railroads, interior airfields, inland waterways, and built-up areas. An area's road and highway system is perhaps its most critical infrastructure as the majority of units and supplies will travel overland. Road networks must allow units to move from their staging bases to the battle area and support the flow of supplies from the port to the foxhole that will sustain the combat units.

Roads must be in good repair and of adequate capacity to handle the volume and weight of traffic necessary to move and sustain the force. Critical points such as bridges and tunnels must be in good repair and of adequate capacity to allow oversized loads such as heavy equipment transporters across or through them. Bridges and tunnels must also be protected as these are likely targets for enemy attacks.

A working railroad system is an invaluable if vulnerable asset for moving heavy equipment and large amounts of supplies. Along with track location, capacity, and level of repair, the analysts must also consider the location and capacity of railheads; the availability and capability of locomotives and rolling stock. The same chokepoint issues are important since bridges and tunnels must have the capacity to support the movement, and be protected from enemy action. Host nation support in providing railroad personnel is critical, as it is unlikely that U.S. forces could efficiently operate or maintain a foreign railroad system.

Inland waterways are another efficient means of transporting oversize equipment and large amounts of cargo within the theater. Transportation analysts must examine the availability and capacity of barges, and other watercraft to transport military cargo. Inland ports are critical in terms of their location, depth, facilities, and available host nation and contract labor. And as always, chokepoints such as sharp turns, locks, or bridges spanning the waterway must be protected from enemy action.

Urban areas that might be used to support the logistics effort are also an important component of infrastructure analysis. While most forces can live in their own shelters, many rear area units and troops in transit benefit from fixed facilities. The availability of facilities to billet troops in warm, dry conditions with good sanitation improves morale, reduces sickness, and speeds their movement forward if they do not have to unpack and repack their own equipment. Garage space with hoists and pits facilitates equipment maintenance, and dry and secure warehousing space is critical for storing the large stocks of supplies necessary to sustain the combat force. A key infrastructure element is storage and transportation of petroleum, oil, and lubricants (POL). Use of existing tank farms for bulk fuel storage and pipelines to transport fuel through the theater reduce the strain on military POL units and speed the flow of this critical commodity to the combat troops. Once again these important but vulnerable facilities must be protected against enemy air attack or sabotage.

Host Nation Support (HNS) is defined as *"civil and military assistance rendered by a nation to foreign forces within its territory during peacetime, crises or emergencies, or war based on agreements mutually concluded between nations"* (Department of the Army, 1999). The availability of HNS reduces the demand on U.S. logistical forces and allows for more combat forces to get to the fight faster. Specific services and facilities that can be partially delegated to HNS include: life support, medical facilities, construction and engineering, local security, transportation assets and infrastructure, labor, emergency services, fuel and power facilities, and communication facilities. In recent years, the use of contractors in MOOTW to provide services equivalent to HNS has become the norm. In Bosnia, contractors provide food services, sanitation, fire fighting, and many other services to U.S. forces, thus reducing the number of logistics troops required. Contract operations are also quite cost effective and add to the local economy.

Case Study: Deployment to Kosovo

Kosovo is a province of southern Yugoslavia surrounded by Serbia, Macedonia, and Albania (see Map 1). Historically it is the cultural hearth of the Serbian people, but it now has a population that is 90% ethnic Albanian Muslim and only 10% Serbian. Kosovo was an autonomous province of Serbia until 1989 when the Serb leader Slobadan Miliosevic brought it under the direct control of Belgrade. Kosovar Albanians opposed this move and ultimately violence erupted between Kosovars and Serbs. During 1998, open conflict between the Serbian forces and the Kosovars resulted in the deaths of 1,500 Kosovars and forced 400,000 people to flee their homes. The United Nations and regional European organizations became concerned that the ethnic cleansing taking place in Kosovo was both illegal and immoral, and that the massive refugee flow into neighboring countries would destabilize the already troubled region and might result in a wider conflict. When diplomacy failed, the North Atlantic Treaty Organization (NATO) launched a 77-day air campaign and prepared for a ground invasion. These military operations eventually forced Miliosevic to withdraw his forces from Kosovo and allow the entry of a NATO-led international peacekeeping force known as KFOR into the province to provide security for the return of almost a million refugees and the subsequent rebuilding of the province (KFOR, 2000).

Kosovo is an excellent case study of the strategic mobility challenges inherent in peacekeeping operations. NATO's strategic mobility challenge involved three elements: supply for the air campaign, movement of heavy U.S.

and allied units into the Balkans, and sustaining the buildup of KFOR and the concurrent operations of the NATO led Stabilization Force (SFOR) beginning its fifth year of peacekeeping duties in Bosnia. Concurrent with the air campaign was the necessity to supply enormous amounts of relief supplies to support the hundreds of thousands of refugees huddled in camps in Albania and Macedonia (Capas, 2000).

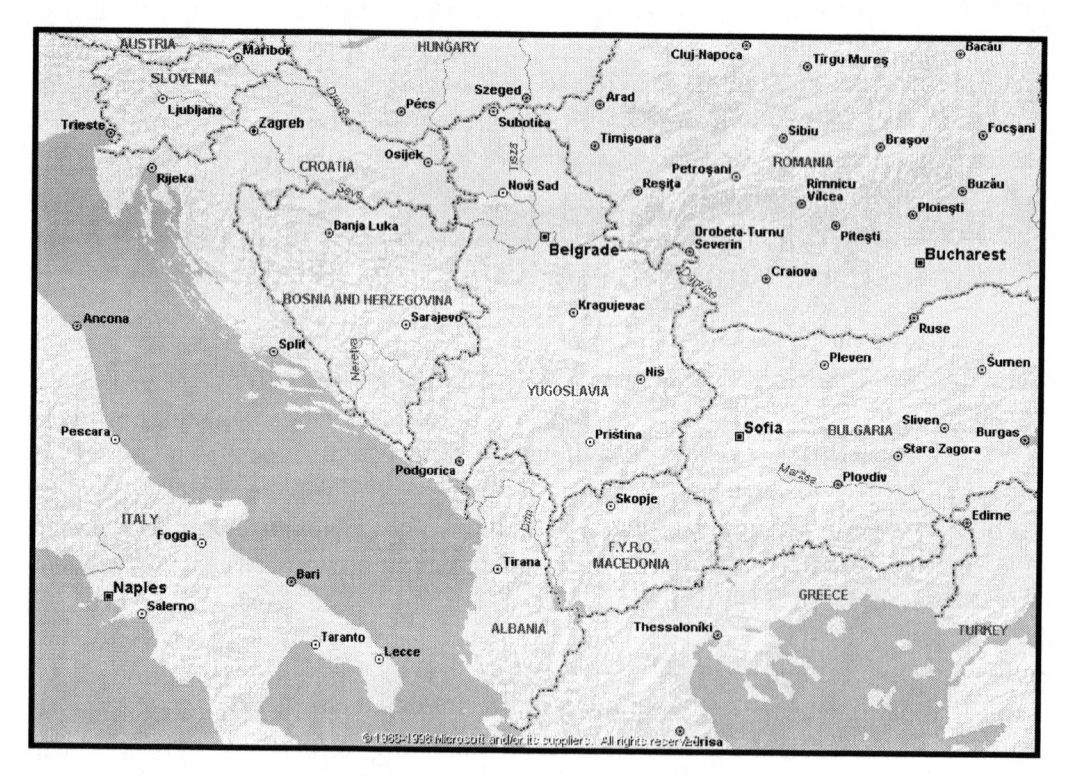

Map 1. Map of the Balkan Peninsula.
Source: Microsoft Encarta Interactive Globe, 2000.

The physical infrastructure, and political geography of the region had a significant impact on strategic mobility. Kosovo and the surrounding countries are part of the Dinaric Alps, which rise 6000-8000 feet above sea level. Kosovo itself is a plateau of rolling hills surrounded on all sides by mountains and accessible by only a few roads that pass through narrow mountain valleys. Kosovo is the poorest and least developed area of the former Yugoslavia, and Albania is the poorest and least developed country in Europe. There are only two major roads accessible to KFOR. In terms of political geography, NATO enjoyed a significant advantage given the proximity of its member countries (including the recent addition of Poland, the Czech Republic, and Hungary), and the full-fledged support of Albania. Macedonia was very hesitant to fully embrace the NATO operation but given the presence of so many refugees on its territory it had little choice but to cooperate. The fact that Montenegro is still a part of the Serb dominated "Rump Yugoslavia" meant that there is no short north-south land route from the NATO countries to Albania and Macedonia.

The early phases of the Kosovo operation were focused on Albania. When it became evident that a ground invasion from Albania was not in the offing, and the Greeks became more politically supportive; the focus changed to Macedonia. Mobility planners made the decision early that most personnel and essential cargo such as sensitive items, special spare parts, and critical supplies would move by air into the APOD at Tirane, Albania. Despite the lack of infrastructure at Tirane's international airport and an abundance of mud, U.S. military engineers made enough improvements to support the deployment of an Apache attack helicopter unit, and vast quantities of humanitarian aid (Bennett, 2000).

The next decision involved whether to move the material and equipment based in Germany using ground routes or sealift (see Map 2). Planners of the 598th Transportation Terminal Group considered three courses of action (COA). COA 1 involved moving all the cargo overland from Germany (primarily by train) through the Czech Republic, Hungary, Italy, Slovenia, to Croatia. The new NATO allies were willing to cooperate; however, difficulties during the 1996 deployment of NATO forces into Bosnia through Hungary surfaced problems with railroad throughput capacity, and there was no acceptable road route from Croatia to Albania. COA 2 involved moving cargo from Germany to a SPOE at Bremerhaven and then sealift to a SPOD at Durres, Albania. Once again, physical geography and infrastructure played a role. Durres is a small port and not deep enough to handle large ships such as LMSRs or Fast Sealift Ships. Additionally, Durres has very little open space for staging operations, poor material handling equipment, and little host nation support capacity. A third COA was to rail and truck the cargo to Italian SPOEs (Anocono, Bari, or Brindisi), then ship the cargo to a SPOD at Durres using small vessels. Brindisi in particular had the necessary staging space so that cargo could move to Durres at a rate that would not choke the city. The highway and rail connections within Western Europe are some of the best in the world and could easily handle the traffic, although the trip through the Alps down the Italian peninsula to Brindisi was lengthy. Another political consideration was that non-NATO countries such as Switzerland and Austria generally refused permission for Kosovo associated cargo to pass through their territory.

The port selection process for the SPOD involved thee ports: Durres, Albania; Thessaloniki, Greece; and Rijeka, Croatia. Durres was most appealing because of the focus on Albania as a base for the Apache attack helicopter task force and its potential as a base from which to launch a ground invasion of Kosovo. However, Durres was not a viable port. Thessaloniki is a world-class port capable of servicing the largest LMSRs. It has excellent infrastructure, and host nation labor. Thessaloniki is 150 road miles (240 kilometers) from Kosovo; however, KFOR would have to cross Macedonia. Macedonia would not allow its territory to be used as a springboard for an invasion, but once an agreement was reached with Serbia to allow KFOR to peacefully enter Kosovo the Macedonians became much more cooperative as did the Greek government. The U.S. also considered Rijeka as a distant third choice. Rijeka is strategically located between Germany and Albania or Macedonia. It is a modern port capable of handling large ships and it has large areas of newly paved staging areas available to NATO at very low cost. The problem remains, however, that there is no acceptable road route from Croatia directly to Albania (Capas 2000).

After careful consideration planners decided to use Bremerhaven as the northern SPOE, and the Italian ports as the southern SPOEs. Two U.S. Army Logistic Support Vessels (giant 272 foot long, ocean going landing ships with ramps at both ends) made over 130 trips from Brindisi across 100 miles of the Adriatic Sea to Durres. The LSVs were supplemented by the use of a chartered commercial ferry. Eventually the SS *Osprey* carefully navigated its way into Durres where the Albanian newspaper *Zer-LiPopuli* immediately dubbed it the "American Titanic" as Albanian waters had never seen such a large vessel (Capas, 2000; Randt, 1999a). As the operation proceeded and NATO began to plan for a ground invasion, it became evident that such an operation launched from Albania was fraught with difficulty. The terrain was difficult, presenting only one avenue of approach that was reduced to a two-lane road through a narrow mountain pass. The APOD in Tirane and the SPOD at Durres were already badly strained and despite the enthusiastic support of the Albanians the infrastructure and host nation support were inadequate.

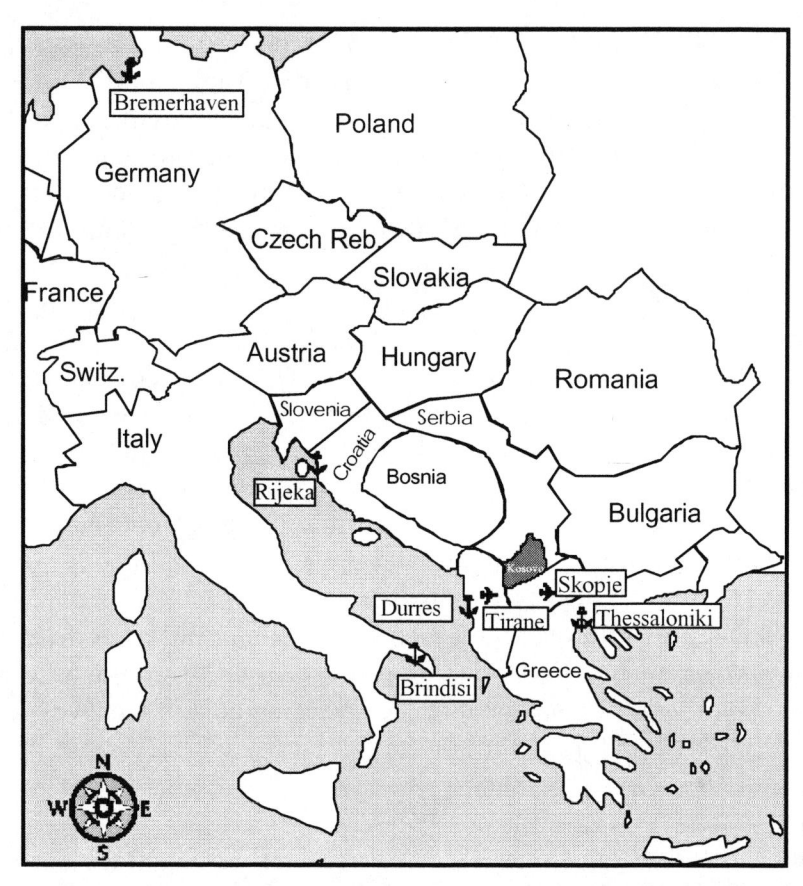

Figure 2. Map of Air and Seaports used for movement of American and other allied KFOR elements into Kosovo.

Fortunately for NATO a diplomatic solution was reached that allowed the peaceful entry of KFOR into Kosovo. The Greeks made the port of Thessaloniki available and soon U.S. Marines were rolling from Greece, through Macedonia, to Kosovo. NATO operations in Macedonia were expanded and the airport at Skopje became the main APOD. The major U.S. component of KFOR consisted of elements of the 1st Infantry Division who deployed from bases in Germany. Their tanks and other heavy equipment were transported by rail to Bremerhaven and shipped to Thessaloniki on the LMSRs USNS *Bob Hope* and USNS *Soderman*. The equipment was off-loaded in Thessaloniki and the tracked vehicles were loaded onto heavy equipment transporters (HET). The HETs and convoys of wheeled vehicles then made the 135-mile trip to Skopje where soldiers at the APOD linked up with their vehicles. From Skopje it is only 15 miles to the Kosovo border. As the focus shifted to Thessaloniki and Skopje, the operations in Albania were curtailed until only humanitarian support activities remained. The Croatian port of Rijeka, while not used to support KFOR was used to rotate the American elements of SFOR in Bosnia demonstrating the necessity to conduct concurrent missions to different areas (Capas, 2000; Randt, 1999b).

In March 2000, the Greeks asked NATO to stop port operations at Thessaloniki during their upcoming national elections. While the Greek government fulfilled its commitment to NATO, many Greek citizens were infuriated at the NATO for bombing their Serbian neighbors and large anti-American demonstrations occurred in major Greek cities including Thessaloniki. Convoy operations from Thessaloniki were actually limited to the hours of darkness to avoid incidents. American military leaders, concerned about the security situation in Greece decided to cease KFOR sustainment operations at Thessaloniki and find an alternate route. Cargo from CONUS is now shipped to Bremerhaven and all cargo then moves on a roundabout rail route from Germany through Austria, Hungary, Romania, Bulgaria, Greece to Camp Able Sentry in Macedonia. From there it convoys into the operating area in Kosovo.

Conclusion

With the end of the Cold War, American military strategy changed from forward defense to a strategy of regional defense relying on CONUS-based forces that can deploy rapidly to trouble spots. Strategic mobility is a critical component of this strategy. The process of strategic movement presents numerous challenges that are fundamentally geographic in nature. Issues of place, space, distance, and movement are core themes in the strategic mobility challenge. In order for our military to successfully implement our contemporary power projection strategy, it is vital that everyone from the President to the lowest private understand the inherent geographic challenges posed by strategic movement, and the systems and processes in place to overcome the challenge.

Three strategic mobility trends are likely in the future. First, strategic mobility will remain pre-dominantly an American endeavor. Secondly, in order to increase our deployment capability the nature of the cargo may well change. And third, America's strategic deployment capability is likely to be tested again and again in the 21st Century.

Readers will have noticed that this chapter deals almost exclusively with American strategic mobility capabilities and doctrine. Frankly, no country but the United States has a major strategic mobility capability. During the Cold War, the Soviet Union maintained a force projection capability in the form of seven airborne divisions and an amphibious assault capability. They also maintained a limited number of RO-RO ships optimized to carry military cargo. The breakup of the Soviet Union means that Russia and other successor states, which can ill afford to maintain such an expensive capability, can no longer move large forces over long distances. Ironically, on the 10th anniversary of the fall of the Berlin Wall, a ship (the Ukrainian RO-RO *ship MV Balakleya*) built to carry Soviet tanks was instead transporting the M1 tanks and Bradley fighting vehicles of the U.S. 1st Infantry Division from Bremerhaven to Thessaloniki for eventual duty in Kosovo.

Another approach to increasing the speed and efficiency of strategic movement is to change the structure of the force. American Army leadership has come to recognize that the heavy tank and mechanized infantry forces designed to defeat the Soviets on the plains of Central Europe are too unwieldy to rapidly deploy in the Post Cold War Era. The American Army is now experimenting with lighter, more mobile units that would be capable of deploying by air anywhere in the world in 96 hours. These new Brigade Combat Teams (BCT) will have approximately 3,700 soldiers (700 less than are currently serving) and will consists of three mechanized infantry battalions and a reconnaissance, intelligence, surveillance, and target acquisition squadron. The units will trade in their heavy Bradley Infantry Fighting Vehicles and Abrams tanks for a lighter, more mobile gun system and wheeled personnel carriers both of which can be transported on C-130 *Hercules* transport planes. Although the BCTs will have fewer soldiers and less firepower, Army leaders feel they will be capable of serving in major theater wars as well as quickly deploying to stabilize regional hot spots throughout the world. Once again the equation of mobility versus protection and firepower.

Kosovo will not be the last trouble spot requiring American military intervention. Crises ranging across the spectrum from major theater wars to humanitarian relief operations will test America's strategic deployment capability. The purpose of this chapter has been to demonstrate how strategic mobility is a fundamentally geographic problem and to provide our current and future leaders with an overview of the challenge and the systems and processes currently in use. Additionally this chapter seeks to educate the citizenry about the inherent challenges posed by distance to our National Military Strategy.

❧ ☙

Bibliography

Baxter, E. and de Jong, B. 1999. Cold War Ship Carries the Weight for NATO-Ukrainian Ship Moves Equipment to Kosovo. *598th Trans Tribune 8*. 2, 4:3.

Bennett. 2000. Interview with CPT L. Perry Bennett who served during the Kosovo Crisis as an operations officer in the operations center of the in the 598th Transportation Terminal Group (headquartered in Rotterdam, Netherlands) which was responsible for the Kosovo deployment..

Capas. 2000. Personal correspondence with LTC Edmund Capas who served as an operations planner with the 598th Transportation Terminal Group during the Kosovo Crisis.

Caprano. 1996. Telephone interviews with LTC Caprano of the Department of the Army Strategic Mobility Division-The Pentagon January 1996.

Department of the Army. 1994. *Army Strategic Mobility Program Brief* Washington DC: Department of the Army.

Department of the Army. 1995a. *Field Manual 100-17-1 Army Pre-Positioned Afloat* Washington DC: Department of the Army.

Department of the Army. 1995b. *Field Manual 55-65 Strategic Deployment.* Washington DC: Department of the Army.

Department of the Army. 1999. *Field Manual 100-17-3 Reception, Staging, Onward Movement & Integration*. Washington DC: Department of the Army.

Department of the Defense. 1992. *Conduct of the Persian Gulf War: A Final Report to Congress*. Washington DC: Department of Defense.

Fletcher, C. 1996. Interview with COL Charles C. Fletcher—Head of the Joint Strategic Deployment Training Center at Fort Eustis, VA in January 1996.

KFOR. 2000. *Background Information* from the KFOR website found at http://www.kforonline.com

Kudalka, L. 1996. Interview with LTC Larry Kudalka—Operations Officer of the 14th Military Airlift Squadron at Charleston AFB, SC in February 1996.

MSC, Webpage. 2000. Military Sealift Command Webpage http://www.msc.navy.mil

OSC Web page. 2000 Operations Support Command (Provisional) of the Army Material Command WebPages at http://www.osc.army.mil

Randt, J. 1999a. MTMC Begins Deliveries to Albanian Docks *TRANSLOG* Summer 1999 page 10-11.

Randt, J. 1999b. *MTMC Lands Kosovo Heavy Force* MTMC Public Information Paper.

Reid, G. 1996. Interview with George Reid of the Joint Strategic Deployment Training Center at Fort Eustis, VA in January 1996.

Taaffe, E. and H. Gauthier and M. O'Kelly. 1996. *Geography of Transportation* (2nd edition) New Jersey: Prentice Hall.

Third U.S. Army/ARCENT. 1995. *Executive Summary Army War Reserve 3: Army Pre-Positioned Afloat* Volume 1.

U.S. Air Force 437th Airlift Wing. 1995. *U.S. Air Force C-17 GLOBEMASTER III: Benefits to the U.S. Army* Fact Sheet.

USTRANSCOM. 1997. *USTRANSCOM Pamphlet 35-1 Today's Vision Leads to Tomorrow's Reality*.

USTRANSCOM. 2000. Security Concerns: MTMC Shipments Travel in a New Route to Kosovo. *USTRANSCOM News Service* Release Number: 000608-1 June 8, 2000.

White House. 2000. *A National Security Strategy for a New Century* (Available at http://www.pub.whitehouse.gov)

Wilson, J. and R. Capote, 1995. Levering Logistics Technology Toward Force XXI *Army Logistician July-August: 14-18.*

The C-17 Globemaster III is the newest cargo aircraft to enter the airlift force. The C-17 is capable of rapid strategic delivery of troops and all types of cargo to main operating bases or directly to forward bases in the deployment area.
Source: U.S. Air Force photograph.

11

Insurgencies and Counter-Insurgencies:
A GEOGRAPHICAL PERSPECTIVE

By

Andrew D. Lohman

"Throughout history unconventional warfare has been affected far more by indigenous political, social, cultural, and economic factors and, of course, by geography than has conventional warfare."

(Laqueur, 1976)

Introduction

Tracing the trends of military geography, Palka (1995) advocates a broader scope for the field of military geography to accommodate the changing missions of modern military forces. Collectively termed Military Operations other than War, (MOOTW) these diverse, non-traditional missions include a number of activities that have never before been considered within the realm of military geography. While these new missions certainly require appropriate geographic study and present a number of challenges and opportunities for geographers, there is also a need for more in-depth geographic analysis on topics within the realm of military geography that have not been fully explored and developed. Insurgencies and counter-insurgent operations are certainly topics that warrant further geographic study and analysis not only because they have received inadequate attention in the past, but also because modern military forces have faced, and will continue to face, such missions within the scope of MOOTW.

Since the end of World War II, insurgencies have become commonplace on the international scene. Numerous definitions of insurgencies exist, but U.S. Army FM 7-98 (*Operations in a Low Intensity Conflict*) provides the most succinct version, *"An organized movement aimed at the overthrow of a constituted government through the use of subversion and armed conflict."* Although armed conflict may take many forms, guerrilla warfare has consistently been the method of waging such a struggle because it is, as Mao (1937) stated, *"A weapon of the weak"* (O'Neill, 1990). There is ample literature on insurgencies, guerrilla warfare, and counter-insurgent operations, but there have been few geographic analyses. Certainly, such diverse and complex subjects cannot be adequately addressed within the scope of one chapter. Nevertheless, I will provide a point of departure for further geographic analysis of these three interrelated topics. First, I will briefly outline and describe what preliminary work has been done on these topics within military geography. I will initially concentrate on the insurgent perspective, before considering counter-insurgent strategies. Second, this chapter will identify areas where geographic analyses can contribute to a more complete and thorough understanding of what modern military forces, especially those of the United States and her allies, are most likely to encounter within the realm of MOOTW. The purpose of this chapter is to enhance understanding of insurgency and counter-insurgency operations from a spatial perspective, and to recognize how they evolve and take shape on the landscape.

Insurgency and Guerrilla Warfare

Examples of guerrilla warfare date to earliest recorded history. It was a method by which indigenous people resisted invading armies and harassed foreign occupation forces. It wasn't until Napoleon's war in Spain between 1807-1813 that the term *Guerrilla* came into common usage, derived from *guerra*, the Spanish word for war. It is, as Huntington (1962) defined, *"a form of war by which the strategically weaker side assumes the tactical offensive in selected forms, times, and places."* Selected forms differ from conventional methods of land warfare because the guerrillas often lack equipment and organization needed to conduct 'regular' warfare. Instead, the guerrillas use what resources they have to conduct raids on key enemy objectives and lines of communications. Characterized by small unit hit-and-run tactics, the guerrilla selects the time and place for these attacks, striking at the enemy's weaknesses while capitalizing on his own strengths. The intent of guerrilla activity is to harass the enemy, keep his units off guard, spread his forces, and ultimately wear down his resistance and resolve.

As a form of war, guerrilla warfare received relatively little attention from military theorists, although von Clauswitz and Jomini addressed the subject briefly in their classical works on strategy and theory. Their discussions, however, described the traditional connotations of guerrilla warfare whereby irregular forces operated in conjunction with conventional armies to resist invading and occupying armies. Clauswitz and Jomini referred to such guerrilla struggles as "Partisan warfare," a term which more accurately describes this complementary method of warfare. Although numerous examples can be found throughout the 18th and 19th centuries, Lawrence of Arabia is often credited with illustrating how effective this form of war could be in the modern era (Wilkins, 1954).

Although Lawrence's direction of the Arab Revolt against the Turks in World War I did not single-handedly win the war for the Allies in the Middle East, his efforts were, without question, an invaluable force multiplier for the British Command in that theater of operations. In the opening days of World War II, the Allies again sought to strengthen their weakening position and counter the initial victories and territorial gains of the Axis powers. The British, and later the Americans, funded, supplied, and assisted many partisan and guerrilla units throughout Europe and Southeast Asia in order to support Allied objectives, but many more were initially formed and operated without any such support. The French Resistance, Russian and Yugoslav Partisans, Filipino and American guerrillas in the Philippines, and the Communist guerrillas under Mao Tse-tung in China are the most recognized and noteworthy examples during the Second World War. Besides contributing to the eventual demise and defeat of the Axis powers, guerrilla operations during World War II are often attributed as the impetus that sparked many of the conflicts of the post-war world (Thompson, 1994).

Following the capitulation of the Axis forces in 1945, guerrilla warfare was used to achieve other political ends. After years of fighting Axis occupation, many indigenous peoples saw guerrilla warfare as a method to oppose continued colonial rule following the war's end (Chaliand, 1982).

Former colonies, most notably in Africa and Southeast Asia, saw this method of war as a means to achieve a political end—independence from colonialism—and this trend prompted Lomperis (1992) to label the period from 1945 to 1975 as the "*Era of People's War*." Although the techniques were the same during the Second World War and post-war years, the latter category has come to be known as an insurgency.

Robert Asprey (1994) noted that, "*For a number of reasons guerrilla warfare has evolved into an ideal instrument for the realization of social-political-economic aspirations of under-privileged peoples.*" This phenomenon is easily understood considering Mao Tse-Tung's proclamation that guerrilla warfare is a "weapon of the weak," to be used by those who are "inferior" to their oppressors (Mao, 1938). Although at the time of his writing Mao was embroiled in the struggle against the Japanese occupation of China, he recognized the significance of the relationship between politics and guerrilla warfare. The long-term goal of his struggle was not just freedom from the Japanese occupation, but the defeat of the Nationalist forces under Chaing Kai-Shek for the ultimate control of China. Thus he asserted, "*Without a political goal, guerrilla warfare must fail.*"

Preliminary Work

The success of the Chinese Communists in ridding themselves of Japanese occupation and Chinese Nationalist control inspired the rise of many other revolutionaries, and prompted them to employ the guerrilla methods to free themselves (Marks, 1996). While Mao's ideas have unquestionably influenced insurgent groups throughout the world, they have also served as the basic foundations for the few works on insurgencies and guerrilla warfare within the literature of military geography. The chapters within *The Selected Military Writings of Mao Tse-Tung* (1963) are replete with geographic concepts that address both the physical and human landscapes. Topography, climate, vegetation, population density, and core-periphery are only a few of the geographic topics and concepts that Mao used to formulate and express his theories on guerrilla warfare. And, while there is a tremendous amount of literature devoted to the topics of guerrilla warfare, insurgencies and counter-insurgent operations, relatively few studies have been conducted by geographers.

Historians, political scientists, and government agencies have traditionally done the "lion's share" of analyses on guerrilla warfare and insurgencies. Most works on counter-insurgency stem from the efforts of former military officers who gained their experiences first-hand fighting guerrillas. Within these works, however, geography, and its many subfields, have received a great deal of attention. Virtually every work on these topics, whether historical accounts or theoretical analyses, discuss the influence of the physical terrain at length, and to a lesser degree, the components of the cultural landscape. Joes (1996) concisely described how these components are intertwined within an insurgency when he wrote, "*the study of guerrilla wars requires attention to the conjunction among political, military, geographical, and sociological factors.*" Such pronouncements are common throughout the literature on insurgencies and guerrilla warfare, yet despite the apparent importance of geography to the birth and evolutions of insurgencies, geographers have contributed little within the published literature. While this undoubtedly reflects the ebb and flow of interest in military geography as a legitimate subfield of geography (Palka, 1995), what little work has been undertaken, I would argue, has been a mere cursory attempt to keep pace with other disciplines.

In the 1960s the United States was involved in a number of small- and large-scale conflicts, that today would be categorized within the MOOTW realm. Certainly our involvement in Vietnam and the struggle against the guerrilla methods of the Viet Cong prompted a great deal of study and volumes of written works on the subject. Peltier's and Pearcy's *Military Geography* (1966), however, discussed insurgencies in a mere two pages. Within these two pages, they attempted to identify the geographic setting in which guerrilla activity would mature. Population density and economic development, they argued, were the overriding factors. A region with 100 to 250 persons per square mile and an urbanization level of about 25%, with two-thirds of the population engaged in agriculture, forestry, hunting, and fishing (the primary sector of the economy) were the ideal geographic conditions for an insurgency to develop and flourish. These conditions, they asserted, allowed for adequate sources of supply for the guerrillas while inaccessible terrain and territorial space enabled freedom of movement. Urban areas, they insisted, were untenable for guerrillas, but were ideally suited for clandestine undergrounds, a concept that will be discussed more fully later in this chapter.

O'Sullivan and Miller (1983) attempted a much greater geographic analysis of guerrilla warfare and insurgency and devoted an entire chapter to the subject in their *Geography of Warfare*. They, of course, drew from Mao's ideas, discussed the influence of the physical landscape and the need for spatial proximity to targets, and also attempted to determine the relationship between such factors as population density, guerrilla-to-regular force ratios and guerrilla as well as regular force to space ratios. Surprisingly, their findings suggest that guerrillas are more likely to be successful in areas with "higher densities where targets, support, and cover are potentially greater." While their work was certainly a long stride towards including guerrilla warfare as a viable subject within the realm of military geography, the components of the cultural landscape associated with this form of warfare were not addressed.

In a subsequent work, O'Sullivan included a chapter devoted to guerrilla warfare in *Terrain and Tactics* (1991), and also included a chapter on revolutions. Within these chapters he continued with the themes that he and Miller presented in their earlier work, namely the relationships between population density and guerrilla-to-regular forces ratios, but he also included geographic aspects of guerrilla warfare that *Geography of Warfare* did not address. The most noteworthy additions included discussions of the cultural landscape, the geopolitical significance of this form of war, and the demographics of revolutionary leadership. In terms of the cultural landscape, O'Sullivan discussed the relationship between revolution and the degree of urban and infrastructure development. From a geopolitical perspective, O'Sullivan explored how geopolitics influences, and is also influenced by, revolutionary change and noted that changes of government within a state are never isolated in a regional or global sense. Revolutionary ideas and effects often permeate boundaries and have lasting effects that are often felt in far off corners of the globe. Both of these ideas will be further explained later in this chapter.

In considering the demographics of revolutionary leadership, O'Sullivan noted that while most of the insurgent manpower is drawn from the lower socio-economic classes within a state, the leaders generally emerge from the upper-middle classes of society. The professional class, with higher education levels, more freedom and mobility, and greater ambitions, serve as the rallying forces behind the masses and are able to effectively initiate and organize an insurgent movement. While O'Sullivan's additions in *Terrain and Tactics* again made significant contributions towards including guerrilla warfare and revolution as important topics within the field of military geography, the most recent texts by Winters (1991) and Collins (1998) fail to include any discussions at all of guerrilla warfare or insurgencies.

At the height of U.S. involvement in Vietnam, Robert McColl wrote what are arguably the most profound and relevant geographic studies of guerrilla warfare to date. *A Political Geography of Revolution: China, Viet Nam, and*

Thailand (1967) detailed the importance of insurgent territorial bases and discussed Mao's conclusions regarding the locational attributes of guerrilla areas. His later article, *The Insurgent State: Territorial Bases of Revolution* (1969), published in the *Annals of the Association of American Geographers*, further developed these locational factors, which he regards as attributes of ideal guerrilla areas, but focused on the spatial dimensions of areas under insurgent territorial control.

Spatial Dimensions of Insurgency

While Mao tied the physical and political attributes of terrain together from a guerrilla's perspective (and his subsequent strategy certainly illustrated the spatial evolution of an insurgency), McColl refined Mao's guiding principles and presented them from a geographical perspective. Mao (1938) described guerrilla warfare as a three-phased process that proceeded in a progressive sequence from one phase to the next, yet lacked clear and distinctive boundaries between phases. Phases may even occur at the same time in different areas within the country. The ultimate objective of this three-phased process is the control of territory and the population. It is this evolution that leads McColl to equate the development of an insurgency with the evolution of a state.

The State, as a political-geographical concept, serves as the foundation of McColl's analysis. The development of an insurgency using guerrilla warfare as a method, which he terms revolutionary movement, parallels the creation of an "insurgent state, complete with the elements of power, a *raison d'etre*, core areas, and administrative units, that is the manifestation of the insurgent's territorial imperative." In order to achieve this territorial imperative, the insurgent's ultimate objective, first voiced by Mao and commonly restated in most other works on guerrilla warfare, is the command of the indigenous population. While this is certainly true, since a lack of popular support will deny an insurgency the manpower, intelligence network, and support necessary for success, O'Sullivan and Miller (1983) argue that the incumbent government aspires to this same objective. However, McColl's idea highlights another critical component of the objectives in this, as in any other form or type of war. Control of territory must coincide with support of the people (McColl 1969). Walt (1996) confirmed this imperative when he wrote, "*The main object of revolutionary struggle is control of the state*," and a critical component of the state is that it controls territory and exerts a degree of sovereignty over that space within the defined boundaries.

Mao (1938) envisioned insurgency as a protracted war, spanning years and potentially even decades, and defined three stages for such a conflict: strategic defensive, stalemate, and counteroffensive. McColl analyzed these three stages and formulated his evolution of an insurgent state along these lines, using Vietnam, Cuba, and Greece as examples to further illustrate and prove the ideas he developed from China's successful revolution. During the initial stages of an insurgency, the insurgents lack control of territory. Described by McColl as the "contention" or "mobile war" stage, the insurgents seek to establish and organize themselves and recruit the support of the people. Without a safe area from which to operate and use as a base to train, rest, and refit, the insurgents must constantly move to elude government forces and survive. A "guerrilla area," as opposed to a "guerrilla base" as described by Mao, is an area, which the guerrillas control only as long as they occupy it. It is not an area that exists firmly within their grasp, and without guerrillas physically on the terrain, control reverts to the government or ruling authority. A guerrilla base, on the other hand, is directly under the control and influence of the insurgents. Government or opposing forces lack control over a guerrilla base and enter only under risk of determined attacks. These bases, however, develop from guerrilla areas as insurgents grow in strength and are able to effectively resist, defeat, and halt government attacks and incursions. Eventually the insurgents' search for a stable and secure "base" is realized and the second phase, termed the "equilibrium" or "guerrilla warfare" stage is achieved. An insurgency has achieved this stage when the guerrillas' strength and degree of popular support increases and the movement is able to gain physical and political control of territory. Once firmly under insurgent rule, these areas, labeled guerrilla bases by Mao, serve as the territorial embodiment of the insurgent state (McColl, 1969).

Guerrilla bases become the "core areas" to insurgents (McColl, 1969). From these locations, insurgents develop the administrative facilities necessary for a state to function and from which to continue the spread, or diffusion, of their ideas and influence. The insurgent movement continues its attacks on government forces and targets from these bases using guerrilla methods and seeks to further expand its territorial control. These bases, then, serve as the initial core areas of the insurgent state and the rudiments of a functioning governing structure evolve. To further solidify their legitimacy and support from the populous, the insurgent state establishes basic social services such as schools, hospitals and clinics, and develops the basic administrative systems required to govern territory and people.

These guerrilla bases, as the core areas of an insurgency, serve as central nodes of insurgent activity and subsequent territorial gains are referred to as "liberated areas" (McColl, 1969). As insurgent influence and control spreads beyond the guerrilla bases, and liberated areas continue to expand, the government is forced to modify its offensive stance and becomes, as Mao (1938) described, fixed to safeguard those areas under its control. This marks the beginning of the third and final stage, the conventional or "regular war" stage, where the guerrillas have transformed their forces into standing regular units and fight the government's forces in open, positional warfare as the former continue to expand their territorial control. Mao realized that guerrilla warfare by itself could rarely, if ever, complete the defeat of government forces. To accomplish this military objective, Mao (1938) foresaw the transformation of guerrilla units into regular, conventional forces which could then, using the material and equipment procured during the protracted war and from outside sources, face and defeat the enemy in conventional, regular warfare.

Another component of this spatial evolution is the need for the insurgents to establish guerrilla areas and bases in a number of locations. While an insurgent movement may initially develop and grow from one location, it must develop and expand it's guerrilla areas and bases through processes analogous to both contagious and relocation diffusion. In essence, the insurgents attempt to spread and disseminate, or diffuse, their revolutionary ideals not only outward from their guerrilla bases and areas but also to new locations in other regions of the state. If an insurgency attempts to develop and expand control from one single location, this presents the government with a tremendous opportunity. With only one area of guerrilla activity with which to contend, the counter-insurgent forces have the opportunity to contain and isolate the insurgent movement, concentrate all its efforts, resources, and manpower on that objective, and slowly tighten the cordon until the insurgents are finally defeated.

Besides preventing this scenario from occurring, Che Guevara (1961) discussed a number of other reasons for the relocation and contagious expansion of insurgent controlled areas. Such methods increase the operational range of the insurgents, spread the ideas of revolution, gather wider support, and place greater pressure on counter-insurgent forces, forcing the government to contend with several dispersed threats simultaneously. They also ease logistical needs as the guerrilla base may find itself incapable of supporting the growing number of the guerrillas. But for such expansion to occur, a "core area" must serve as the focal point. These areas are the guerrilla bases.

While this insurgent state model describes the importance of guerrilla bases and the spatial evolution of an insurgency, an often-critical component of an insurgents' preliminary successes are the locational factors of the initial guerrilla areas (McColl, 1969). From Mao's original deductions, and verified by examining other insurgencies, McColl elaborated on the seven geographical factors that constitute what he regards as the attributes for an ideal guerrilla area, an area conducive to the birth and evolution of an insurgency. McColl's necessary attributes include the following: (1) If possible, an area should be chosen that has had previous experience in revolution or political opposition to the central government. (2) Political stability at both the national and local levels should be weak, or actually lacking. (3) The location must provide access to important military and political objectives, such as provincial capitals, regional cities, and critical resources and transport services. (4) Areas of weak or confused political authority, such as borders between police or military areas or even along international boundaries are ideal locations. (5) Terrain should be favorable for military operations and personal security. (6) Insofar as possible, the area should be economically self-sufficient. (7) Once established, the base should never be abandoned except under the most critical circumstances.

These attributes constitute geographic criteria, reflecting both the physical and human landscapes, which the insurgent should seek to gain an advantage over an adversary during the initial phases of an insurgency. David Galula (1964), a French army officer with extensive counter-insurgency experience stated the need for such an advantage when he wrote, *"If the insurgent, with his initial weakness, cannot get any help from geography, he may well be condemned to failure before he starts."* While McColl certainly understood the importance of these locational factors, he acknowledged that these factors alone will not ensure the success of an insurgency but their relevance and validity cannot be disputed. While these attributes are fairly straight forward, there is need to discuss them further. This is necessary not only to elaborate on the more salient features of each, but to also suggest refinements based on further research. O'Sullivan and Miller (1983) noted that insurgencies are difficult to study as current events because of the lack and incompleteness of information that is available during such conflicts. Both the insurgents and the government are often secretive about aspects of such a conflict and it is only after victory has been won that information becomes available. For these reasons, it is certainly worthwhile, indeed necessary, to consider new information and assess the validity of previous conclusions.

These seven criteria essentially describe the ideal site and situation for guerrilla areas and perhaps it may prove more useful to consider these attributes in those terms. Site, refers to the internal attributes of a location, including its spatial organization and its physical setting. Clearly, the attributes which describe previous experience in revolution or political opposition, weak or absent political stability and presence, terrain favorable for military operations and personal security, and economic self-sufficiency characterize the internal attributes, or site, of such a location. Situation refers to location relative to other areas and features of the landscape. The attributes that address weak or confused political authority along boundaries, important military and political objectives relative to an area are examples. While all of these criteria certainly influence the initial stages of an insurgency, it is the situation, which is regarded as the most critical.

The location must provide access to important military and political objectives, such as provincial capitals, regional cities, and critical resources and transport services. This is certainly the most critical attribute of a guerrilla area according to Mao and, in geographic terms, reflects the situation of the guerrilla area. The relative location of a guerrilla area and its relationship and proximity to other places, namely targets, is considered paramount because isolation, as McColl has illustrated, can doom an insurgency to failure. Guerrilla activity can certainly survive in remote areas away from government control, but they can only be effective in challenging the government if they are within reach of critical objectives. In comparing the physical security of rugged terrain versus the proximity of insurgent objectives, Liddell Hart (1991) wrote, "*A guerrilla movement that puts safety first will soon wither.*"

Although critical resources and transport services are included in this category, these features should be considered sub-parts of a broader category under economic targets. Examples of economic objectives throughout the history of warfare are easily identified. Scorched earth policies and targeting of an enemy's manufacturing industries exemplify the relevance and importance of these objectives during war. A state's economic activities, especially those of the basic sector of the economy, are certainly critical targets in an insurgency since they represent not only the economic means by which a state earns its revenues, and hence contributes to funding the counter-insurgent campaign, but also because of their related political importance. Although it may seem redundant to include economic objectives as critical targets since such activities are concentrated in large urban areas, which McColl addressed, economic centers are worth including as vital targets within this description. Recent attacks on tourists in Egypt illustrate the need to include this aspect along with military and political targets. As one of Egypt's largest industries, tourism contributes tremendous revenues to the economy and national treasury. Insurgent attacks on foreign tourists during the 1990s can be viewed to have the dual effect of weakening the Egyptian economy and discrediting the government.

Another important aspect of the situation of a guerrilla area is its location astride borders, especially international boundaries. Such locations provide a number of opportunities for guerrillas, as a lack of coordination or cooperation between adjacent provincial agencies may prevent a concerted and coordinated drive by government forces to defeat the insurgents. International boundaries provide even greater assets to the insurgents, especially if the neighboring state is sympathetic to the insurgent cause, and have long been recognized as critical locations that afford the insurgent a crucial form of external support known as sanctuary. Such proximity to the sovereign territory of another state provides the insurgents with an area safe from government reprisals and also provides a line of communications through which external support in the form of equipment and material may be funneled. While any state friendly to the insurgent cause provides a potential asset to the guerrillas, a neighboring state that itself has a weak or unstable regime in power and is unable to prevent the flow of support or presence of the guerrillas may prove especially beneficial to the insurgents.

Although considered less important as a locational attribute, the site of a guerrilla area is also an important consideration. Terrain suitable for military operations of course refers to the physical landscape of the area in question. This attribute, by far, has received the most attention within the literature on insurgencies and guerrilla warfare. Rugged, difficult terrain such as mountains, forest, jungles, and swamps has traditionally been ideal for guerrilla operations and were labeled as "favorable terrain" by Che Guevara (1961). Guevara likewise identified deserts, plains, and other types of relatively flat, open terrain, devoid of vegetation as "unfavorable" despite successes by past insurgents in these environments. Although a number of successful guerrilla campaigns have been waged in these environments, doubts exist whether these victories could be achieved today because of the improvements in technology; namely aircraft and reconnaissance satellites (O'Sullivan and Miller, 1983). However, aircraft and satellite systems can only assist a counter-insurgency campaign if a state that has these resources at her disposal or readily available from allies and friendly governments. A "marginal air force capability" as O'Neill (1980) described, enabled the Polisario insurgents in Spanish Sahara to survive in a desert environment. Likewise, Barnard (1985) found McColl's insurgent state model and the locational attributes of guerrilla areas applied to the South West African People's Organization (SWAPO) struggle to gain independence from South Africa. Despite the disadvantages of its site in southern Africa, an area composed mostly of flat arid Savannah, SWAPO eventually gained independence for Namibia in 1990.

Although not discussed as extensively as the physical landscape, climate is also a factor, which has been considered in the conduct of guerrilla warfare. Gallula (1964) contended that the ideal climate for guerrilla activity is a temperate zone, whereas tropical or arctic conditions favored counter-insurgent forces since harsher climates require greater logistical support in terms of food, clothing, and shelter. O'Neill (1990), however, argued that climate is neutral, but noted that bad weather offered greater advantages to the insurgents since such conditions eliminate the technological superiority of the government's forces. He cited as examples the Sultan's Armed Forces in Dhofar, Oman, who suspended operations against the insurgents during the monsoon months, and the Soviet propensity to cease operations in Afghanistan during the winter. Barnard (1985) also described the effects of the monsoon on guerrilla activity during SWAPO's struggle for independence as the insurgent operations there increased during the wet seasons when the rains limited aircraft support and washed away insurgent tracks, making it more difficult for South African units to tack insurgent movements. Clearly, insurgents of the future will attempt to reduce their enemies' advantages in the same ways and counter-insurgent forces must understand the effects that weather, climate, and terrain will have on their operations.

While not necessary, if possible an area should be chosen that has had previous experience in revolution or political opposition to the central government. This attribute, of course, enables an insurgency to take root in an area in which such a struggle is not a new concept. A legacy of war and conflict may provide the insurgents with an experienced pool of battle-hardened fighters if the previous rebellions were relatively recent. While McColl illustrated that initial guerrilla areas in the examples he cited certainly met this criteria, the impression from reading these descriptions suggests that this attribute is considered in an historical context. I would argue, based on more contemporary works on this subject, that such an attribute should be considered in a terms of culture. More recent and in-depth studies, such as Laqueur (1976), O'Neill (1990), and Marks (1996) to name but a few, discuss the effects of culture on insurgencies and described specific instances where cultural components affected the course of particular cases. Mao himself acknowledged that China's was a unique situation in terms of physical characteristics and its political situation as well its cultural dynamics. When he considered guerrilla wars in other states he noted differences and wrote, *"these differences express the characteristics of different peoples in different periods"* (Mao 1937).

A few short years after McColl's article, Gurr (1970) wrote an extensive work entitled *Why Men Rebel*. Just one of the theories among the many potential causes that incite men to fight, Gurr wrote, *'Some of men's perspectives on violence are psychocultural in origin, the result of socialization patterns that encourage or discourage outward displays of aggression, and of cultural traditions which sanction violent collective responses to various kinds of deprivation."* While McColl's premise about the historical nature of resistance against the government certainly holds true, this historical recurrence is arguably the manifestation of culture traits that define a people as fiercely proud, stubborn, independent, and intolerant of outside intervention and influence. The past experience in revolution or opposition to the government, then, should arguably serve as an indicator to analyze the cultural traditions of an area of potential insurgent activity.

Laqueur (1976) hypothesized *"that beyond a certain stage of cultural development it is difficult for a guerrilla movement to gain mass support."* He proclaims *"civilized countries"* are not fertile grounds for these types of movements but fails to clarify his definition of *"civilized."* But considering how culture, in conjunction with the political, social, and economic elements of a society are related, a geographer could attempt to characterize a "civilized" society based on several factors. Literacy rates, education levels, levels of infrastructure development, and economic indicators, to name but a few, should be able to provide some indicator of "civility." While this question deserves further geographic analysis, a possible indicator of "civility" may be the degree of political representation. Marks (1996) contended that democracy was the answer to defeating and preventing insurgent threats and Gurr (1970) discussed the idea that political violence erupts when the political voice of the people is revoked or their representation in government is declared illegal. Every realm of the world has experienced an insurgency at one time or another, yet several regions have proven more susceptible to this phenomenon than others, and these regional differences should spark the interest of geographers.

Mao arrived at the list of locational attributes, summarized by McColl, after years of continuous struggle against both the Chinese Nationalist forces and Japanese occupation. Driven from the cities by the Nationalist forces under Chiang Kai-shek, the communists found that the rural hinterlands provided the setting where Mao and his followers could develop their insurgent movement. Within the articles of his *Selected Military Writings*, Mao (1963) emphasized the Chinese communists' weaknesses when compared to the industrial, economic, and political strengths of his adversaries, yet he never questioned Red China's ability to ultimately win those conflicts. He described his rationale for this argument as his enemies' inability to effectively control anything beyond the large cities of the country. The inability to control the hinterland, of course, was due to the limited resources and manpower of Japan and the lack of popular support for the Koumingtang. Since his enemies' control would subsequently be limited to cities and the crucial lines of communications, Mao saw the vast rural areas beyond the urban areas, or hinterlands, as the regions that would facilitate the evolution and growth of guerrilla movements.

The rural areas were also attractive for the conduct of guerrilla warfare because these areas contained China's greatest asset—the people. Mao (1938) voiced this idea when he wrote, "*the richest source of power to wage war lies in the masses of the people*" and he emphasized this belief throughout all his writings. China was, and still is, a predominately agrarian society where the vast majorities of people live and work the land. Even today, not withstanding the rising urbanization levels throughout the world, China is only 30% urbanized by most estimates. Although Mao's conduct of guerrilla warfare is often regarded as a rural approach, he never lost sight of the economic, cultural, and political importance of cities and his strategy revolved around "surrounding the cities with the countryside" (McColl, 1969). His method, then, was the development of a guerrilla movement in the rural hinterlands, developing from the initial guerrilla areas selected with regards to the locational attributes mentioned above, into guerrilla bases which slowly and steadily expanded the liberated areas toward the cities. Control of the principle cities and regional capitals, the core areas of the state, are the ultimate objectives of such a revolutionary movement and signify the maturation of the insurgent state. Figure 1 illustrates this rural approach as the insurgents expand their control from rural guerrilla areas and bases towards their ultimate and final objective, the cities.

The legacy of Mao's success in China has been numerous insurgencies attempting this same rural approach, especially among communist movements. While Marks (1996) concluded that democratic reform was the key to successful counter-insurgent campaigns against communist insurgents following Mao's model, he also determined that many failures resulted from misunderstanding Mao's intent: that revolutionary struggle must be adapted to meet the unique circumstances within each society. He further contended that insurgencies succeeded in Vietnam and Cuba because those two cases deviated from the classic rural approach, which proved successful in China and undertook a rural-urban approach. While both insurgencies developed guerrilla areas and bases in the rural hinterlands, both movements also established networks, or clandestine undergrounds, within the major cities and population centers (Marks, 1996). While many believe that Fidel Castro landed in Oriente Province, Cuba with only twelve men and eventually grew into the powerful force that overthrew the Batista regime, many in-depth studies of the Cuban revolution acknowledge that an extensive underground network was in place, mostly in the urban areas, that provided Castro's fledgling guerrilla force with invaluable support and assistance which contributed greatly to his ultimate victory (Anderson, 1997).

Urban areas are the settings for establishing the clandestine undergrounds discussed by Peltier and Pearcy (1966). Although it is difficult for insurgents to fight using classic guerrilla methods in cities, undergrounds serve as vital sources of intelligence information, develop and build a base of popular support among the population, and conduct acts of sabotage and subversion against political, military, and economic targets that support overall insurgent strategy. These activities, of course, force the government to divert their efforts from the rural counter-insurgent campaign and commit more resources, manpower, and effort to maintain their control over these important objectives. While Mao's rural approach advocated control of cities as the final stage of protracted war, the rural-urban approach described by Marks is more readily suited to states that have much higher levels of urbanization. And, as urbanization levels around the world steadily increase, this phenomenon will surely continue (O'Neill, 1990).

Of course, the notion of urban guerrillas is not a new concept and there have been several works devoted entirely to this subject. Within the literature there are varying opinions on whether the term 'urban guerrilla' or 'terrorist' should be used to describe these insurgents, chiefly because their methods include bombings, kidnapping, hijackings, and assassinations. Further discussion of this subject is beyond the scope of this chapter, but this does highlight another subject area where military geographers can certainly contribute. O'Sullivan (1991) touched on this subject when he wrote, *"Revolution will have a greater chance of success in a nation with an overwhelming dominant capital, a strong radial component in its networks and constitutional arrangements and a minimum of circumferential linkages."*

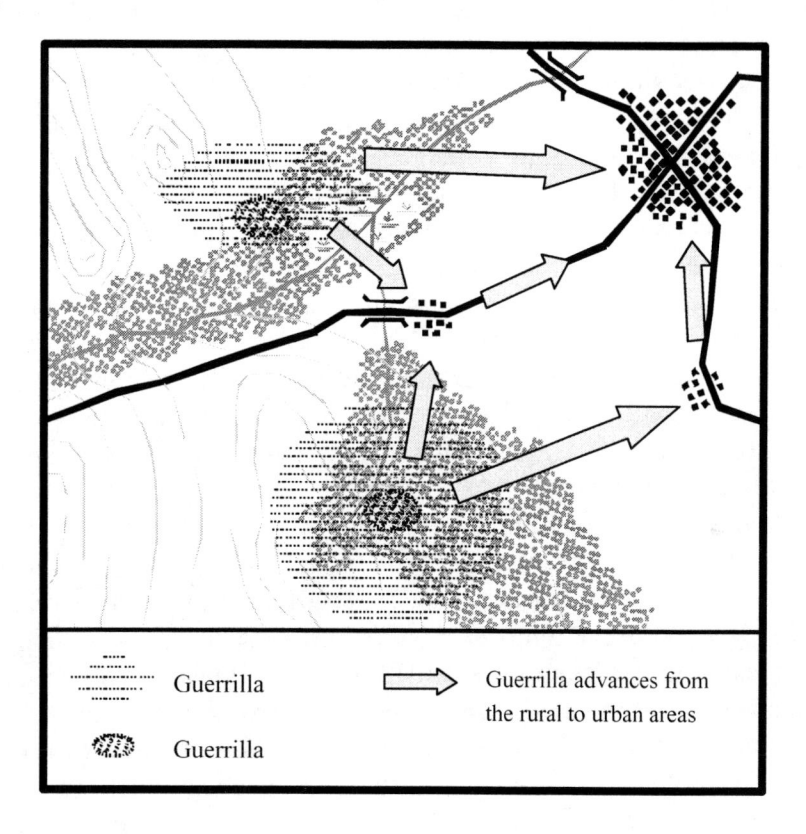

Figure 1. Rural Insurgent Strategy.

Counter-Insurgency Operations

While insurgent success following the strictly rural approach used by Mao is questionable today, urban areas undoubtedly remain the most important objectives to insurgent and the counter-insurgent forces. In protecting these areas and seeking to defeat guerrillas, the state, in essence, becomes the Counter-Insurgent State (Rich and Stubbs, 1997). The ultimate objectives for a government seeking to rid itself of an insurgent threat is to defeat the insurgents and retain control of the population and territory. Two strategic approaches exist. Although either can be used, the most effective strategies have proved to be a combination of political and military approaches (Paret and Shy, 1962; Paget, 1967; Rich and Stubbs, 1997). The literature on counter-insurgency operations advocates many of the same principles of guerrilla warfare, such as control of the population (O'Sullivan, 1991). However, another critical component of these two approaches is controlling the terrain. In simple terms, the military approach attempts control of the territory while the political approach seeks control and support of the population.

Paget (1967) acknowledged that the defeat of guerrilla forces is a *"long and laborious process."* He concluded that it is *"far more effective, and also more economical, to defeat them by making it impossible for them to fight on."* This defeat is possible, Paget continued, by denying insurgents the "essentials" needed to continue an insurgency. These essentials include support of the local population, bases, mobility, supplies, information, and the will to win. Denial of these essentials requires a combination of political and military approaches, but an overriding factor in this relationship is the need for the military to be subordinate to the political solution (Paret and Shy, 1962; Paget, 1967).

One of the more notorious methods, which blend political and military aspects of a counter-insurgent campaign, is resettlement. The British first used this technique to move Boer families away from guerrilla areas to deprive Boer fighters of local popular support (Paret and Shy, 1962). Additionally, this technique has been used in subsequent successful counter-insurgent campaigns in Malaya, Kenya, and Oman. While this technique is intended to remove the population from potential insurgent influence, reduce insurgent sources of supply, and effectively create a free-fire zone without fear of civilian casualties, it does present a significant challenge to undertake. In order to properly and effectively gain the support of the people who are being moved to new areas, a number of questions arise. One of the most obvious and important issues is deciding where to resettle these people. Logistical questions also arise in terms of funding, transportation, housing, food, and new employment. While resettlement has proved successful in several cases, it certainly presents a number of challenges in implementation.

One of the most effective solutions during counter-insurgency operations has been the offering of amnesty to insurgents who surrender. Unconditional surrender has no place in this form of war, Paget (1967) proclaimed. Amnesty, coupled with promises of opportunity and assistance, has proved successful in many counter-insurgency campaigns such as in the Malayan peninsula in the 1950s and in Dhofar, Oman in the late 1960s and early '70s. These opportunities can be even greater for former insurgents who join the counter-insurgency campaign. While this is considered to be a political solution to an insurgent problem, it does offer significant military advantages as well. Former guerrillas are often enticed to serve the government and can effectively harass their former comrades and lead counter-insurgent forces to guerrilla bases.

Amnesty, while an effective incentive to some insurgents who doubt the validity or success for their cause, does not address the larger issue of support of the population. On a larger scale, the most ideal political approaches address the sources of grievances among the people by developing and implementing social, economic, and sometimes political reform (Paret and Shy 1962). By thus eliminating the causes, the state can erode the popular support for the insurgents and bring it to their side. Institution of social welfare programs, including education and health care, and economic restructuring to relieve unemployment and poverty should be the objective of any government hoping to avoid the development of an insurgency. Such measures effectively eliminate the *raison d'etre* of the insurgent state. However, many times these options are not employed.

Some regimes proved to be so corrupt that the needs of the people are ignored. Gurr (1970) acknowledged that this has a far greater potential to lead to political violence than simple incompetence by the government. In many cases, governments in lessor developed countries lack the expertise, structure, or resources to adequately address the social, political, or economic problems they encounter. In such situations, external support is almost as vital to the state as it is to the insurgents. Unfortunately, this external support often comes in the form of military aid. The effects of political solutions are often not immediately evident so military solutions to insurgency development are frequently adopted as the first line of defense (Rich and Stubbs, 1997). Counter-insurgent forces recognize the importance of terrain in the battle against guerrillas and aim to deprive them of the safety that guerrilla bases offer. Unless such operations are conducted in the very early stages of an insurgency, military approaches are rarely successful in and of themselves. A combination with political solutions, and a careful mixture and balancing act between the two are critical to successful counterinsurgency (Rich and Stubbs, 1997).

Galula's (1964) strategy for counter-insurgency operations includes a combination of political and military approaches in order to deny guerrillas of the essentials they need for success. The strategy involves the concentration of government forces to destroy or expel the guerrillas from an area, gain local support from the people to isolate the guerrillas, guard this area through the development of local constabularies and militia, establish new provisional local authority, erode confidence of the insurgents in their cause, and hence win them to the government's side. This, of course, is only possible if the government forces are proactive instead of reactive.

Paget (1967) believed that concentration of government forces in an area is analogous to the guerrilla's development of a base area. According to this theory, once a base is firmly established, government forces must expand their control outward, clearing the guerrillas from the area. These actions must be based on a systematic analysis of the insurgent situation. The locations of guerrilla bases and areas must be identified and the government must develop an overall plan to systematically wrest control from the insurgents (Thompson, 1966). The site and situation of the government base is somewhat similar to that of the guerrilla base. However, as the guerrillas establish rural bases and expand control towards the cities and population centers from the hinterlands to the core areas, government strategy operates in reverse. The government's initial concern is to maintain and consolidate their control over these core areas and population centers and isolate these areas from insurgent influence (Thompson, 1966). With these areas firmly in their grasp, the government attempts to extend this control outward, systematically regaining control of the hinterlands. Figure 2 illustrates the counter-insurgent method of consolidating control of cities and population centers and slowly, but systematically, expanding their presence and control further away from these areas into the rural hinterlands.

The strategy of counter-insurgency, as Paret and Shy (1962) proposed, becomes a battle for territorial control. From the areas under its control, the government will concentrate its forces and expand outward, leaving behind the local constabularies and police to maintain control while the military continues its outward expansion. As this process continues, the government seeks to isolate the guerrilla bases and areas, closing them in and separating them from one another. By eliminating these lines of communications among guerrilla forces, which reduces their mobility and infringes on their supply flow, it becomes more difficult for the insurgents to retain the initiative (Paget, 1967).

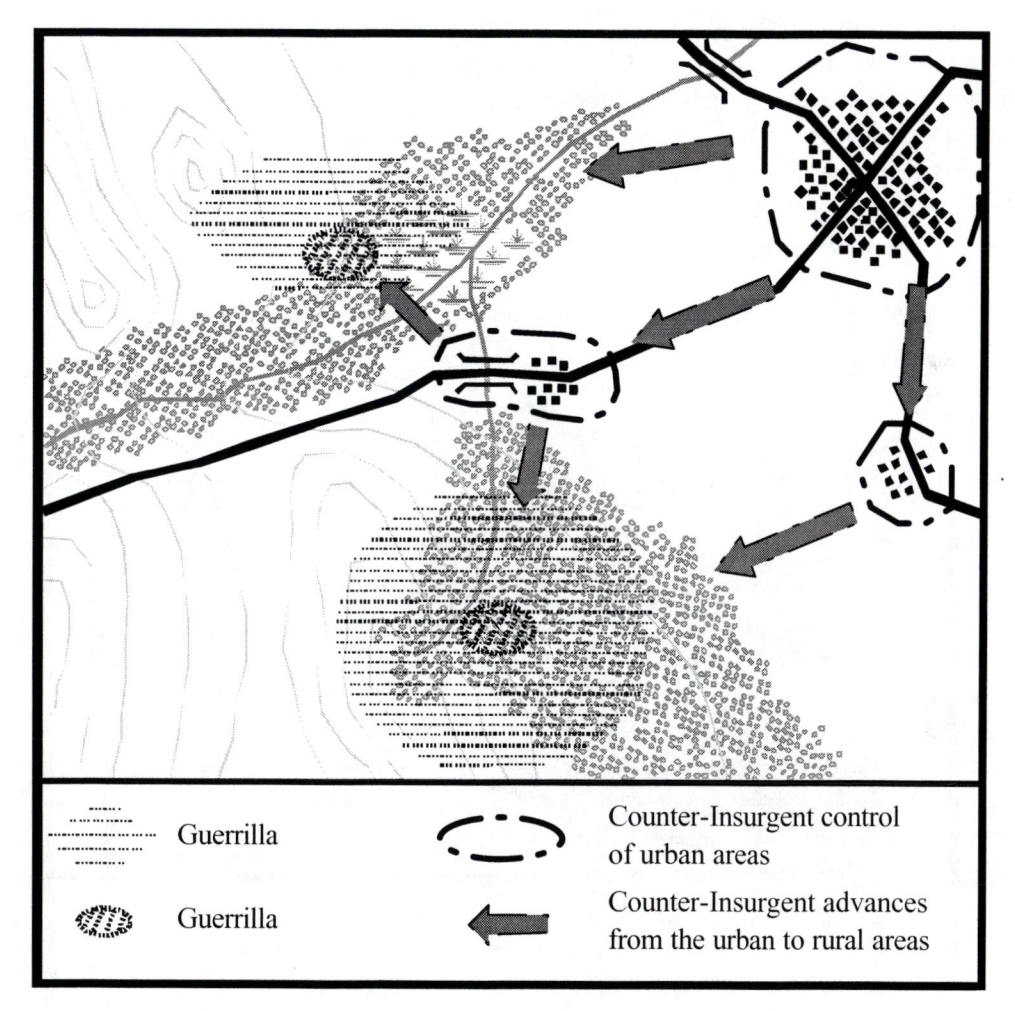

Figure 2. Urban-to-rural counter-insurgent strategy.

Another government imperative is to isolate the guerrillas not only from the people, but also from external support. O'Neill (1990) identified four broad categories of external support: moral, political, material, and sanctuary, and there is agreement within the literature that such outside support is necessary at some time during the course of a conflict. A proper assessment of both the physical and human components of the landscape is necessary if government forces are to succeed in eliminating this effort. The Dhofar Insurgency in Oman between 1965 and 1975 illustrates how the government, assisted by British advisors, was able to effectively use terrain to limit and eventually stop the flow of support to the Popular Front for the Liberation of Oman (PFLO) (see Figure 3). Omani and British forces were able to isolate the PFLO insurgent movement by blockading the coast with naval forces and land and aerial patrols on the northern slopes of the Dhofar mountain range. Not only did this limit the flow of supplies and assistance to the guerrillas, it allowed the government to systematically regain territorial control from the insurgents by pushing them westward towards their primary base in Yemen.

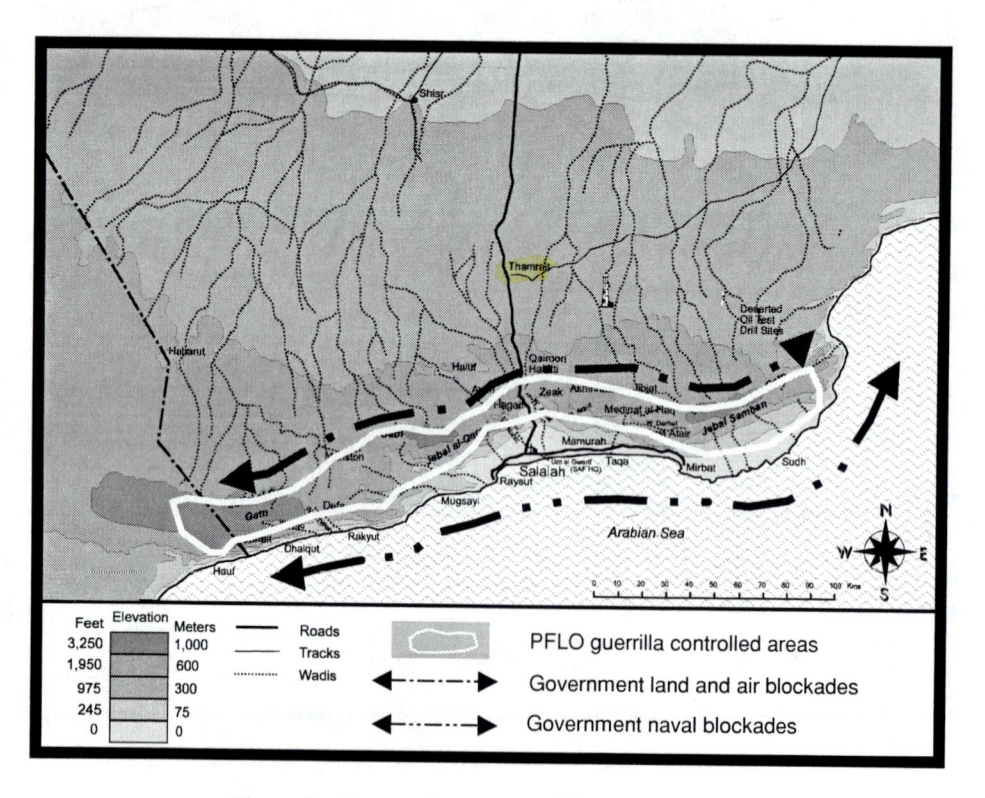

Figure 3. Counter-insurgent strategy in Oman.

Control of the territory, however, does not necessarily guarantee support of the people, and so the government must enact political approaches to deny this support to the insurgents. Military defeat of armed insurgency does not always end the dilemma (Thompson, 1966). If the seeds of discontent are still alive, the insurgency has the basis from which to rise again. To eliminate this threat, the reestablishment of authority becomes imperative as the government reclaims areas. The responsibilities, which accompany this authority, however, must be met. The social and economic reforms designed to eliminate the grievances of the people must be instituted to gain and maintain local support, thus denying assistance to the guerrillas for a potential return to the area.

The Future of Counter-Insurgency

At the time he wrote *The Insurgent State*, McColl describe that period of time as the "Insurgent Era" and his study was devoted to the subject of mass revolutions that attempt to solicit widespread support from the population in an overthrow of the existing regime. He noted a significant difference between these revolutions and secessionist movements, a region seeking to break away from a state and gain independence, which he did not consider applicable to his study. While there are certainly examples following the precepts of the insurgent state model seeking to overthrow existing regimes today, with Colombia, Sri Lanka, and the People's Democratic Republic of the Congo as the most current examples discussed by Deblij and Muller (2000), there are currently far more secessionist movements attempting to gain independence. Eritrea gained its independence from Ethiopia in 1993 and Kosovo, Chechnya, East Timor; several islands in the Philippines, and the Caprivi Strip in Namibia are among others attempting the same. Perhaps this current period should be referred to as the "Secessionist Era."

While these scenarios may be difficult to truly understand due to the scarcity and incompleteness of adequate information, I would argue these secessionist movements also form and develop using the locational attributes described by McColl (1969) and essentially evolve along the same lines as the insurgent state model. This is, however, one area within the field of military geography that certainly warrants further study. Undoubtedly, this phenomenon will continue in the near future and as U.S. and United Nations forces will continue to be committed to these areas as peacekeepers or peace enforcers. It is essential to have an intimate understanding of these conflicts if we hope to resolve them and prevent being drawn into potentially bloody campaigns. Additionally, in the event of another large, conventional conflict, the same locational attributes may reasonably indicate where partisan units will seek to establish their bases since Mao did arrive at these conclusions while fighting against Japanese occupation as well as against the incumbent Nationalist government.

As Laqueur (1976) stated, the insurgent state is a very complex and dynamic phenomenon that includes a great many subjective and objective factors. His attempt has been to summarize the guerrilla experience and determine common factors and threads by analyzing numerous insurgent movements and studying the works of renowned practitioners of guerrilla warfare. The importance of his work is that he attempted to discover patterns rather than empirical laws within guerrilla movements. Although the latter is impossible, he asserted, because the complexity and diversity associated with such movements, the analysis of patterns can lead to a greater understanding of why some insurgencies succeed and others fail. The focus of this chapter has been the spatial patterns of insurgencies and counter-insurgent strategy. While I contend that this is by far the most critical dimension of these conflicts, there are a great number of other considerations, which deserve further geographical study. Geographers are, without question, uniquely qualified to contribute to a greater understanding of these scenarios that military forces will undoubtedly continue to face in the future.

The review of preliminary work by geographers was intended to identify the sources to others interested in this topic and draw attention to a number of hypotheses that were proposed in those works. And, considering newly released information and more recent analyses by other sources, further testing is certainly warranted. This will not only test their validity in light of more detailed information but may also allow new theories to be developed. Certainly, more contemporary studies of the influence and impact of the physical and cultural landscapes on the course of insurgencies and counter-insurgencies can only serve to enhance a greater understanding of these conflicts.

Conclusion

Under the umbrella of Military Operations Other Than War, the United States and her allies will continue to face insurgent threats in the global arena. With such a specter on the horizon, an understanding of the birth and evolution of guerrilla insurgencies is crucial to maintaining and projecting political and military strength. The true essence of counter-insurgency was probably best stated by McColl (1966) when he described his intent behind identifying the locational attributes of guerrilla areas. It was, in his words, to *"predict, at least the general areas where preemptive efforts at control and development should be concentrated.*" The key to counter-insurgency is to stop it before it begins and by focusing development in the areas ripe for insurgency a government can make significant strides in preempting such a conflict.

Much study has been devoted to the causes that incite insurgencies and escalate into guerrilla war. In a general context, failure by the state to meet the needs or desires of the people are widely recognized as causes that lead to the birth of an insurgency but an insurgent movement is far more likely to develop where political, economic, or social oppression occur (Gurr, 1970). Political, economic, social, and cultural causes are often attributed to the birth and development of insurgencies (Campbell, 1967; Gurr, 1970; Joes, 1992; and O'Neill, 1990), and while one may play a more significant role in inciting the people, rarely do they operate independently. More accurately, the causes of an insurgency result from the combination of these factors, which encourages people to rebel. This requires an intimate understanding of people and places so as to eliminate the grievances that so often lead to political violence. Examine these topics together, consider the physical landscape, and we've described the regional approach in geography. Certainly this discussion illustrates the significance geographers can make towards understanding, and predicting, where such conflicts may erupt.

It is the interesting mix of patterns and dissimilarities Laqueur (1976) discussed that should draw the interest of geographers. It stands to reason that every aspect of an insurgency and guerrilla war is heavily dependent upon and influenced by geography. It is the combination of these factors, the dimensions of the human and physical geography, at every scale of analysis, which influence the birth, evolution, direction, and future of an insurgency. Although it is often an unattractive and even ugly venture, often with high casualty figures, insurgency is a topic worthy of study. B.H. Liddle-Hart (1967) wrote *"War is always a matter of doing evil in the hope that something good may come from it."* This sentiment has been echoed by many others (Laqueur, 1976; Asprey, 1994; Paget, 1967; Galula, 1964) who claim the study of insurgencies and guerrilla wars will shed light and provide insight, which, hopefully, will prevent their recurrence. Considering the significance geography plays in all aspects of such wars, it stands to reason geographers have much to contribute towards this end.

Bibliography

Anderson, Jon Lee. 1997. *Che Guevara: A Revolutionary Life*. New York: Grove Press.

Asprey, Robert B. 1994. *War in the Shadows, The Guerrilla in History*. New York: William Morrow and Company, Inc.

Barnard, W.S. 1985. The Border War: After 19 Years. In *Kompas Op Suidwes-Afrika/Namibie*, Dennesig, RSA: Society for Geography Special Publication No. 5.

Deblij, Harm and Peter O. Muller. 2000. *Geography: Realms, Regions, and Concepts 2000*. 9th ed. New York: John Wiley and Sons.

Campbell, Arthur. 1967. *Guerrillas, A History and Analysis*, London: Arthur Barker Limited.

Chaliand, Gerard. 1982. *Guerrilla Strategies*. Los Angeles: University of California Press.

Ellis, John. 1995. *From the Barrel of a Gun: A History of Guerrilla, Revolutionary, and Counter-Insurgency Warfare*, from the Romans to the Present, London: Greenhill Books.

Galula, David. 1964. *Counter-Insurgency Warfare: Theory and Practice*, New York: Frederick Praeger, Publisher.

Guevara, Che. 1961. *Guerrilla Warfare*, New York: Monthly Review Press. Reprint. 1985. Lincoln: University of Nebraska Press.

Gurr, Ted Robert. 1970. *Why Men Rebel*. New Jersey: Princeton University Press.

Huntington, Samuel P. 1962. Introduction. In *Modern Guerrilla Warfare*, ed. Franklin M. Osanka. New York: The Free Press of Glencoe.

Joes, Anthony James. 1992. *Modern Guerrilla Insurgency*. London: Praeger Publications.

Laqueur, Walter. 1976. *Guerrilla, A historical and Critical Study.* Boston: Little, Brown and Company.

_____. 1997. *Guerrilla Reader, A Historical Anthology*. Philadelphia: Temple University Press.

Liddle Hart, B.H. 1991. *Strategy.* New York: Penguin Books.

Lomperis, Timothy J. 1996. *From people's War to People's Rule.* Chapel Hill: University of North Carolina Press.

Mao Tse-Tung. 1937. *On Guerrilla Warfare.* Trans. Samuel B. Griffith. 1961. Reprint. New York: Praeger Publications.

_____. On Protracted War. 1938. In *Selected Military Writings of Mao Tse-Tung.* 1963. Peking: Foreign Language Press.

Marks, Thomas A. 1996. *Maoist Insurgency Since Vietnam.* London: Frank Cass & Co. LTD.

McColl, Robert A. 1967. Political Geography of Revolution: China, Viet Nam, and Thailand. *The Journal of Conflict Resolution.* 11: pp. 153-167.

_____. 1969. The Insurgent State: Territorial Bases of Revolution. *Annals of the Association of American Geographers.* 59: 613-631.

O'Neill, Bard E. 1990. *Insurgency and Terrorism, Inside Modern Revolutionary Warfare.* New York: Brassey's.

_____. ed. 1980. *Insurgency in the Modern World.* Boulder: Westview Press.

O'Sullivan, Patrick and Jesse W. Miller Jr. 1983. *The Geography of Warfare.* New York: St. Martin's Press.

O'Sullivan, Patrick. 1991. *Terrain and Tactics.* New York: Greenwood Press.

Paget, Julian. 1967. *Counter-Insurgency Operations: Techniques of Guerrilla Warfare.* New York: Walker and Company.

Palka, LTC Eugene J. 1995. The US Army in Operations other than War: A Time to Revive Military Geography. *GeoJournal.* 37 (2): 201-208.

Paret, Peter and John W. Shy. 1962. *Guerrillas in the 1960s.* New York: Frederick Praeger, Publisher.

Peltier, Louis C. and G. Etzel Pearcy. 1966. *Military Geography.* Princeton: D. Van Nostrand Company, Inc.

Rich, Paul B. and Richard Stubbs, eds. 1997. *The Counter-Insurgent State, Guerrilla Warfare and State Building in the Twentieth Century.* New York: St. Martin's Press.

Thompson, Leroy. 1994. *Ragged War, The Story of Unconventional and Counter-Revolutionary Warfare.* London: Arms and Armour Press.

Thompson, Robert. 1966. *Defeating Communist Insurgency.* New York: Frederick Praeger, Publisher.

U.S. Army. Field Manual 7-98: *Operations in a Low-Intensity Conflict.* 1992. Washington, D.C.: Headquarters, Department of the Army.

Walt, Stephen M. 1996. *Revolution and War.* Ithaca, New York: Cornell University Press.

Wilkins, Frederick. 1954. Guerrilla Warfare. In *Modern Guerrilla Warfare.* 1962. ed. Franklin M. Osanka. New York: The Free Press of Glencoe.

12

The Case of Bosnia:
MILITARY AND POLITICAL GEOGRAPHY IN MOOTW

By

Mark W. Corson and Julian V. Minghi

Introduction

Political and military geography are inter-related just as military operations are driven by political policies. The interface between political and military geography is especially evident in MOOTW where soldiers are focused on preventing fighting and rebuilding countries. Soldiers serving as peacekeepers must combine military and diplomatic skills. These skills include knowledge of the military and political geography of their area of operation. This chapter focuses on the inter-relationship between military and political geography in MOOTW by presenting a case study focusing on the on-going peacekeeping mission in Bosnia. Two articles are presented that address the political geography of the Dayton Accords. The first article was written in March of 1996 and sought to apply three political geography perspectives (i.e., boundary studies, peace geography, and Hartshorne's Functional Approach) to the Dayton Accords to make policy recommendations. The second article was written two years later and analyzed the author's recommendations from their first effort. The student of *military geography* will note the important role of the military in bringing and keeping peace and Bosnia, and how this role is inseparable from the political geography of the Dayton Accords.

I. Political Geography of the Dayton Accords

During the first twenty days of November 1995 at Wright-Patterson Air Force Base near Dayton, Ohio, the U.S. Government virtually incarcerated three foreign presidents—Izetbegovic of Bosnia-Herzegovina, Milosevic of Serbia, and Tudjman of Croatia—in an attempt, to reach an agreement on a peaceful settlement and a permanent solution to the civil war that had raged in Bosnia for the past three or more years

The purpose of this analysis is to apply political geographic approaches to a discussion of the Dayton Accord. Both in the manner in which it was reached and in its initial implementation in order to provide analytical insights, constructive criticism, and policy recommendations. As political geographers we have a conceptual tool-kit of relevant analytical approaches that include: boundary and frontier studies; core areas and capital studies; electoral geography; Hartshorne's functional approach; Jone's unified field theory; partition and reunification studies; state power analysis; state structure studies, such as federalism as a state structure; territoriality and perception studies that include mental maps, cognitive distance and direction, and iconography; and peace geography, i.e. association and disassociation (Glassner, 1996). Because of space constraints, we will limit ourselves to applying boundary and frontier studies, peace geography, and Hartshorne's functional approach.

Background

The events that led to this extraordinary three-week event—officially called "The Dayton Proximity Talks"—are history and not our focus, although some background is relevant. Emerging from W.W. I after the collapse of Austro-Hungary and the Ottoman Empire as the Kingdom of the South Slavs, Yugoslavia survived W.W. II, scarred by German occupation and internal ethnic conflict including widespread genocide. Under Tito, it turned into a Communist federation of several ethnically based republics. Following Tito's death in 1980, the central control of Belgrade and the Communist party declined. The earth-shaking events of the late 1980's—the end of the Cold War, the collapse of the Soviet Union, and the end of Communist party domination in the states of Central and Eastern Europe—hastened the breakup of the federated Yugoslav state, with declarations of independence, first in 1991 by Slovenia and Croatia, and then in 1992 by Macedonia and Bosnia. Attempts by Serbia to preserve the Federation failed as the new states won diplomatic recognition from Europe and the United States.

Bosnia, however, was different from the others. Critical factors in its newly elected government's effort to retain central power and territorial integrity were its ethnically-mixed population—44% Moslem, 32% Serb, and 18% Croat according to the 1990 Yugoslav Census—and its relative location, virtually land-locked and sharing a long boundary with only two neighbors, Serbia and Croatia. A bloody civil war developed between poorly armed Bosnian government forces and Bosnian Serbs and Bosnian Croats, armed, supplied and helped directly by their neighbor sponsors. Despite international support for the idea of a united Bosnia—including an embargo on most goods to Serbia—Serbs and Croats sought to avoid any possibility of being ruled within a Moslem-dominated Bosnia by a policy of systematic ethnic cleansing of Moslem and/or the other's populations in towns and villages falling under their control (by forced migration or genocide) thus expanding territorial control over cleansed self-declared Serb and Croat "republics" within Bosnian territory. The core of the capital, Sarajevo, remained in Bosnian government control but for the most part was surrounded by Serb held territory—a city center without most of its suburbs and bereft of a hinterland - and periodically subjected to artillery and sniper fire from Serb forces. Despite the deployment of UN peacekeeping forces and several attempts at reaching agreement among the combatants on a reshaping of Bosnia, such as the Vance-Owen Cantonal Plan of early 1993 and the Owen-Stoltenberg Mini-States Plan of August 1993, the conflict continued into the Fall of 1995 with a rising military and civilian dead and wounded and estimates of over 2 million refugees. Finally diplomatic pressure, the squeeze of sanctions and the embargo, and NATO airpower forced a cease-fire in early October 1995, which in turn led to what has become known as the "Dayton Peace Agreement on Bosnia-Herzegovina", initialed on November 21 and implemented in Paris on December 14, 1995 when it was signed by Croatia, Bosnia and Serbia, and also by the "contact group" representing the international community—Russia, Great Britain, France, Germany, and the United States.

The Dayton Peace Agreement

The general framework of the agreement retains the state of Bosnia as a federated republic with Sarajevo as its reunified capital, and consists of two distinct entities, a Moslem-Croat Federation (officially the Bosnian-Croat Federation) and a Bosnian Serb Republic (Republika Srpska) (see Figure 1). The complex agreement is composed of thirteen annexes running to over 100 pages (see Figure 2). Its four major points are summarized below:

Figure 1. Bosnia and Herzegovina showing the entities of Republika Srpska (dark grey) and the Bosnian-Croat Federation (medium grey).
Source: U.S. Department of Defense Map.

1. Full mutual respect for sovereign equality among the three republics of Croatia, Bosnia and Yugoslavia (FRY or Serbia).

2. Mutual recognition between Bosnia and FRY.

3. Full respect for and promotion of the details of the agreement, including an obligation to respect human rights and the rights of refugees and displaced persons.

4. Full cooperation with all entities involved in implementing the peace settlement - the UN, NATO Implementation Force (IFOR), and others - and investigation and prosecution of war crimes (US State Department, 1995a).

The Dayton Agreement

General Framework Agreement
Annex 1-A Military Aspects
Annex 1-B Regional Stabilization
Annex 2 Inter-Entity Boundary
Annex 3 Elections
Annex 4 Constitution
Annex 5 Arbitration
Annex 6 Human Rights
Annex 7 Refugees and Displaced Persons
Annex 8 Commission to Preserve National Monuments
Annex 9 Bosnia and Herzegovina Public Corporations
Annex 10 Civilian Implementation
Annex 11 International Police Task Force

Figure 2. Components of the "Dayton Peace Agreement on Bosnia and Herzegovina".
Source: U.S. State Department, 1995a.

Facts and Assumptions

Before beginning our analysis, we should state some basic facts and assumptions that are valid in the months following the agreement.

- Fact one is the international community and the directly involved parties publicly accept and support the idea of an independent and sovereign federal state of Bosnia-Herzegovina as evidenced by their signing of the Dayton Peace Agreement and supporting annexes.

- Fact two is that at least over the first five months of the agreement the parties have fulfilled the major requirements of the military annex and that organized violence has ceased.

- Fact three is that political components of the agreement are slowly being implemented.

- Assumption one is that the ultimate goal of the current leadership of the Bosnian Serbs is to prevent ethnic reintegration and ultimately to unite the Republika Srpska with Serbia.

- Assumption two is that the current Serbian and Croatian leadership want to avoid further sanctions and grudgingly support the Dayton Agreement but would, given the right conditions, absorb their ethnic kinsmen with their territory.

- Assumption three is that Bosnian Croat leadership is uneasy with the alliance with the Bosnian Muslims and feels they have inadequate say in the federation. They continue to look to patrons in Croatia for support for eventual unification with Croatia.

- Assumption four is that the Bosnian Muslims remain politically split with hard-line factions pushing to dominate the federation thus diminishing the commitment of the Bosnian Muslims to a truly multi-ethnic, secular, democratic state.

- Assumption five is that the international community is committed to the creation of a sovereign multi-ethnic Bosnia-Herzegovina so as to stabilize the Balkans, to undo ethnic cleansing and to punish its most violent perpetrators, and to set a precedent by which war and ethnic cleansing in Europe will not work and will not be tolerated.

Boundaries and Frontiers

Traditionally, the generally accepted boundary-making model recognizes four related but distinct stages: definition, delimitation, demarcation and administration. Commonly the process of moving from one stage to the other is subject to delay and can be so slow as to be influenced by policy shifts and by changes on the ground, that is in the newly forming borderlands themselves (Glassner, 1996). Armed conflict and migrations in the intended borderland change the facts upon which original definitions and delimitations were based and hence, can make the demarcation step an anachronistic exercise.

The Dayton Accords have challenged, perhaps even shattered, this traditional model. By the use of high-tech geographic information processing systems integrating GIS, remote sensing, and terrain visualization, boundary-making stages are compressed, with definition and delimitation merged into one. The rapid determination and implementation of the inter-entity boundary in Bosnia was a necessary precursor for peace. The boundary negotiations that occurred as part of the Dayton peace accord in November 1995 saw the advent of a new geographic technology that has the potential to revolutionize boundary making so that it is faster, accurate, and less liable to future dispute. The three "proximity talks" presidents were subjected to the virtual reality experience of *Powerscene* as they were "flown" down the intended zones of separation between the inter-entity boundaries of the future two halves of Bosnia.

Powerscene is a computer based terrain visualization system that provides a three-dimensional landscape scene of an entire country using digital cartographic data overlaid with remote sensing imagery allowing users to "fly" over, explore, and examine a virtual country. *Powerscene* was developed by the United States Defense Mapping Agency for military purposes such as target selection, mission planning, and rehearsals. In 1995, U.S. Air Force aircrews used *Powerscene* to plan many of the missions launched against the Bosnian Serbs during NATO's air campaign.

In Dayton, differences of opinion as to the location of the boundary or the width of the separation zone were settled immediately at the conference table. After the agreement was signed representatives of the three presidents spent nine hours "flying" the 660 mile-long boundary to determine the exact demarcation of the boundary. In some cases this involved determining along which side of a road the boundary would run—a critical point as a boundary line on a 1:50,000 scale map is 50 meters wide on the ground.

As the convenor and gatekeeper of the Dayton talks and as the eventual leader for any consequent implementation of decisions reached, the United States ensured its interests were paramount so that priorities were placed on successful achievement of the military aspects and regional stabilization during the first year. Within days of the agreement, the demarcation and administration stages were jointly initiated through the deployment of IFOR troops, equipped with global positioning systems. In the case of any positional dispute, the IFOR commander has the right to determine the exact delimitation of all lines and zones of separation, and then to demarcate them (Defense Mapping Agency, 1996).

While the success of the first four months of this process does not guarantee the viability of the reconstituted Bosnian republic, it does serve as a remarkable example of time/space convergence in political geography by the application of high technology in a pressure cooker diplomatic environment. Working from a general agreement on the basis of definition, negotiators can move to decisions on boundary delimitations on the firm basis of indisputable contemporary border landscape realities, as if they were being periodically "beamed" into the field, indeed, into the future, as potential demarcators. The use of *Powerscene* eliminated the normal delays in boundary making involving particular delimitation disagreements to be eventually evaluated by demarcation teams in the field, who then report back to the reconvened negotiators thousands of miles away. This complex boundary making process normally spread over several years during which governments change, new conflicts develop and people migrate, has, in the Bosnian case, been simplified and has produced an unambiguous line on the ground so crucial in the peace process for Bosnia.

The new "inter-entity" boundary separating the Moslem-Croat Federation and the Republika Srpska is 1,053 kilometers long. The boundary involves an "inter-entity zone of separation" (ZOS) averaging 4 kilometers in width, producing a built-in unproductive and hostile borderland of over 4,200 square kilometers. In addition, several exceptional circumstances further complicate this internal Bosnian boundary. The reunified Sarajevo metro area now lies totally within the Federation but is to be "open to all people in the country". Gorazde is virtually surrounded by the Republika Srpska but has assurances for remaining "secure and accessible" by a land corridor link with the Federation, and the disputed Brcko area is to go to arbitration for a decision before the end of 1996.

Clearly, with the likely end of the IFOR deployment after 1996, this internal boundary is likely to take on a function of increasing separation in the flow of peoples, goods, capital and ideas, creating a marginalized border zone on either side. On the other hand, the international border between the Republika Srpska and FRY, and the border between the Federation and Croatia are both likely to decline as barriers to interaction, and hence create a dynamic borderland working to blur the territorial distinction between the Bosnian state and its two neighbors.

Given the above, certain policy changes, which would slow down but hardly reverse these processes, would seem desirable especially in the context of the local scale. IFOR should encourage, perhaps on an experimental basis at selected points, an increase in socio-economic interchange across the zone of separation, and place a priority on returning some of the abandoned borderland to original owners for productive use. Any policies which do not endanger the peace but which serve to lessen the image and the fact of this borderland as a separator are to be welcomed. At the same time, the international community should pressure Croatia and Serbia to respect fully the territorial sovereignty of the Bosnian state in administering their boundary with Bosnia, and also to control the activities of their own citizens regarding violations of international law, which could undermine Bosnian authority in the border zone.

Peace Geography

Peace geography is the application of spatial concepts to a conflict situation in order to stop violence and bring about harmony such that the conditions leading to the violence will not recur. Peace Geography has the two major components of disassociation and association (O'Loughlin and van der Wusten, 1986).

Disassociation involves separating states that might engage in violence. It includes partition, peacekeeping, political\economic isolation, and deterrence. In the case of Bosnia, Annex 2 established the "Inter-Entity Boundary" and resulted in de facto partitioning (i.e., the creation of two entities with a buffer zone) of the Bosnian/Croat Federation and Republika Srpska entities separated by the four-kilometer wide "Zone of Separation." In the near term, this is good since it is necessary to separate the military forces of the warring sides.

Annex 1A (Military Aspects) addresses the issue of peacekeeping by introducing IFOR as a peace enforcement element that guarantees that the armed forces of the warring parties will remain in their respective territories and serves to build confidence in their security and safety (U.S. State Department, 1995b). Notice that we characterized IFOR's mission as peace-enforcement rather than peacekeeping. In military doctrine there are three types of peace operations. Peacekeeping occurs when the warring factions have requested or accepted the presence of a lightly armed force that monitors the "zone of separation" and the warring sides in order to build the confidence in their security. Peacekeepers are not authorized to prevent further conflict and are typically withdrawn if hostilities resume. Peace-making is at the opposite end of this spectrum where the warring parties are forced by a heavily armed outside force to cease hostilities. This approach is theoretical since it has never been attempted. Peace-enforcement falls between the two previous approaches in that the warring sides have agreed to cease hostilities and to the deployment of the outside force. The difference peace-keeping and peace-enforcement is that, in the latter's case, the force is larger, more heavily armed and authorized to use overwhelming force to prevent or punish violations of the agreement (Department of the Army, 1993). IFOR is the first application of the peace-enforcement approach on an international scale.

IFOR and the U.N. High Representative have another tool of disassociation at their disposal in the form of sanctions and political isolation. While not specifically addressed in the Agreement, the threat of sanctions (e.g. selective withholding of international reconstruction funds) to coerce uncooperative parties into complying with the agreement has been effectively used by the UN High Representative Carl Bildt.

Annex 1B (Regional Stabilization) uses disassociation techniques of deterrence. The intent of Annex 1B is to create a balance of power through arms control negotiations sponsored by the Organization for Security and Cooperation in Europe (OSCE). In the event these negotiations fail, there is a provision for creating a balance of power by arming and training the Bosnian/Croatian Federation force so that it is equal to Bosnian Serb forces. Additionally, Annex 1B calls for the negotiations to implement confidence building measures within 45 days and a moratorium on arms purchases within the former Yugoslavia (excluding Slovenia and Macedonia) for 90 days and heavy weapons for 180 days or until an arms agreement comes into effect (U.S. State Department, 1995b).

Essentially, there are three problems that we would like to highlight and make tentative recommendations. First, the zone of separation coupled with the results of ethnic cleansing has resulted in the *de facto* partitioning of the country into territories with relatively homogenous ethnic groups, which hard-line factions on all sides want to maintain. Annex 7 on Refugees and Displaced Persons states, "*The agreement grants refugees and displaced persons the right to safely return home and regain lost property, or to obtain just compensation.*"

(State Department, 1995a). People who want to return home to areas in other entities may have difficulties recovering their homes. In many cases other displaced persons have occupied those homes. Who will ensure the security of returnees, and how long will it take them to be accepted and reintegrated into their old communities? This problem has the potential to seriously increase tensions and result in further violence. While we agree that it is morally correct that people should have the right of return, we suggest that in the near-term, people should be encouraged to accept compensation or lease arrangements for their property, at least until stability has returned and some of the cultural wounds have healed.

Second, creating a balance of power through the "Arm and Train" program currently being pushed by the U.S. is problematic. In our estimation to create deterrence and a balance of power through local and regional arms control negotiations is preferable. The "Arm and Train" program is problematic because it: 1) offends the Bosnian Serbs thus strengthening the position of their hard-line factions; 2) it concentrates greater military power in the hands of the Bosnian Muslims who have dominated the Bosnian/Croat Federation force by ousting non-Muslim leaders and reducing he power of other ethnic elements; 3) assumes force stability of the Bosnian Serb Army when in fact it could result in a local and regional arms race; and 4) it makes the Europeans anxious because renewed hostilities will be at a higher level of violence that could draw in other countries resulting in the worst case scenario of an expanded regional war.

Our recommendation is that the United States should back away from the "Arm and Train" program and the international community should put its full diplomatic efforts into bringing about a local and regional arms control agreement that will result in deterrence without increasing the potential for greater violence and wider conflict. This agreement would include a sanctions regime for those who would violate the agreement as well as appropriate security guarantees backed by NATO and the international community.

A third problem is that while IFOR and the military agreement have been very successful to date, IFOR's one-year mandate will soon expire [editor's note: international presence continues in Bosnia today via the Stabilization Force (SFOR)]. While we understand the political constraints in United States, it would be a tragedy to withdraw the force prematurely and allow Bosnia to slip back into war and anarchy. We recommend that IFOR's mandate be extended for another year and that efforts to establish arms control be expedited, confidence-building measures be rapidly initiated, the threat of further sanctions be articulated to those who would violate the accord, and the association components of peace geography be implemented and expedited.

Association is based on interaction, sharing, and cooperation. Once the violence has ceased and some measure of stability restored, association approaches should be the focus in order to ensure a lasting peace. Association approaches include: application of international law, economic integration, regional cooperation, efforts to facilitate intercultural understanding, and confidence building measures.

The General Framework portion of the agreement establishes a basis for understanding by articulating that Bosnia-Herzegovina, Croatia, and the Federal Republic of Yugoslavia will respect one another's sovereignty and settle all disputes by peaceful means. It additionally states that all the parties agree to cooperate with all entities associated with the agreement and to cooperate with war crimes investigations. Annex 6 guarantees respect for internationally recognized human rights within Bosnia-Herzegovina and establishes a Human Rights Commission authorized to investigate violations of human rights and initiate appropriate proceedings.

Annex 4 (Constitution) and Annex 9 (Public Corporations) address economic integration. The constitution will provide for a central bank and monetary system. Annex 9 provides for a national transportation corporation to organize and operate transportation facilities such as roads, railways, and ports. It also creates a Commission on Public Corporations to examine the establishment of other public corporations to operate joint public facilities such as utilities and postal services. Given the size and the historically integrated infrastructure and economy of Bosnia and Herzegovina, economic integration measures financed in large part by the international community provide a means to bring the entities together through reliance on shared resources such as transportation facilities, the power grid, postal services (especially international outlets), and the monetary system. We recommend that the High Representative use all of his influence over policy and finance to ensure that international reconstruction money be used to create an infrastructure and economy that make the entities mutually reliant.

Regional cooperation is built into the agreement by including Croatia and the Federal Republic of Yugoslavia as signatories. Additionally, there is a special provision for the entities to have special parallel relationships with their neighboring patrons (e.g., Republika Srpska with Serbia). As we will see later, this arrangement has pros and cons. Given their signatures on the Dayton Agreement, Croatia and Serbia are publicly committed to supporting the idea of a sovereign Bosnian Federation. The international community will now have to hold them accountable through diplomacy.

An aspect of association not covered specifically in the agreement is that of intercultural understanding. We recognize that historical animosities exist as well as the experience of four years of brutal warfare; and this has radicalized much of the population. However, we must also remember that Bosnians also share the experience of having lived peacefully with members of other ethnic groups. We believe that there are people on all sides who want to see peace and a return to civility. Therefore, our recommendation is that the High Representative, the IFOR Commander, and other representatives of the international community should endeavor to politically and economically isolate radical elements and bolster the position of moderates and bring them together in and effort to rebuild that intercultural understanding.

Annex 10 (Civilian Implementation) and Annex 5 (Arbitration) address confidence building measures and dispute resolution mechanisms. Annex 10 establishes a High Representative who coordinates and facilitates civilian aspects of the peace settlement, such as humanitarian aid, economic reconstruction, protection of human rights, and the holding of free elections. The High Representative also chairs a Joint Civilian Commission comprised of senior political representatives of the parties, the IFOR Commander, and representatives of civilian organizations. Annex 5 states that the parties agree to enter reciprocal commitments to engage in binding arbitration to resolve disputes between them, and they agree to design and implement the system of arbitration.

An essential confidence building measure is that of bringing war criminals to justice. This is a problem in the short term as the international community tries to bring often-influential people to justice. However, in the long-term, bringing such people to justice is essential to demonstrate that the world community will not tolerate crimes against humanity. Additionally, by precluding indicted war criminals from dealing with the international community, these radicals are marginalized within their communities. This ultimately diminishes the power of radical factions. Finally, the vigorous prosecution and punishment of war criminals is necessary to break the cycle of violence that has plagued this land for centuries.

Hartshorne's Functional Approach

Hartshorne's functional approach involves state function and structure. Essentially he said that a viable state must have a *raison d'être*. Additionally, he said that for a state to function effectively its centripetal forces (i.e., those that hold it together) must overcome its centrifugal forces (i.e., those that would tear it apart) (Hartshorne, 1950).

Thus, the first question we must ask is what is Bosnia and Herzegovina's *raison d'être*? There is no clear answer to this question in the accord. Is the reason for existence of a Bosnian state the international community's desire not to reward war and ethnic cleansing? Is it the Bosnian's desire for a multi-ethnic, secular, democratic and federal state? This is an important question for informing the future path of the international community as well as the Bosnian parties and in our estimation it has been inadequately answered.

The second issue is that of centripetal versus centrifugal forces. Centrifugal forces in Bosnia include: ethnic nationalism, religion, relations with external patrons, historical animosities, memory of the recent conflict, and power sharing and distribution within the Bosnian/Croat Federation. Centripetal forces include: external (i.e., international community pressures), war weariness, experience of having lived together peacefully, moderate elements in the population who want a multi-ethnic and secular society, small size of the territory, and traditional economic integration and interdependence.

Using Hartshorne's approach, we are perplexed to see how a viable Bosnian Federation will emerge under the current conditions. We see no *raison d'être* that would be acceptable to all entity leaders, and we believe that once the overwhelming centripetal force of the will of the international community is removed the balance will change such that centrifugal forces dominate and the state collapses.

We therefore offer recommendations. First, we believe that the parties involved must develop and accept a *raison d'être*. This may take time and may not be possible under current circumstances. Thus, the international community should remain involved with the internal governance of Bosnia until changes and healing to take place. The Dayton agreement already allows for this. During this grace period the High Representative and his agents should have the goal of bolstering centripetal forces while minimizing or marginalizing centrifugal elements. The following are possible approaches to counter centrifugal forces:

1. Marginalize radical nationalist by isolating them from international community support and funding. Additionally, prosecute war criminals to prevent retaliatory violence.

2. Work on inter-religious understanding through the local and denominational church leadership.

3. Monitor relations between parties and their external patrons to ensure that these relations foster rather than undercut the development of a Bosnian state. Careful intelligence work must ensure that covert arms transfers, training, funding, and other support does not occur.

4. Historical animosities and memory of the recent conflict will take time to heal. The fair prosecution of war criminals will show that justice is being done. Parties who would use historical animosities stir unrest should be politically and economically marginalized in order to diminish their power and appeal.

The following are possible approaches to supporting centripetal forces:

1. Maintain international involvement and commitment to a multi-ethnic Bosnian state.

2. Capitalize on war weariness by working to rebuild the country and improve the quality of life. Publicize rebuilding efforts and quality of life improvements.

3. Identify and support moderate elements in the population. Place such people in positions of power where possible and support moderate politicians in all factions while being careful not to taint these people as puppets of foreign masters.

4. Use the small size of the territory and traditional economic integration and dependence to make the factions economically interdependent and develop an integrated economy.

Summary and Conclusions

We have shown that the application of traditional political geographic concepts to the Dayton Agreement provide useful analytical insights into the process through which Bosnia is likely to evolve as a restructured political entity. Serious questions about Bosnia's longer-term viability are raised. This constructive criticism in turn gives rise to speculation about recommendations for policy changes that would enhance the chances of the successful survival of Bosnia as an independent state, precisely the basic aim of the Dayton Agreement.

The "boundaries and frontiers" approach yields two noteworthy findings. The first extensive use of a computer-based terrain visualization system in negotiations among civil war combatants enabled an external power to force acceptance of an agreement by all parties, moving directly from a fragile month-long cease-fire to a complete blueprint for a restructured state in a matter of a few days. The reality of such time/space convergence stands as a unique aspect of Dayton, and augurs well for effective and efficient negotiated settlements of similar disputes in the future. Second, the impermeability of the new inter-entity borderland will inevitably place an increasing strain on the Bosnian State and its ability to function as a viable, independent unit in the Balkans. This problem will be exacerbated when seen in combination with the likely increase in cross-boundary ethnically based circulation between each Bosnian entity created by Dayton and its neighbor.

The association/disassociation analysis applied within the peace geography context also raises serious questions about the workings of the various aspects of the Dayton Agreement, and hence about Bosnia's future. The disassociation approaches of partition, peacekeeping, and deterrence are being applied successfully as evidenced by the establishment of the inter-entity boundary and the deployment of IFOR. Political/economic isolation has been used to a limited extent by the UN High Representative but deserves consideration as a major policy tool for isolating and marginalizing the hard-line factions resisting the peace process. There are three major problems, however, that must be addressed: the return of refugees to hostile areas; the potential of the "Arm and Train" Program to fuel a regional arms race; and the prospect of a premature IFOR pullout. In the near term, refugees should be encouraged to accept compensation or lease arrangements until a modicum of stability and civility returns. In regards to deterrence and security, the U.S. should back away form the "Arm and Train" Program and instead foster security through regional arms control and Western security guarantees. Additionally, IFOR's mandate should be extended to prevent a return to war and anarchy.

Association approaches are a necessary complement to disassociation techniques and include actions to bring about interaction, sharing, and cooperation. The application of international law is critical in holding the parties to their agreement to respect one another's sovereignty and to their obligation to cooperate with the International War Crimes Tribunal. Bringing war criminals to justice is essential to breaking the cycle of violence and vengeance. Economic integration initiatives are critical to knitting the entities together and improving the quality of life of the Bosnian people. International funding of Bosnian reconstruction must proceed immediately. Finally, the difficult but not impossible task of restoring inter-cultural understanding must be addressed. The issue of scale again plays an important role in that efforts at the local level (oftentimes by local and non-governmental organizations) are often the most successful approach. The international community should work at all levels from the regional to the local to overcome the scars of war and build good faith between the communities.

Lastly, the almost half-century old Hartshornian functional model of centripetal and centrifugal forces also yields some important findings when applied to the Dayton Agreement. However defined, the *raison d'être* of the current Bosnia remains inadequate. Centrifugal forces dominate while the few centripetal forces are fragile at best and largely products of the international community rather that indigenous to Bosnia itself. Policies to bolster identified centripetal forces and to counteract the many centrifugal forces are suggested. All these require sufficient time and appropriate conditions on the ground that the short-term aims of Dayton - permanent separation of the combatants - are unlikely to produce.

While not all findings about Bosnia's future are negative, the unavoidable conclusion remains that through the application of these selected political-geographic concepts, a prognosis at this early stage for an eventually healthy post-Dayton Bosnia - even if suggested policy changes are enacted - is not particularly encouraging.

Bibliography

Defense Mapping Agency. 1996. *Powerscene Briefing*. Fort Belvoir: Defense Mapping Agency.

Department of the Army. 1993. *Field Manual 100-5 Operations*. Washington DC: U.S. Government Printing Office.

Glassner, M.I. 1996. *Political Geography.* 2nd edition. New York: Wiley.

Hartshorne, R.T. 1950. "The Functional Approach in Political Geography." *Annals of the American Association of Geographers,* 40 (2): 95-130.

O'Loughlin, J. and H. van der Wusten. 1986. "Geography, War and Peace: Notes for a Contribution to a Revived Political Geography." *Progress in Human Geography*, 10-4: 484-510.

United States State Department. 1995a. *Summary of the Dayton Peace Agreement on Bosnia-Herzegovina.* Fact sheet released by the Office of the Spokesman, November 30, 1995.

United States State Department. 1995b. Full text of the Dayton Peace Agreement on Bosnia-Herzegovina. Treaty text released by the Office of the Spokesman.

II. "Prediction and Reality in Bosnia: Two Years after Dayton"

Geographers have long been active in publishing their assessments of the future impact of major ongoing events in changing the political geographic structure of a region. Existing models and concepts have been applied frequently to simulate the likely processes generated by such events, allowing for prediction concerning future structural changes. It seems only fair that such predictions should be put to the test of reality after sufficient time has passed. This analysis will focus on a contemporary study of this kind—an assessment of the likely impact of the Dayton Agreement on Bosnia, originally written in March 1996 (Corson and Minghi, 1996). Now, two years later, in an attempt at unbiased self-criticism, the original authors assess the veracity of their original findings. However, we need to first briefly establish the context of our 1996 predictions and then see what has actually happened so far in Bosnia, to assess their veracity.

Our purpose was to provide and analytical insight, constructive criticism, and policy recommendations by applying selected traditional political geographic approaches to the Dayton Accords. By such a study, we hoped that colleagues might be inspired to apply their own analytical perspectives to studying the impact that this unique agreement would have on Bosnia's human landscape. We presented a litany of relevant political geography approaches, but chose to focus on what we felt were the three most appropriate: boundary and frontier studies, peace geography, and Hartshorne's functional approach. We concluded that the application of these related concepts did indeed provide useful insights into the process through which Bosnia was likely to evolve as a restructured political entity. Serious questions about the longer-term viability of a unified Bosnian state were raised. We speculated about the future, identified problems and recognized the need for policy shifts by the forces overseeing peacekeeping and reconstruction.

Much has happened since the original IFOR (Peace Implementation Force) troops were deployed in early 1996 (Klememcic and Schofield, 1996). Among the more significant events have been: national elections in September 1996 and municipal elections a year later; the new Bosnian Parliamentary Assembly meeting once in October 1996; a smaller but more international follow-on force deployed under the new title SFOR (Stabilization Force) to provide a security framework for general reconstruction; and the International Criminal Tribunal for the Former Yugoslavia (ICTY) based in The Hague in May 1997 bringing 17 indictments against 74 individuals, only a handful of whom are in custody (Landay, 1997).

Boundaries and Frontiers

Given the short-term priority for disengaging and disarming warring militias, especially along the zone of separation (ZOS) between the two newly-created republics we felt that Dayton guaranteed little of any long-term viability for the Bosnian state (see Figure 1). A key to this obvious paradox—likely success in the short-term helping to undermine the longer-term aim of a united stable state of Bosnia functioning without external military presence—would be found in the policies followed in administering this 1,053 kilometer-long borderland. Should these policies lead to a permanent decrease in human interchange across this line, regional stabilization would be difficult to achieve, and already shattered borderland communities would become even further marginalized.

Furthermore, we predicted that Bosnia's international boundaries would likely begin to blur as separators of distinctive political landscapes, especially as they encouraged interaction and interdependence among ethnic Croats and Serbs. Dynamic borderlands of economic and cultural exchange and mutual dependency that such a process would create, could serve to tie much of the Republika Srpska irrevocably to Serbia, and the Bosnian Croat Federation to Croatia, which in turn would undermine still further the territorial integrity of the Bosnian state.

The IEBL

The reality of the past two years supports our predictions regarding the Inter Entity Boundary Line (IEBL) and Bosnia's international boundaries. IFOR/SFOR is set up in distinct zones of operations, or Multi National Divisions (MNDs), which allow for a sharing of commands among American (North), French (Southeast), and British (Southwest) leaders (see Figure 2). Each MND is at right angles to the IEBL so that the same command is in charge on both sides of the line (CIA, 1997). In this way, the deployment and functioning of occupation troops at least do not reinforce the IEBL as a permanent separator. In all divisions, however, the bulk of active troops available are stationed in the

IEBL borderland itself. Even so, this has not prevented attempts to build the IEBL into a more permanent and effective political separator. For example, the Republika Srpska has attempted to establish a visa requirement for non-residents entering across the line. Such a requirement would give it greater legitimacy as a separate "state", control over immigration, and much needed extra funds. Although IFOR/SFOR has successfully blocked these attempts, there is strong anecdotal evidence from traveling academics and journalists that such visas exist.

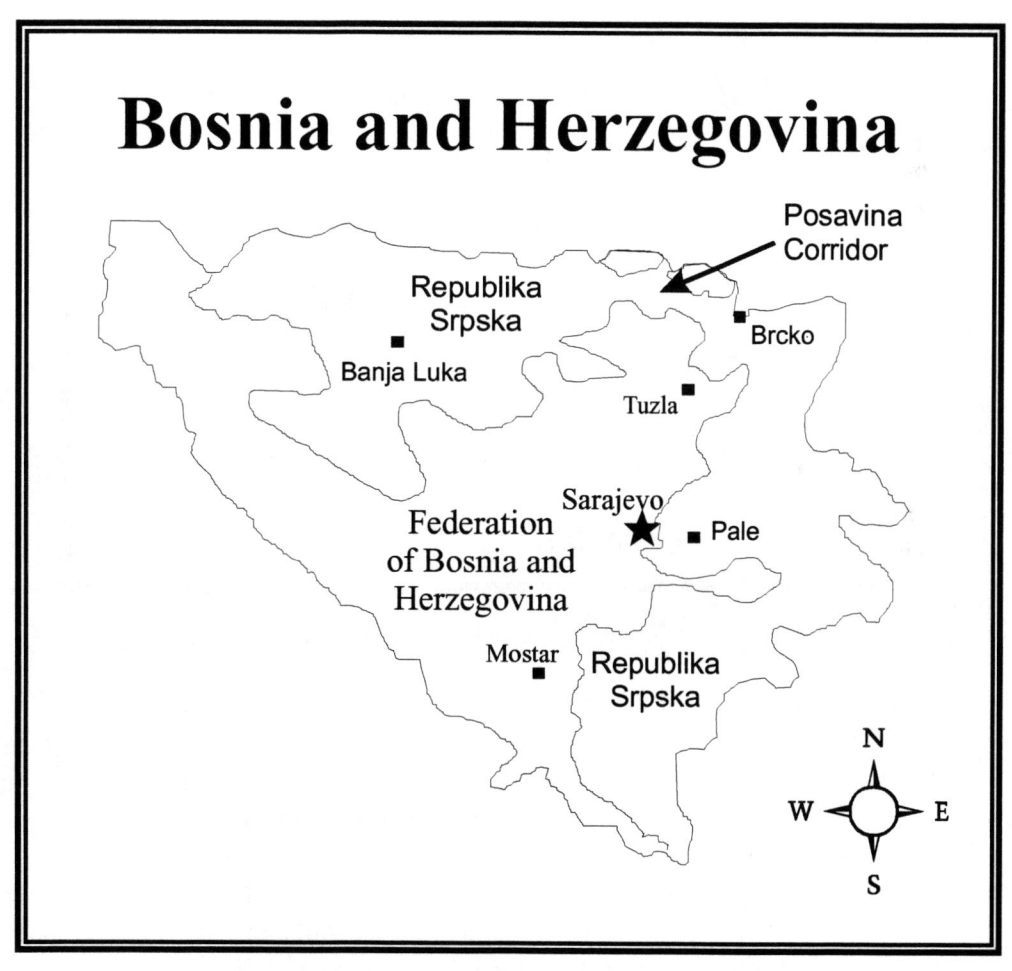

Figure 1. Bosnia and Herzegovina showing the two entities of the Federation of Bosnia and Herzegovina (Bosniac-Croat Federation) and Republika Srpska.

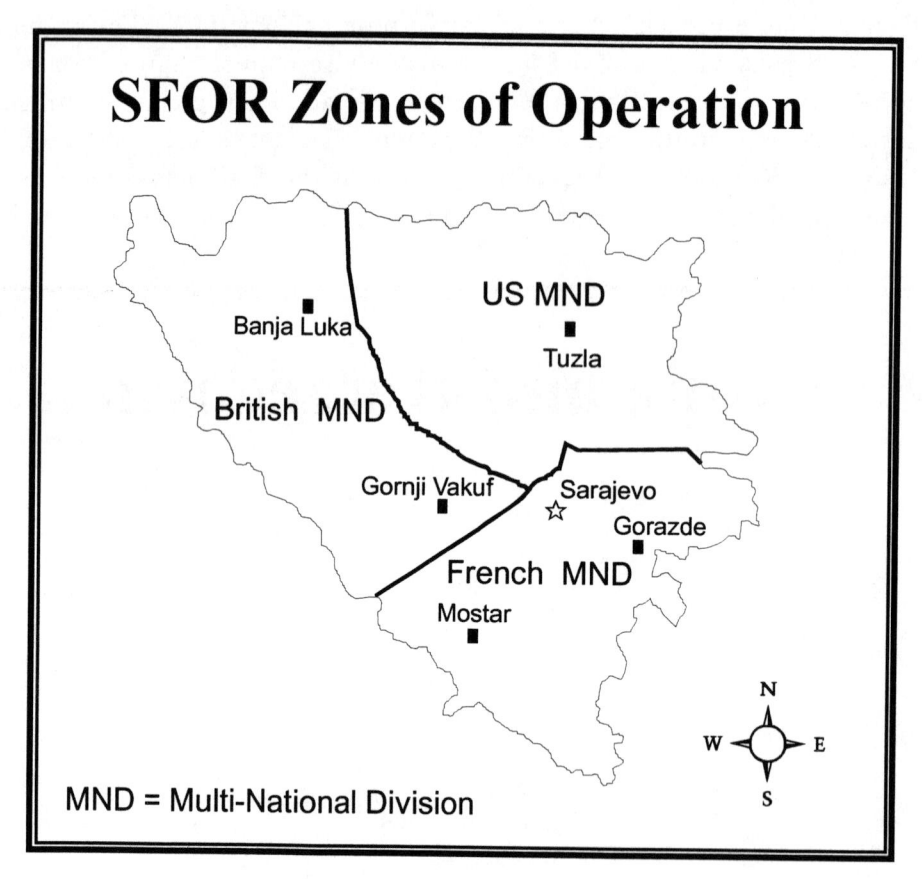

Figure 2. SFOR Zones of Operation showing the three Multi-National Division sectors.

It can be argued that as the months and years pass, the three MNDs are adding a unique overlay to the already complex human landscape as differing military priorities and cultures evolve and are reflected in regional contrasts. For example, one of the articles in Dayton is the requirement to arrest and bring to trial all war criminals indicted by the ICTY in The Hague. The vast majority of these criminals reside in and around Pale in the eastern portion of the Republika Srpska which falls under the French MND, yet the only arrests made thus far have been in the Southwest division under British command.

Recognition of our plea not to allow the IEBL to become permanently entrenched is seen in the recent regional priorities given to resettlement. In

1997, UNHCR targeted 22 locations for high priority Muslim resettlement, most of which were in and around the ZOS, with the specific aim of overcoming the growing marginalisation of the borderland. The right to return of families forced to migrate during the war, however, is still severely circumscribed by the presence of the IEBL.

International Boundaries

All evidence supports the case that Bosnia's international boundaries are porous, and hardly function as international separators. Multiple crossing points with little or no Bosnian state control have developed as centers of economic exchange and daily cross migration. Television stations viewed and currencies used are Serbian or Croatian. The flow of mail and common international telephone codes tie these areas to the cross-border neighbor. Such Serb and Croat borderland communities clearly intrude on the sovereignty of the Bosnian state. This trend is seen most dramatically in the Posavina Corridor that links the two wings of the Republika Srpska. Here the Sava River port of Brcko is the epitome of this process. From the start of the civil war, the Serbs (20% of Brcko's population) understood the town's strategic significance. They seized and ethnically cleansed it, and still hold power there six years later. Dayton's territorial arrangements confirmed its strategic importance. Under the Accord, a solution to Brcko's dilemma was delayed for six months. Since, the deadline for a decision has been repeatedly extended and no solution is yet in sight. Brcko remains a Bosnian border town in name only (Cigar and Williams, 1998). In short, realities on the ground regarding the human geography of Bosnia's internal and national boundaries fit very closely to the scenarios suggested in our prediction of early 1996.

Peace Geography

Peace geography, as advanced by O'Loughlin and van der Wusten (1998), involves the application of spatial concepts to a conflict in order to bring about peace and to prevent a resumption of hostilities. Peace geography consists of the two major components of disassociation and association. Disassociation involves efforts to separate the warring parties through partition, peacekeeping, political/economic isolation, and deterrence. Association, on the other hand, seeks to facilitate interaction and mutual inter-dependence. Association techniques include: the application of international law, economic integration, regional cooperation, intercultural understanding, and confidence building measures.

Our initial analysis using the perspective of peace geography was correct in the three areas of peacekeeping, intercultural understanding, and the pursuit of war criminals. Our analysis was incorrect, however, in addressing the issues of refugees and the "Arm and Train" Program. Additionally we failed to foresee the geographically significant split within the Bosnian Serb leadership, the unrealistic effect of absentee voting provisions, and the enormous economic reconstruction problems.

The deployment of IFOR and SFOR has succeeded in preventing a return to hostilities and creating a stable environment in which further progress towards peace and normality can occur. IFOR/SFOR has been successful in establishing the IEBL and ZOS, causing the parties to place their heavy weapons and forces in cantonment areas, and generally de-militarizing the Bosnian landscape. SFOR has also established a center to coordinate and facilitate the efforts of all parties to remove the millions of landmines dotting the Bosnian countryside (NATO, 1998). We suggested that a NATO military presence would be essential well past IFOR's one-year mandate and we applaud the open-ended deployment policy recently adopted by NATO. We note, however, that SFOR is a substantially smaller force and we caution that future versions of SFOR must be large and well enough armed to deter any aggression and if necessary, to enforce the peace.

Efforts at intercultural understanding have had limited success to date but show great promise in the end. Again, the geographic perspective of scale has proven critical. Macro efforts at promoting intercultural understanding have had very little success, but, as we predicted, local efforts, especially in border regions, have had some limited success that bodes well for the future. In Gornji Vakuf, a multi-ethnic knitting group of 140 women sets the example for other townsfolk. This initiative was started in 1995 by international NGO volunteers and demonstrates how humanitarian organizations can help locals take the first steps in rebuilding a multi-ethnic society. Bosnians who mix with the other side often risk the disapproval of their friends and neighbors. It is significant that in Gornji Vakuf it is the women taking this risk. As Bosnian Serb Pavo Koralic noted, *"the women here are very strong...They are the bearers of the family. When the woman crosses the line, the men will follow."* (Mertens, 1997). While local initiatives have shown that progress in creating intercultural understanding is possible, it is sad to note that such projects receive little of the one billion dollars being spent on reconstruction.

The indictment and pursuit of war criminals has had the desired effect of signaling that justice is possible without extra-judicial retaliation. The pursuit of war criminals, many of whom are radical nationalists who oppose the implementation of the Dayton Accords, has also served the purpose of marginalizing and isolating many potential spoilers. Progress in apprehending alleged war criminals, however, has been slow. As of this writing, the ICTY has handed down 17 indictments charging 74 individuals. Ten people are in custody, one person was killed resisting arrest, and only one person has been convicted. What is surprising is how much progress has been made given the burden under which the ICTY labors. Despite their pledges, Serbia and the Republika Srpska have failed to apprehend and to turn over indicted persons that live openly in their territory. Croatia has cooperated to a limited extent while the Bosnian/Croat Federation has cooperated as well.

Perhaps most troubling is SFOR's failure to apprehend indicted persons who live openly in their areas of operation. There is an interesting geography here in that the British and Americans have been most proactive while the French have been the least enthusiastic. The British and Americans control the northern portion of Republika Srpska including Banja Luka (home of the most moderate Bosnian Serbs), while the French control southern Republika Srpska including Pale (home of the radical nationalists such as Radovan Karadzic). Apprehending indicted war criminals especially in French-controlled Pale would go a long way in suppressing the radical nationalists movement and bolstering the moderates. SFOR's failure to apprehend war criminals residing in plain sight can only be cause for the Bosnian and international publics to question the seriousness of UN/NATO efforts.

As mentioned previously, our recommendations concerning refugee return proved to be misguided. We previously suggested that, while it was morally correct that people had the right of return, they should in the near term accept compensation or lease arrangements until stability and civility return. As of November 1996, 2.3 million Bosnians remained displaced with one million of these in Bosnia itself. Only ≈250,000 have returned home but another ≈90,000 were displaced during the same period. The refugee problem has essentially three elements: there are still limitations on freedom of movement; the personal security of returnees cannot be assured; and, with the destruction of many dwellings, there is a lack of shelter and infrastructure. US Secretary of State Albright noted that individual returnees are vulnerable and thus people must be allowed to return in groups for their own security (Department of State, 1997). What our initial recommendation failed to foresee was the *de facto* partition that is solidifying in the landscape. The international community representative in Bosnia, seeing this dilemma, has initiated an effort to resettle groups in targeted areas where security can be assured and resources made available to overcome the problems of freedom of movement, personal security, and shelter.

A second area where we were incorrect (at least in the short term) concerns the "Arm and Train" program. We suggested that the program of arming and training the Bosnian Croat Federation so as to ensure a balance of power was less desirable than a regional arms control regime that would have limited the power of all sides. The "Agreement on Sub-Regional Arms Control in the Former Yugoslavia" signed in Florence on 14 June 1996 was apparently inadequate to assure the Federation of its security. The "Train and Equip" Program, as it is now known, is designed to: integrate the Bosniac and Croat militaries, orient the Federation force to a western model, reduce destabilizing foreign influences (e.g. Iran), provide continued compliance with the Dayton Accords, and, most importantly from a NATO perspective, enable an SFOR withdrawal. The "Train and Equip" Program involved $100 million in US arms and $140 million (from Saudi Arabia, Kuwait, UAE, and Malaysia) for training (Department of State, 1997). Some U.S. Congressman believe that the program is shifting the balance of power to favor the Federation and that the it increasingly sees an opportunity to reclaim all of Bosnia by force. Federation Defense Minister Ante Jelavic (a Croat) stated, "*Achieving a military balance is most vital to ensuring security in Bosnia once SFOR leaves. But we have not achieved that balance, and to do so we need continued material assistance from the outside world.*" American military trainer Dick Edward (head of the simulation center in Hadzici) said, "*All of our training operations center on a 'deter and defend' strategy.*" (Woodard, 1997). We believe that the weapons mix of the "Train and Equip" program should provide for a balance of power in that it does not provide the Federation with major offensive weapons systems. Additionally the "Train and Equip" Program's focus on a deter and defend strategy as well as the emphasis on western military values, training, and doctrine should serve to limit the offensive capabilities of the Federation Army.

We still contend, however, that the perceived need for the "Train and Equip" Program is unfortunate and that regional arms control and disarmament would have been a preferable course.

Our previous analysis did not foresee the split of the Bosnian Serbs into a moderate element based in Banja Luka (falling within the British MND) and controlling the north of Republika Srpska, and a radical element based in Pale (in the French MND) and controlling the south. As we suggested, efforts to engage and reward moderate elements while marginalizing and isolating radical elements brought the moderates into power. What we did not anticipate was that geography, economics, and SFOR would allow the radical nationalists to survive. The Republika Srpska consists of two elongated territories held together by a narrow corridor and separated from each other by hostile Federation territory. Communication between the two is difficult and thus the weak government in Banja Luka cannot extend state control in the southern part of the country. Radical nationalists have been able to maintain and finance their local power through their central control of the black market and other criminal activities. Their armed followers protect the nationalists and maintain their control in Pale and the southern countryside. Some of the nationalists leaders are indicted war criminals but the French SFOR contingent seems disinclined to apprehend these men. The good news is that the split among the Bosnian Serbs weakens the potential for renewed hostilities. The bad news is that the spatial division of Republika Srpska allows the radical nationalists a haven with direct contact along an open boundary with Serbia from which they may someday emerge.

Another problem our initial analysis failed to anticipate was the effect of absentee voting in local elections. In order to overcome the effect of ethnic cleansing people were given the option to vote for candidates in either their current location or original hometowns. In several cases displaced persons cast absentee ballots that elected a Muslim or Croat mayor in a Bosnian Serb occupied town. For example, in Brcko absentee Muslims voted in a Muslim mayor in September 1997, but the town is actually run by the Serbs who are now a dominant majority. The outsider, while nominally the duly elected leader, gets no cooperation from the hostile local inhabitants thus creating an ungovernable situation. The *de facto* local leadership continues to run things and thus the elections in some cases have proven farcical.

Yet, another circumstance our analysis did not foresee were the problems of economic reconstruction. Twenty-eight months after the war ended Bosnia's economic recovery has moved from a slow start to paralysis. Unemployment in the Federation is 50% with an average monthly wage of $145 while the Republika Srpska suffers 70% unemployment and average monthly wages of $40 a month. Foreign aid has been evident in the rebuilding of housing, roads, bridges, and airports, but most of the aid has gone to the Federation. The Bosnian Serbs' refusal to cooperate has furthered their isolation and caused them to receive only 3% of the $1.4 billion in foreign aid. Again, showing the critical importance of scale, the position of the moderate Serbs improved when the International Community changed the scale of their aid focus from entities to localities. Now Serbian towns in the north that cooperate with the international community receive aid while those that insist on a nationalist's path are further isolated. Bosnia, unlike other Eastern European countries, is establishing a market economy while at the same time rebuilding a war-ravaged country.

The Functional Approach

The Functional Approach, as advanced by Richard Hartshorne (1950), addresses state function and structure. He suggested that viable states must meet two minimum political criteria. First, the state must have a *raison d'être* that is accepted by the majority of the citizens. Secondly, both centripetal forces and centrifugal forces are constantly at work within states. States remain viable as long as centripetal forces overcome centrifugal forces. Once centrifugal forces overcome centripetal forces the state will start to disintegrate. Our previous analysis resulted in eight specific recommendations on supporting centripetal forces and countering centrifugal forces. Our recommendations concerning ethnic nationalists and support of moderates, international community involvement, and patron relations proved to have greatest veracity.

Perhaps the single greatest centripetal force in Bosnia is international community determination to see a unified Bosnian state as manifested in the continued presence of SFOR and work of various UN agencies and non-governmental organizations. To our surprise, the international community has shown remarkable staying power in Bosnia. The stability provided by the SFOR/UN presence remains essential if centripetal forces are to dominant centrifugal forces. Simply stated, the international community must maintain a strong presence to ensure the dominance of this essential centripetal force.

As indicated in the peace geography analysis, efforts to support moderate elements and marginalize ethnic nationalists on all sides have proven successful and served to enhance centripetal tendencies while diluting centrifugal forces. The success of more moderate Bosnian Serbs at the expense of Serbian nationalists in recent elections as well as the removal of a hard-line Bosnian Muslim Defense Minister with ties to Iran is examples of this policy in action. This limited success to date could be further enhanced by vigorously pursuing ethnic nationalists leaders who are also indicted war criminals.

We were also unfortunately correct in our observation that patron relations would play a critical role. While it appears that Serbia and Croatia are generally abiding by the Dayton Accords, a combination of geography and open borderlands is acting as a centrifugal force in a fashion that we did not anticipate. An economic rather than a military dependency has arisen between clients and patrons that undermine the economic integration of Bosnia. The continued use of Croatian and Serbian currencies, postage stamps, and other instruments of state sovereignty in lieu of Bosnian instruments undermine efforts to create Bosnian institutions that might weld the two entities together into a viable state. Additionally, we felt that the small size of the two entity territories as well as their traditional economic integration would force them to work together to rebuild a viable economy. The ability of the entities to rely upon their patrons economically, however, acts as a dis-incentive for further economic integration between the Federation and Republika Srpska. The Dayton Accords call for the creation of Bosnian federal institutions and public corporations that will provide necessary services. This is easier said than done, however, as on a day-to-day basis people need a viable currency, postal and telephone systems, and other necessities of modern life. The challenge for the international community representatives will be to strongly encourage both entities to cooperate in the creation of effective federal institutions while gently weaning the population away from currencies and services associated with the patron states. This is no small task but necessary in the end if federal institutions are to become centripetal forces.

As with many observers, our view of the role of historical animosities as a centrifugal force was somewhat skewed. During the war, it was widely held in government and media circles that Serbs, Croats, and Muslims had been fighting and hating each other continuously for hundreds of years and would never get along. Post-war analysis by many knowledgeable commentators has, however, disputed this view pointing out that with a few exceptions modern Bosnians have gotten along well for decades even to the extent that 13% of marriages in Bosnia as part of Yugoslavia were ethnically mixed Maass, 1996). The destruction of Yugoslavia and the war in Bosnia did not occur because these people are incapable of living together, but rather were the result of unscrupulous communists leaders who, seeing they had no future in a democratic and free market society, used ethnic nationalism to spark a conflict that ensured their continued power and privilege. We over-rated the power of historical animosities as a centrifugal force. The real centrifugal force were the agendas of cynical leaders such as Slobadan Milosevic of Serbia and Franjo Tudjman of Croatia (Silber and Little, 1996). Warren Zimmerman (the last U.S. Ambassador to Yugoslavia) notes with sorrow that the view that historical animosities made it impossible for Yugoslavs to live together was used as justification in western capitals to take a hands-off approach that let the tragedy of Bosnia occur (Zimmerman, 1996). The belated recognition that the true centrifugal force is the exploitation of ethnic nationalism by radical leaders makes it evident that the marginalisation and suppression of such leaders is essential if centripetal forces are to prevail.

The most troubling question concerning the future of a unified Bosnian State remains. What is Bosnia's *raison d'être*? The international community's desire to see a unified Bosnian state will ultimately prove to be inadequate. At some point a majority of Bosnians in both entities will have to accept that living in a unified Bosnian state is to their advantage and the state is thus deserving of their allegiance. Given the current balance of centripetal and centrifugal forces in Bosnia, there is no indication that this will happen any time soon.

Conclusion

The events in Bosnia over the past six years have been an exception to the general rule regarding conflict in Europe since World War II. The Cold War and the attendant threat of nuclear annihilation imposed a political geographic framework in which European conflicts were resolved through peaceful negotiation and compromise. The violent breakup of Yugoslavia and the war in Bosnia demonstrated both how this system has broken down in the post Cold War era and the potential for further violence and unrest in other parts of Eastern Europe and the former Soviet Union. Bosnia is a warning to the international community that further conflict again threatens Europe and that a new political geographic framework is essential to resolve future conflicts. Bosnia and the Dayton Accords have also provided a laboratory in which political geographers can apply their perspectives in order to contribute to the development of a new European political geographic framework that prevents the exploitation of ethnic nationalism and resolves future conflicts through compromise and negotiation.

A review of our initial analysis and recommendations shows that the application of this limited set of political geographic concepts had real utility in understanding the problems in Bosnia and making useful policy recommendations. Most of our predictions and recommendations were correct, although a number of unforeseen situations arose. It is also evident that two years or so is too short a time frame in which to draw any long-term conclusions, but still time enough to see the direction taken by the processes spawned by Dayton. We thus offer this critical analysis as an interim report on the short-term validity of our analysis and predictions.

Finally, we would note that we used only three approaches from the political geographers large and varied analytical toolkit. We would once again suggest to our colleagues that there is great utility in applying varied political geographic approaches in the analysis of real-world problems and challenge them to undertake such work with an eye toward making policy recommendations which make a contribution toward the maintenance of peace and to improving people's quality of life.

Bibliography

CIA/DID. 1997. *Report on the SFOR Mission.* (Fall, 1997).

Cigar, N. and Williams, P.R. 1998. 'Rewards Serbs with Town of Brcko? Don't Do It.' *Christian Science Monitor,* (11 March 1998): 19.

Corson, M.W. and Minghi, J.V. 1996. 'The Political Geography of the Dayton Accords.' *Geopolitics and International Boundaries,* 1/1 (Summer, 1996): 77-92.

Hartshorne, R. 1950. 'The Functional Approach in Geography.' *Annals of the Association of American Geographers,* 40/2: 95-130.

Klemencic, M. and Schofield, C. 1996. 'Mostar: Make or Break for the Federation.' *IBRU Boundary and Security Bulletin,* (Summer, 1996): 72-76.

Landay, J.S. 1997. 'Why Troops Won't be Coming Home.' *Christian Science Monitor* (19 Dec 1997): 1 & 8.

Maass, P. 1996. *Love Thy Neighbor: A Story of War.* New York, New York: Alfred A. Knopf.

Mertens, R. 1997. 'Grass Roots Groups Knit Bosnia Together Again.' *Christian Science Monitor,* (8 Sep 1997): 6.

NATO/SFOR Website (1998).

New York Times. (1996-98).

O'Loughlin, J. and van der Wusten, H. 1986. 'Geography, War and Peace: Notes for a Contribution to a Revived Political Geography.' *Progress in Human Geography,* 10/4 (1986): 484-510.

Silber, L. and Little, A. 1996. *Yugoslavia: Death of a Nation.* New York, New York: TV Books Inc.

US Department of State Website: Statements of the Secretary (1997).

Woodard, C. 1997. 'US Arms Weaker Side for Bosnia After NATO.' *Christian Science Monitor,* (15 Sep 1997): 6.

Zimmermann, W. 1996. *Origins of a Catastrophe.* New York, New York: Random House.

Editor's Note:

Article 1. Political Geography of the Dayton Accords was adapted from Corson, M.W. and Minghi, J.V., 1996, "The Political Geography of the Dayton Accords," in *Geopolitics and International Boundaries,* Vol. 1 No. 1, pp. 77-93 by Frank Cass Publishers, London. Reprinted by permission.

Article 2. "Prediction and Reality in Bosnia: Two Years after Dayton" was adapted from Corson, M.W. and Minghi, J.V., 1998. "Prediction and Reality in Bosnia: Two Years After Dayton" published in *Geopolitics and International Boundaries,* Vol. 2 No. 3, pp 14-28 by Frank Cass Publishers, London. Reprinted by permission.

The Peacetime Context

Introduction

The value of military geography within a theater of war can hardly be disputed. The subfield has also been important during peacetime, however, providing an important forum for the continuing discourse among geographers, military planners, political officials, and government agencies, as each relies upon geographic tools and information to address a wide range of problems within the national security and defense arenas.

One of the most significant paradigm shifts that is necessary for military geography to maintain momentum well into the twenty-first century involves the fundamental requirement to move beyond the traditional, narrowly construed focus on wartime military problems, and take into account peacetime concerns. Ironically, peacetime concerns have always consumed a far greater percentage of the Military's time and effort, yet military geographers have generally ignored these areas. Notable exceptions include McColl's (1993) detailed examination of international refugees and their political and military implications. More recently, Shaw et al. (2000) employed an ecological classification scheme to better manage the US Army's training lands, and Malinowski and Brockhaus (1999) applied spatial analytical techniques to address the Army's recruiting problem, currently one of the Military's most perplexing issues. These examples epitomize the diverse opportunities that exist in the peacetime arena.

A monumental challenge that is routinely overlooked involves the Department of Defense (DoD) responsibility for managing and maintaining installations and training lands all over the globe. DoD is one of the most important environmental resource managers in the world, overseeing 25 million acres of domestic holdings and more than 2 million acres of land abroad. Thus, environmental stewardship is a fundamental military concern during peacetime (Shaw et al., 2000).

Domestically, DoD's formal environmental restoration programs have been underway since 1984. A wide range of multifaceted programs have been administered to remedy various age-old pollution problems, and specific measures have been implemented to better manage natural resources, prevent contamination, protect endangered species, and comply with environmental laws and legislation. Complex environmental problems frequently demand multi-disciplinary approaches, and so it is not surprising that many non-geographers have addressed the Military's peacetime environmental concerns. Geographers, however, have been conspicuously absent from peacetime military geography, a surprising trend, given the discipline's traditional interest in land-use planning.

Aside from the abundance of challenges that surface in the areas of land use planning and resource management are the plethora of issues directly related to training. Climate, terrain, and vegetation, for example, provide constraints and restraints within the training arena. Where soldiers train in terms of climate conditions and weather hazards is an important issue for military geographers to investigate and illuminate for the larger military audience. Additionally, the management and restoration of degraded training and testing lands has become a highly sensitive topic in military circles.

One of the pervasive dilemmas facing the nation is the problem of recruiting military manpower to serve in the armed forces. This topic has received little attention in academic literature to date. Jon Malinowski investigates this unique military geographic problem in Chapter 13. His research examines the locational attributes of where our military recruits come from and illustrates interesting geographic correlations. Recruits are trained in rugged, intemperate conditions. Richard Dixon's chapter explores the geography of initial entry training in terms of bioclimatological considerations. This longstanding concern recently surfaced when several training fatalities occurred during severe climatic conditions. In Chapter 15 Robert Shaw, Bill Doe, Gene Palka and Tom Macia examine the geographic distribution of training lands in the United States and their correlation to potential combat environments. Considering the shrinking number and diversity of training areas available to the U.S. armed forces, despite its global mission, this important geographic problem has received little attention in the literature.

Environmental stewardship and land management are "hot-button" themes in peacetime military circles. Land-use planning has a strong geographic foundation and military geographers can certainly offer valuable insights to this significant peacetime concern. In that regard, three chapters provide a unique look at land management issues. Doe et al., employ an "ecoregional" approach to the management of military training lands on the basis of their climate and associated recoverability in Chapter 16. Similarly, in Chapter 17, Wiley Thompson addresses the issue of restoration of military lands damaged by training activities and he examines the methodologies available to land managers. This is an important topic because the Army must preserve and regenerate their training lands in view of future limitations on their use. Finally, Brandon Herl employs a warfighters' perspective to terrain management by looking at the use of land from the perspective of cover, concealment, avenues of approach and fields of fire. This refreshing insight illustrates that not all training land is alike and it is possible to predict and manage the most heavily used portions of training lands based on their "value" to the warfighter.

ಞ ೕ

Bibiliography

Malinowski, Jon, and Brockhaus, John. 199). Correlates of U.S. Army Recruiting Success in Texas. Paper presented at the annual conference of the AAG, 23-28 March 1999. Honolulu, HI.

McColl, Robert W. 1993. The Creation and Consequences of International Refugees: Politics, Military, and Geography. *GeoJournal* 31/2: 169-177.

Shaw, Robert B., Doe, William W. III, Palka, Colonel Eugene J., and Macia, Thomas E. 2000. Where Does the U.S. Army Train to Fight? Sustaining Army Lands for Readiness in the 21st Century. *Military Review*, LXXX/5, (September-October 2000).

An M2 Bradley Fighting Vehicle in an Army training area. The impact of heavy vehicle use on the landscape is clearly evident in this photograph.
Source: Center for Ecological Management of Military Lands (CEMML), Colorado State University, Fort Collins, Colorado.

13

Manning the Force:
GEOGRAPHICAL PERSPECTIVES ON RECRUITING

By

Jon C. Malinowski

Introduction

Headlines in recent years have brought considerable attention to the problems that the United States Armed Forces face in recruiting qualified men and women. In the era of an all-volunteer military, considerable time and effort needs to be expended to identify, recruit, and process new soldiers. Together, the Army, Navy, Air Force, and Marine Corps must recruit 200,000 people for active duty each year, and another 150,000 for duty in the Reserves. In 1999 the Army fell short of its goals by over 6,000 soldiers, and the Air Force was nearly 2,000 recruits short. During the same time period the Army Reserve missed their enlistment goals by over 10,000, and the Navy Reserve by about 4,700 (Cohen, 2000). Clearly, recruiting has become a challenge.

The reasons given for recent recruiting difficulties include economics, attitudes and values, and changing demographics. The economy is cited because some feel that when the economy is good, young men and women have options other than military service. Advocates of a changing values explanations point to surveys that show that young people do not view military service in the same way as they did in past. Military figures are not heroes anymore, and military service not as attractive as it once was. Others look at the changing demographic structure of the United States and note that some racial or ethnic groups that traditionally have relatively high or low enlistment rates are changing in number within the U.S. population structure.

An understanding of geography strengthens each of above explanations of recruiting difficulties, because each reason is inherently spatial or geographic. The overall national economy might be good, but regions of the country are struggling. Values might be changing with time, but there are clear regional differences in teen attitudes towards military service. Finally, while demographic changes are taking place in the United States, they are more concentrated in some cities, states, and regions. Thus, recruiting theorists and practitioners must be able to understand how geographic differences affect their mission.

To examine the linkages between geography and military recruiting, it will be helpful to first look at some historical examples of how military leaders have used geographic differences to help fill their ranks. Some of these examples highlight dark periods in recruiting theory that combined geography, ethnicity, and race to support colonial policies. Other examples will reveal how socioeconomic and ethnic divisions within a country can complicate conscription or recruiting efforts. Finally, current trends in recruiting research will be examined with a geographer's lens to suggest how a spatial perspective can improve recruiting efforts in the United States.

Historical Perspectives

Over the past 150 years there are numerous examples of geographic factors in recruiting or drafting soldiers into armed forces around the world. The examples that follow show that spatial differences in recruiting can be grounded in racism, ethnicity, politics, religion, necessity, and socioeconomic class. While some may seem unique to a particular time and place, and thus not relevant to today's personnel problems, all are necessary to an educated formulation of recruiting theory. Failure to understand the past may cloud our judgement as recruiters view the future.

Northern Conscription and the American Civil War

During the Civil War in the United States, the government of the North enacted conscription laws to draft soldiers into the Union Army. As a result of the Enrollment Act of 1863, over 776,000 men were drafted between 1863 and 1865. Because of volunteer enlistment, only 46,000 of those drafted were actually called to service. Nevertheless, the prospects of being called to serve were real for northern males. Draftees could opt out of service by paying a fee of $300 or by sending another person in their place. Neither option was available for many who sought to escape from service, and about 160,000 men failed to report to their draft boards (Levine, 1981).

The geographic pattern of evasion is interesting. Rates were initially highest in parts of Wisconsin, Michigan, New York, and Pennsylvania, but later spread to all areas. Levine (1981) found that the 27 congressional districts with the highest evasion rates tended to share certain characteristics: a tendency to vote non-Republican (the party in power); relatively high Catholic populations; greater numbers of foreign-born citizens; and generally were poor. Levine suggests that the reasons for the geographic concentration of evaders is rooted in a few related factors. The Republican Party generally stood for and promoted rural, Protestant values and thus was not as appealing to Catholic voters. Furthermore, urban populations tended to be more interested in social-class issues that urbanization was thrusting upon parts of the North. The Republican Party was not seen by many urban residents as being sensitive to urban class issues. Thus, there was, in some parts of the country, an urban-rural difference in attitudes towards the war that became reflected in evasion rates (Levine, 1981).

The British and the Concept of "Martial Races"

One the most controversial approaches ever taken towards recruiting was the British concept of favoring what they described as "martial races" in their colonial armies. This policy conceptualized certain ethnic groups, such as the Gurkhas, as being more appropriate for military service than others. While this aspect of colonial history is often discussed as an example of racism, it included a decidedly geographic component because favoring certain ethnicities generally required a focused recruiting effort in those regions of a colony where the martial race was concentrated. Examples from the British experience in India and Africa highlight the spatial nature and ramifications of this recruiting policy.

British recruiting policies in India developed slowly (Barua, 1995). In the 18th and early 19th century, the British recruited soldiers into the Army of India regardless of caste or ethnic group. As the 19th century progressed, however, British writers began to write about differences among certain regions of India. Lowland areas along the major rivers, where many British administrators lived with their families, were hot and very difficult for Europeans to adjust to climatically. When they traveled into the mountains they found cooler climates more like their own homes in Europe. They also found ethnic groups that had different physical and cultural characteristics than people in the valleys. A few Europeans began to write about these differences, some making connections between the climate and physical features of a region and the characteristics of the people who lived there (Barua, 1995). There was some experience with this concept from the homefront, where Scottish Highland soldiers had developed the reputation as a martial people during the 18th Century.

Students of geography will recognize this argument, that climate and physical geography create differences between people, as examples of **environmental determinism,** i.e. that environment determines the level of development of the culture. This concept has its roots in Darwin's *Origin of Species*, which raised the possibility that time and distance could lead to racial differentiation. Combined with ideas such as "survival of the fittest" some came to believe that some races were superior to others. The eugenics movement of the late 19th century warned "superior" races (the colonizers) from mixing with "inferior" races at the risk of weakening the superior race. Within geography, writers such as Ellen Churchill Semple and Ellsworth Huntington carried environmental determinism into the 20th century (Semple, 1911; Huntington 1915). Huntington wrote extensively on the issue of climate and race, generally concluding that tropical climates make people lazy, stupid, and morally bankrupt.

These deterministic ideas became written into British recruiting policies around the world for their overseas armies. In India, officials began to classify superior and inferior races. Many used the 1858 Indian Mutiny (a.k.a. Sepoy Rebellion or Great Mutiny) as justification for racial classification based on climate and physical appearance (Barua, 1995). In the aftermath of the rebellion, upper-caste Brahmins, mostly from the lowland, riparian state of Bihar were labeled as instigators in the conflict. Gurkhas, Sikhs, and other groups had remained loyal to the British, and thus their position rose in British eyes. Mountainous terrain, such as the areas where the Gurkhas lived, was thought to produce ferocity and stamina hot, low areas created laziness and deceitfulness.

The recruiting consequences of this policy are obvious. At first, recruiters were obliged to recruit only certain ethnic groups from certain regions. Later restrictions were made more stringent by limiting selection to only certain classes within ethnic groups, and to individuals with certain body dimensions. A Gurkha was supposed to have a sturdy, compact build, etc. The proper body types were determined by pseudo-scientific studies and individual perceptions of the ideal. No doubt thousands of competent Indians were turned away from service because of ethnicity, appearance, or class.

The British carried the concept of the martial races to other colonies as well. In east Africa the Kamba were favored over other groups, leading them to feel a sense of entitlement towards Army service after independence ended British rule (Parsons, 1999). In West Africa, northern Nigerians, such as the Hausa, were favored over the Yoruba and Ibo in the south (Barua, 1995; Adams, 1998). This may have been an attempt on the part of the British to insure that the southern groups, close to the center of political power, were kept in balance. These regions possess considerable geographic differences. Northerners live in a semi-arid climate and tend to practice Islam. Traditional religions and Christianity were more common in the semi-tropical coastal areas.

After independence, Nigeria attempted to maintain an ethnic balance by instituting a quota system. The plan sought to maintain a balance of 50% of the army coming from the Northern Region and 25% from each of the Eastern and Western Regions. Instead of creating a harmonious balance, these divisions caused cleavages that led to northern soldiers massacring Ibo soldiers from the Eastern Region. (Barua, 1995).

While the British were not the only power to buy into the concept of martial races (see Lunn, 1999 for a discussion of the French in Africa), many of the attitudes of the colonial British persist to this day. Partly because of their status and partly because of their distinguished record of service, Gurkha soldiers are still honored for their strength and tenaciousness. Reports surfaced during the Falklands War that Argentine troops fled the battlefield upon hearing that Gurkhas in service to the British were about to attack. More recently, in the aftermath of clashes in East Timor, Gurkhas were sent in to patrol the areas because they were deemed to be the most fit to handle the tough climate and conditions.

Modern Perspectives on Recruiting

Today's thinking on recruiting is certainly more scientific than the British notions of martial races, but geography can still enlighten our understanding of military personnel acquisition. Modern research on military recruiting is surprisingly sparse, and there remains almost limitless studies awaiting the eager student or researcher. From the research that has been conducted, three broad foci seem to emerge: propensity to serve, the effect of the economy, and demographic change. Each of these areas has geographic variations that shed light on the constant, difficult job of filling the ranks of modern militaries.

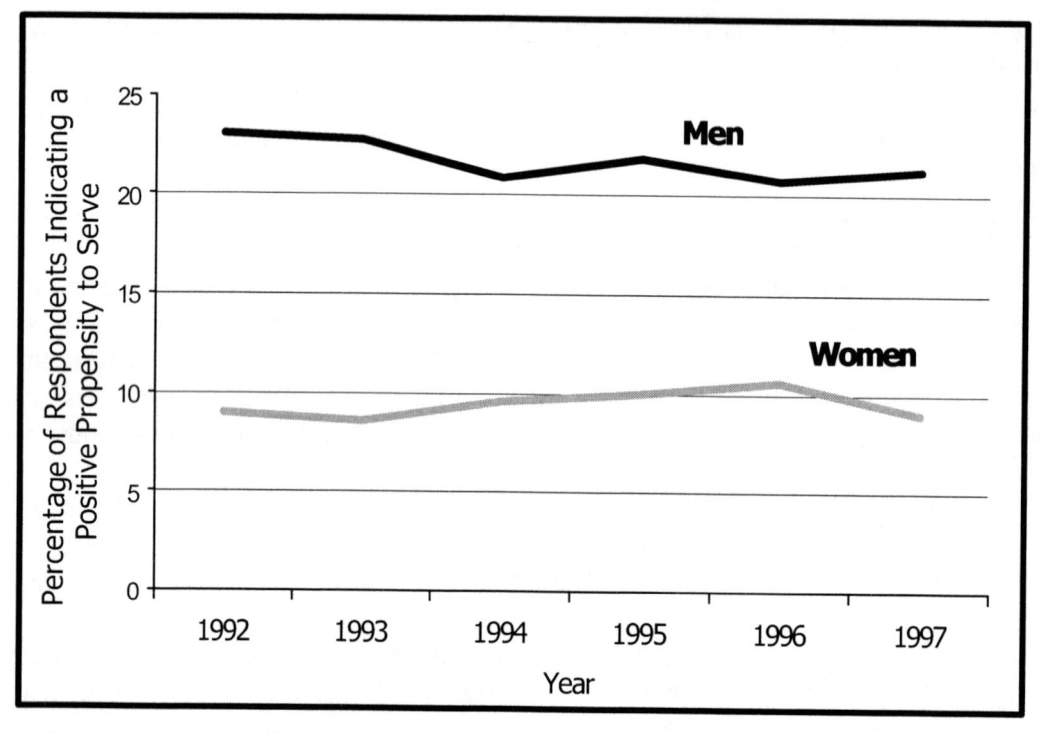

Figure 1: Propensity to Serve by Sex, 1992-1997.
Source: Youth Attitude Tracking Survey, 1997.

Propensity to Serve

A teenager's willingness to consider a career in the military (i.e. "propensity to serve") has been a fertile source of research on recruiting issues. The reason for this comes not from not only a logical connection between teenage attitudes and enlistment, but also from the availability of the data. Two readily available databases exist with propensity data in the United States. The first is the Youth Attitude Tracking Survey (YATS) and the second the Monitoring the Future (MtF) project.

The Department of Defense has collected the Youth Attitude Tracking Survey since 1975. The survey is given to about 10,000 16-24 year-olds each year. One of the questions asked is, "How likely is it that you will be serving on active duty in the Army, Navy, Marine Corps, Air Force, or Coast Guard?" Because of the long history of the survey, the results of the Study are given considerable weight in the recruiting community. YATS data highlights the fact that there are real differences among the population that recruiters have to draw upon based on race, gender, age, and geography

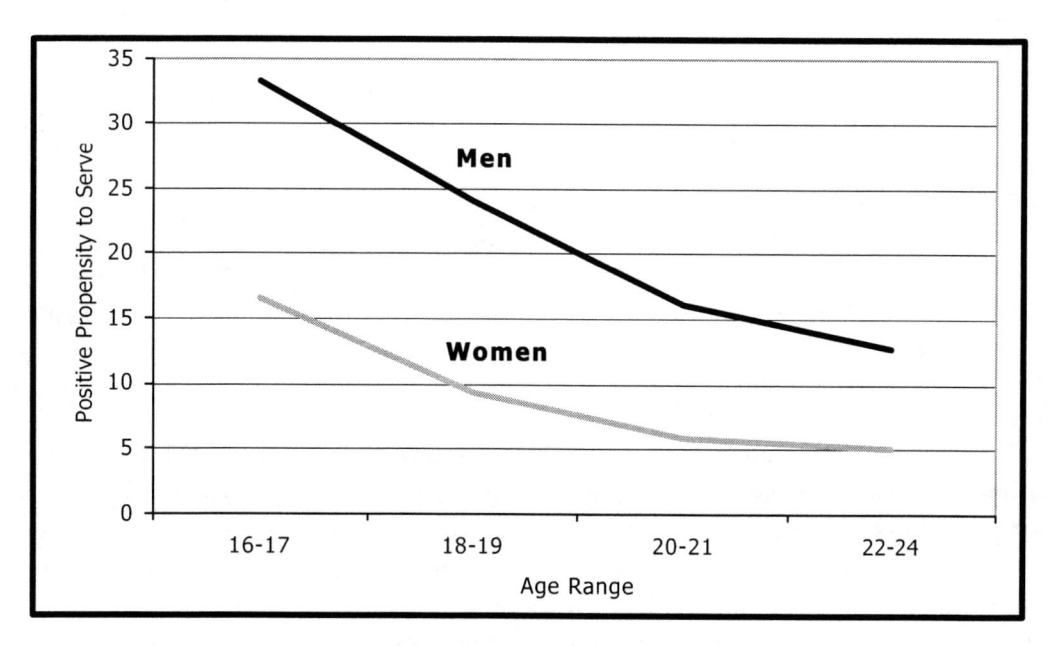

Figure 2: Propensity to Serve by Age, 1997.
Source: Youth Attitude Tracking Survey, 199.7

Figure 1 shows differences in positive propensity to serve between men and women for the six-year period from 1992-1997. It's clear that women are much less likely to see themselves on active duty. This makes sense given the fewer job openings for women in the military today. In terms of branches, the Air Force emerges as the most popular branch choice for YATS participants, followed by the Army. The Coast Guard generally receives the lowest positive propensity scores.

Age is also an important factor. As Figure 2 shows, propensity drops off with the age of survey respondents. Clearly a recruiter's best opportunity is among high school students. This does not mean, however, that recruiting older men and women is fruitless. Many in this age range have already committed to college and other careers and thus would be less likely to see themselves in active service. This is, of course, one of the dangers of using propensity; it does not necessarily predict enlistment numbers.

Ethnicity is also an important determinant of propensity. Figure 3 shows positive propensity by the race or ethnicity of the survey respondent. Clearly there is a greater propensity towards military service among Black and Hispanic youth. Twice as many Black and Hispanic men indicated a positive propensity, and nearly three times as many women. Given the great differences in the ethnic and racial geography of the United States at the national, regional, and local level, it would seem that geographers would have countless opportunities for examining recruiting patterns at various scales. To date, however, the number of studies has been few.

The YATS does provide some rough geographic variables. The U.S. is divided into four regions, Northeast, North Central, South, and West, for basis of comparison. Positive propensity percentages are not the same in each region. This is clear from Figure 4, showing lower propensity levels for males in the North Central region than in the South or West. What factors might explain these regional differences?

The Monitoring the Future (MtF) project provides similar types of propensity data and has been used by a number of researchers. The MtF has been given to between 14,000 and 19,000 teens since the mid 1970s ("Monitoring", 1999). Participants are asked whether they "definitely will", "probably will", "probably won't" serve, and "definitely won't" serve in the military. A 1999 study by Segal et al. found numerous temporal changes in propensity numbers between the early 1970s and the late 1990s. First, they found that propensity generally declined as grade-level increased for teens during the years 1991-1997. This matches the YATS findings discussed above showing declining propensity for age. An unanswered question is why propensity drops off so dramatically and what might be done to slow the decline?

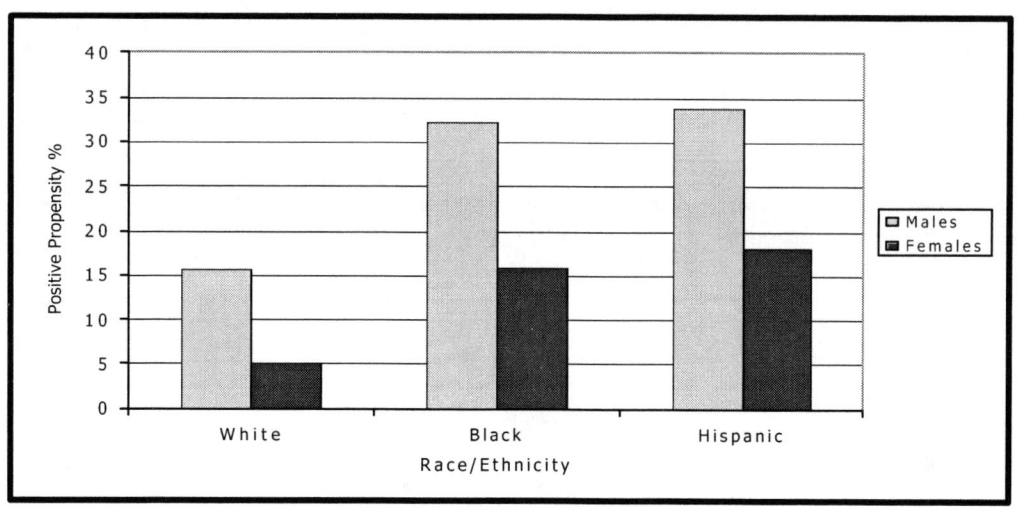

Figure 3: Positive Propensity by Race/Ethnicity, 1997.
Source: Youth Attitude Tracking Survey, 1997.

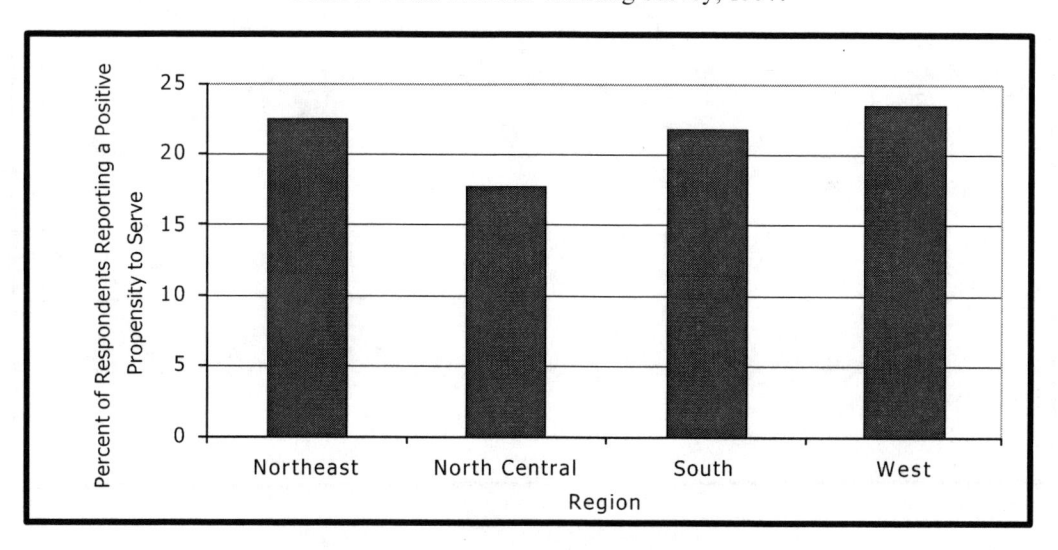

Figure 4: Positive Propensity by Region for Males, 1997
Source: Youth Attitude Tracking Survey, 1997.

Segal et al. (1999) also found changes in propensity over the past two decades. Although "definitely will" respondents remained relatively constant, "probably won't" serve respondents dropped from about 30% in 1976 to about 16% in 1997. At the same time, the percentage of respondents in the "definitely won't" category rose from just below 60% to over 70%. This certainly doesn't bode well for recruiting efforts.

The same study also found changes in propensity by race over time (Segal et al., 1999). During the 1970s and 1980s African American men had much higher propensities to serve than Hispanic or White men. This changed dramatically in the early 1990s as the propensity to serve among African American men dropped to about the same level as Whites and Hispanics.

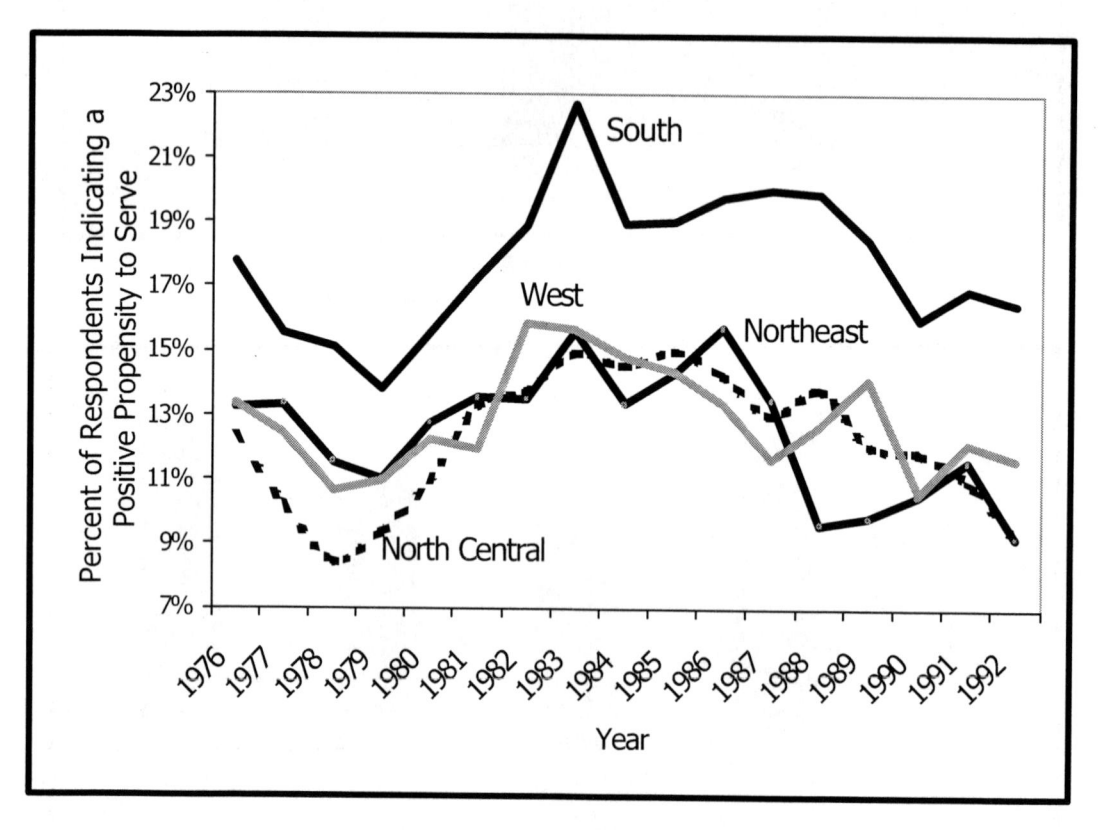

Figure 5: Positive Propensity by Region, 1976-1992.
Source: Monitoring the Future, 1976-1992.

Like the YATS database, MtF data has only basic geographic information. Again, a Northeast, North Central, South, and West division of the United States is used. Figure 5 summarizes regional changes in positive propensity between 1976 and 1992. It's clear that not all parts of the country see military service in the same way. Youth in the South have long indicated a higher interest in serving in one of the armed forces. Although there is some similarity among the regions, such that propensity is generally going up or down at the same time, it is not always the case. Notice, for example, the dramatic decrease in propensity in the Northeast during the late 1980s while at the same time propensity in the South was increasing. It was another year or two before a similar large drop in the South. The point to be taken away from this data is not that differences have existed in the past, but simply that differences exist! A national recruiting policy must take into account that attitudes may be considerably different in some parts of the country than in others.

When dealing with the issue of propensity, a logical question arises, "Does propensity predict enlistment?" A 1998 study by Bachman et al. looked at just this question using Monitoring the Future data. Among high school seniors, they found that about 62% of males who had said they "definitely" would serve had actually enlisted within two years of graduation. By 6 years after graduation the percentage climbs to near 70%. Men who said they "probably" would serve had lower numbers, just 21% enlisted within two years of graduation, 30% by the end of the sixth year after high school. The numbers for women are lower. Only 37% of the women who reported they "definitely would" serve actually enlisted within two years of graduation. The number only climbs to 40% by the end of the 6^h year. This discrepancy between men and women need to be further researched. Bachman et al. (1998) conclude, however, that propensity, i.e. expectation of service, is a good predictor of actual enlistment, especially in men.

The Economy

Many newspaper articles about recruiting problems site a strong economy in the late 1990s as a fundamental reason why the military has struggled to attract recruits. The assumption, which has considerable face validity, is that a strong economy affords an abundance of good-paying jobs that pull potential recruits to other occupations.

But the relationship between economic variables and recruiting is still poorly understood. Even a cursory understanding of American economic geography reveals that while some parts of the United States, such as the Southeast, boomed during the 1990s, other regions, such as parts of the northern Rust Belt were suffering. Yet research and journalistic accounts of recruiting tend to talk about national statistics only. If economics are important, what about rural-urban differences, or urban-suburban?

Research suggests a mixed importance for economic variables. For example, unemployment is often considered as a key indicator of economic well-being. Low unemployment might signal trouble for military recruiting because other jobs are readily available. Research conducted by the author in Texas with over 30,000 Army recruiting contracts found that recruiting did fall over time as unemployment fell, but the relationship was very hard to measure and only accounted for a small proportion of the variance in recruiting contracts over several years (Malinowski and Brockhaus, 1999). In other words, unemployment may be related to recruiting but is not that important overall.

In fact, there may be a danger in relying on monthly unemployment figures. For example, in many areas of the country unemployment increases in the spring after employees are let go following the holiday rush, and decreases in the summer as seasonal employment increases at amusement parks, summer restaurants, campgrounds, etc. But a typical recruiting season is just the opposite. Recruitment numbers are higher in the late spring and summer because of the pool of new graduates and lower during the school year when unemployment numbers might be higher. So analysis of recruiting using unemployment is complicated.

Geographers in the United Kingdom found that although the unemployed accounted for about 25% of applicants for military positions, over 40% were rejected because of criminal of substance abuse problems in their past (Dandeker and Strachan, 1993). Many also withdrew their applications during the process as they found other jobs. The same study also found that certain types of workers account for very few of the applicants for military positions. Namely, skilled manual laborers and junior non-manual workers (mostly administrative or secretarial workers). These young men and women have an easy time finding other work. This study suggests that the types of workers in a particular area or region may affect recruiting. The aforementioned study in Texas, however, found almost no relationship between percentage of blue-collar or service sector workers and recruiting (see Figure 6). As indicated by Figure 7, per capita incomes also seemed to have little predictive value on recruiting success, indicating that wealthier areas may be just as good for recruiters as poor areas. So the picture is cloudy and highlights the importance of considering geography when analyzing recruiting. Every area needs to be understood and analyzed both separately and in the context of state, regional, and national economies.

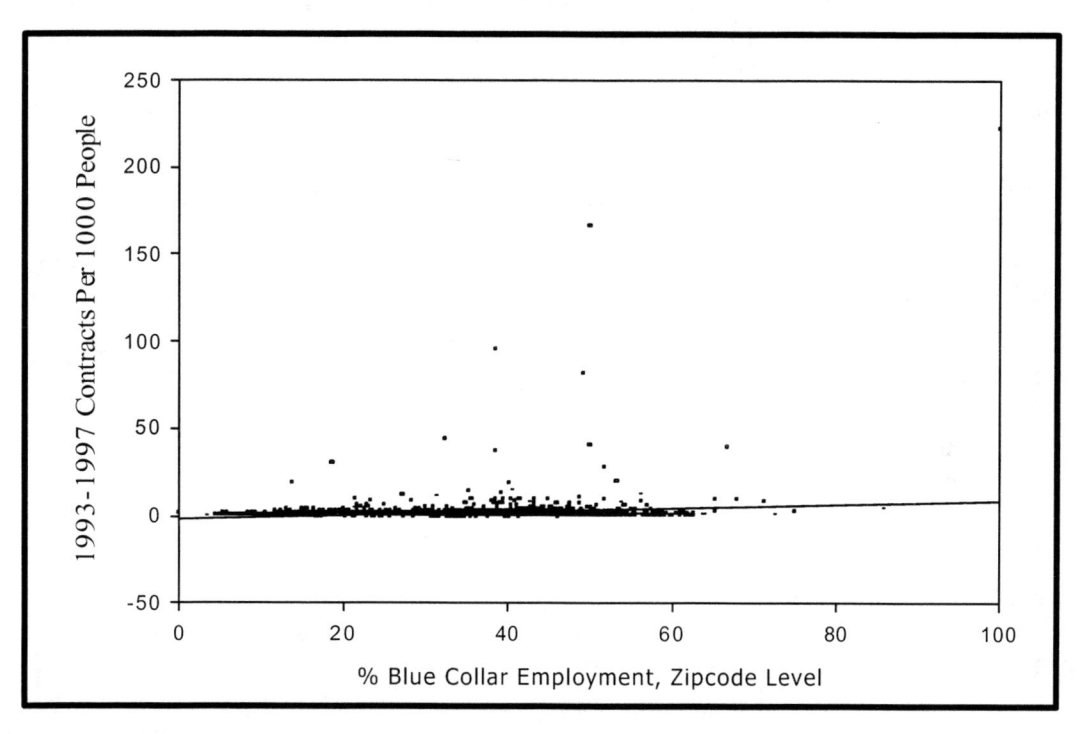

Figure 6: Contracts Per 1,000 People by Blue Collar Employment, Texas 1993-1997.
Source: Malinowski and Brockhaus, 1999.

Another way the economy may affect recruiting is through an indirect effect on recruiting standards. During the early 1990s a lull and the economy and a successful Gulf War created an increase in interest in military service. Because the military had more people to pick from, standards were stiffened. Now in times of recruiting difficulties, discussions have turned to loosening some of these previously imposed restrictions. This, of course, can lead to charges that recruiters are "lowering standards" and somehow weakening national security.

Economic geography needs to be considered on a local scale as well. The location of recruiting stations are made based on economic factors such as rent, space, and proximity to other stations. This is a classic least-cost location strategy as put forth by Alfred Weber. Weber argued that industries should locate at a place between the market and raw materials where overall production costs are lowest. For recruiters, the market is teens and young adults, but no money comes in from them, so all costs really boil down to rent and utility costs. Because of this, many end up in older strip malls that are not heavily frequented by the populations likely to enlist. Perhaps recruiters need to adopt a profit-maximization approach that recognizes that the location of the recruiting stations might actually be at the highest cost site, with expensive rents, but that will create more face-to-face meetings with the market. The high costs would be reasonable if recruiting increased significantly. Along these lines, some researchers are now exploring the idea of putting recruiting stations in shopping malls to gather more attention from young people.

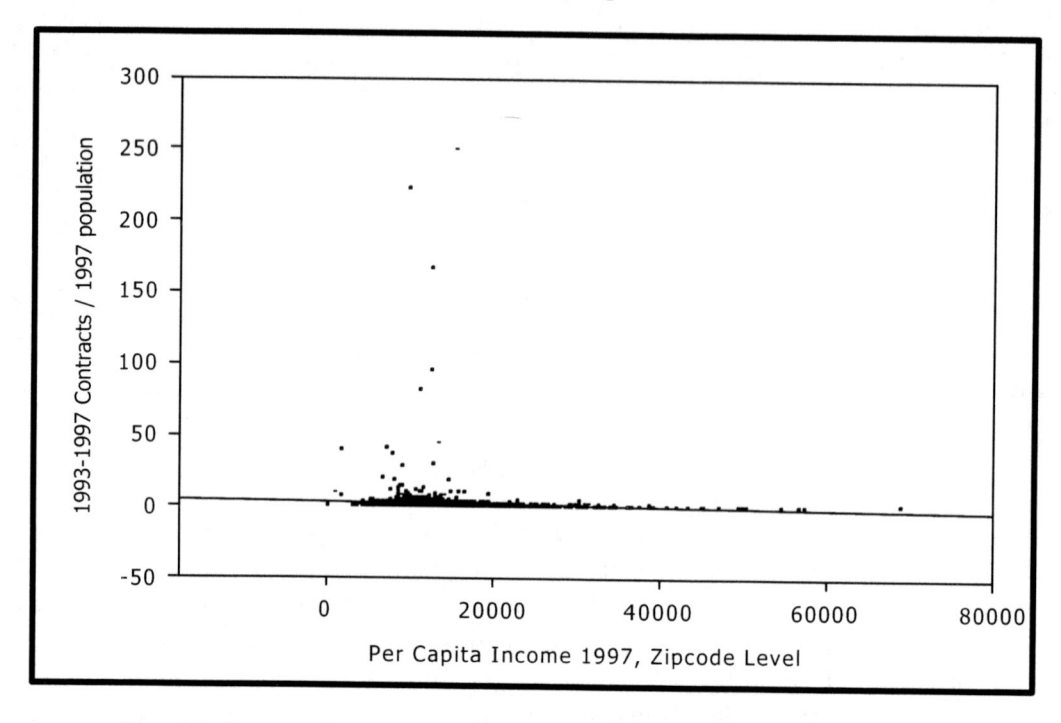

Figure 7: Contracts Per 1,000 People by Per Capita Income, Texas 1993-1997.
Source: Malinowski and Brockhaus, 1999.

Demographics

Finally, changing demographics highlight an important area for the geographic analysis of military recruiting. The 'Baby Boom' of the post-World War II era created ripple effects that the economy still feels today. An increase in births, clearly evident in Figure 8, during the 1950s meant that the 1970s had a large number of men and women of the appropriate age to serve in the military. Their children represent an increase in 20 year-olds in the 1990s. So demographic trends can affect the number of eligible youth for recruiters to approach.

There are also dramatic changes occurring in the ethnic and racial composition of the U.S. population. According to the Census Bureau, by 2020 Hispanics will be adding more people to the population than any other ethnic group in the country. Hispanic growth rates may even exceed 2 percent, higher rates than seen during the Baby Boom. The increasing proportion of Hispanics in the population will likely change the composition of U.S. Armed Forces. White, non-Hispanic populations will actually begin to decline by 2030 or so and will account for smaller and smaller proportions of the American population. Black populations will also account for smaller percentages of the U.S. total by 2050. Asian populations will increase, but not as dramatically as the Hispanic population.

As Map 1 shows, however, population does not increase uniformly. Hispanic populations will increase faster in some areas more than others. How this will affect recruiting is obviously unknown at this time. Demographic change is not aspatial. Regional and even local differences in growth must be considered. Recruiting regions and stations need to be ready to adjust with trends. Many see Hispanics as filling the gaps in recruiting numbers so prevalent in recent years. But the realities of economic and social conditions in some Hispanic areas means that large numbers of Hispanic youth drop out of school, disqualifying them from consideration for service. National drop out rates in late 1990s hung at about 30% for Hispanic teenagers and reached to over 50% in Texas. Although the Hispanic proportion of the recruitment-age population nationally is about 12%, they account for less than 5% of Air Force enlisted personnel, 7% of Army enlisted personnel, and 8.5% of the Navy's enlisted ranks. Only the Marine Corps, with about 11.5%, comes close to an expected representation.

This raises an important fact about recruiting. One might expect that the ethnic composition of recruits would mirror the ethnic composition of the labor force. It does not. Black youth are more greatly represented in the Army than in the national population of the same ages. During the late 1970s well over 30% of all Army accessions were of black soldiers although only about 12% of the labor force was black. Black accessions, however, declined steadily during the first half of the 1980s, rose slightly in the late 80s, and then dropped again in the early 1990s. Seagal and Verdugo (1994) note that during the 1980s, there was a shift among black soldiers from combat to non-combat units. For example, the percentage of blacks in the 82nd Airborne Division infantry battalions dropped from 21 to 16% between 1980 and 1990. Seagal and Verdugo (1994) argue that the non-combat jobs offer more skills training that appeals to black enlistees. White soldiers, they argue, are more interested in the educational benefits, such as the G.I. Bill, and are not as worried about using their time in the military to acquire specific skills. From a geographer's standpoint, the question arises as to how the changing demographics of the United States will affect future interest in the Army. If certain groups whose populations are increasing are not considering combat positions, will these positions become increasingly difficult to fill? There is also the reality of having greatly different racial and ethnic compositions in different branches of service, which might lead to internal social problems.

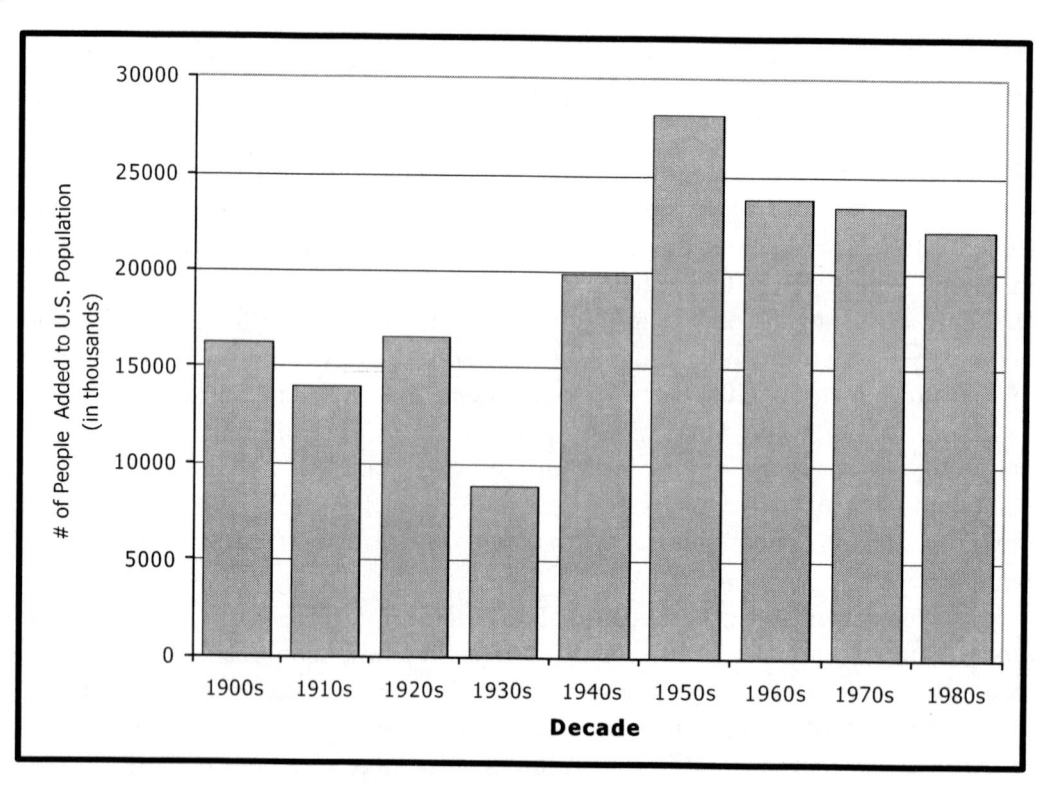

Figure 8: United States Population Added by Decade.
Source: U.S. Census Bureau.

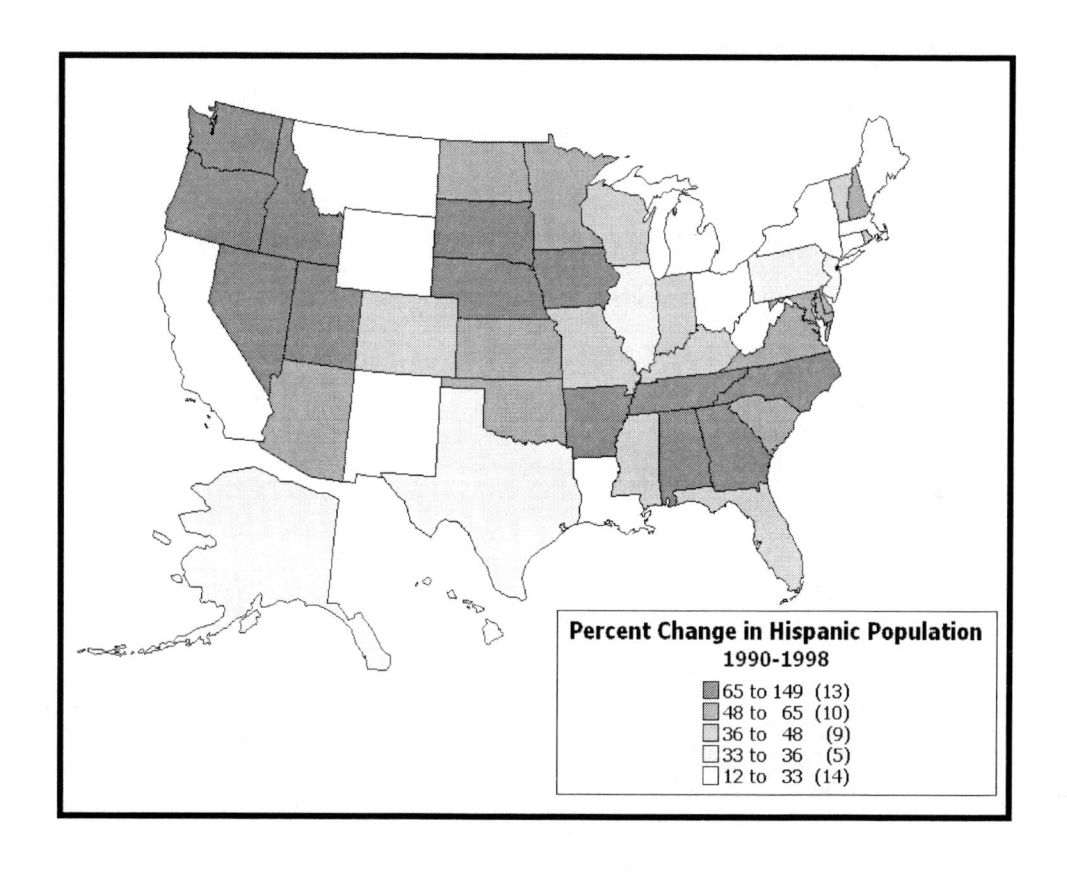

Map 1: Percent Change in Hispanic Population, 1990-1998.
Source: U.S. Census Bureau.

Conclusion

With only 25 years of voluntary service in the modern era, the United States Military and military researchers still do not fully understand the dynamics that lead to recruiting success or failure. More disturbing is the lack of interest in approaching recruiting from the geographer's perspective. A country as large as the United States cannot be considered one large, homogenous market readily to be approached in exactly the same way by every recruiter. Economic and demographic changes occur on local, regional, and national levels. Attitudes towards service will vary greatly within the same city. An old adage states that in business it's "location, location, location". Recruiting is a business and needs to be approached in that way. Failure to adequately address location and geography may worsen a situation already considered critical.

Bibliography

Adams, Paul. 1998. "The Military View of the Empire 1870-1899: As Seen Through the Journal of the Royal United Services Institution." *RUSI Journal* 143:58-64.

Bachman, Jerald G., Segal, David R., Freedman-Doan, Peter, and O'Malley, Patrick 1998. "Does Enlistment Propensity Predict Accession? High School Seniors' Plans and Subsequent Behavior." *Armed Forces & Society* 25:59-80.

Barua, Pradeep P. 1995. "Inventing Race: The British and India's Martial Races." *Historian*, 58:123-137.

Cohen, William S. 2000. Annual Report to the President and Congress. Washington DC: Department of Defense.

Huntington, Ellsworth. 1915. *Civilization and Climate*. New Haven: Yale University Press.

Levine, Peter. 1981. "Draft Evasion in the North during the Civil War, 1863-1865". *Journal of American History* 67: 833-.

Lunn, Joe. 1999. 'Les Races Guerrières': Racial Preconceptions in the French Military about West African Soldiers During the First World War. *Journal of Contemporary History,* 34:517-536.

Malinowski, Jon C. and Brockhaus, John. 1999. Correlates of U.S. Army Recruiting Success in Texas. Paper presented at the Association of American Geographers' Annual Meeting, Honolulu, HI, March 1999.

Monitoring the Future 1999. University of Michigan: Institute for Social Research.

Parsons, Timothy H. 1999. "Wakamba Warriors Are Soldiers of the Queen": The Evolution of the Kamba as a Martial Race. *Ethnohistory*, 46: 671-701.

Segal, David R., Bachman, Jerald G, Freedman-Doan, and O'Malley, Patrick. 1999. "Propensity to Serve in the U.S. Military: Temporal Trends and Subgroup Differences." *Armed Forces & Society* 25:407-427.

Semple, Ellen Churchill. 1911. *Influences of Geographic Environment: On the Basis of Ratzel's System of Anthropo-Geography*. London: Constable.

Youth Attitude Tracking Survey 1997. Washington, DC: DO Defense Manpower Data Center.

A new recruit low crawls through an obstacle course at Fort Benning, Georgia.
Source: United States Army photograph.

14

Training the Force:
BIOCLIMATOLOGICAL CONSIDERATIONS FOR MILITARY BASIC TRAINING

By

Richard W. Dixon

Introduction

The mission of military basic training is to take a group of civilians and turn them into members of the military. This is typically accomplished by rigorous physical training combined with instruction in discipline, military customs, duties, and responsibilities. Recruits are typically exposed to a variety of weather conditions during their basic training and so a basic understanding of thermoregulation and the human response to extreme heat and cold is required. This knowledge will also assist in provisioning/sheltering of troops during advanced training or operational deployments.

Humans are homeotherms; they use active mechanisms to maintain a core body temperature near 98.6°F except in conditions of extreme environmental temperature. In addition to extreme heat or cold, wind, humidity, and excess radiation can also impact the ability of the body to maintain its temperature balance. Loss of temperature balance degrades performance and in extreme cases can lead to serious injury or even death.

Climatologists track changes in extreme heat or cold by using apparent temperatures[1]. An apparent temperature is the environmental temperature adjusted for the effect of wind, high humidity, or excess radiation. It is not a sensible temperature; you cannot use a thermometer to measure the wind chill, but it is customarily reported in temperature units. The two most common apparent temperature scales are the Heat Index and the Wind Chill. Apparent temperatures should be thought of as measures of the efficiency with which the environment interacts with the body to prevent or enable heat loss.

Impact of Temperature on Human Comfort

The body begins to suffer from thermal stress when the heat gain or loss exceeds the ability of the body to shed or conserve heat. When the body is in thermal equilibrium, the heat lost balances the heat gained. This may be expressed in equation form (Griffiths, 1976) as:

$$M + R + C + P + E = 0 \qquad \text{[Equation 1]}$$

where M is metabolic heat, R is net radiation, C is net convection, P is net conduction, and E is evaporation. M will always be positive (leading to warming), while E will always be negative (associated with cooling). The other three terms may be positive or negative depending on environmental conditions. Radiation gain comes from solar short wave and long wave from surrounding objects. Radiation loss from the body occurs via long wave emission. If the surrounding air is warmer than the body a convective heat gain occurs as air is advected across its surface. Similarly, convective heat loss can occur when the wind advects cool air past the body. Typically the smallest component of the heat balance is the conduction term. Conduction occurs when the body is in contact with another surface. If the surface is colder, heat will flow from the body. Heat flow into the body occurs when the object is warmer than the skin. This is a familiar sensation to anyone who has burned their fingers on a hot pot. Activity level is another concern in assessing risk of thermal morbidity or mortality. Long periods of relative inactivity during extreme cold or physical exertion during extreme heat increases the strain on the thermoregulatory system.

The body reacts to changes in its thermal environment by a series of physiological responses coordinated by the hypothalamus, that section of the brain, which regulates thermal balance (Morgan and Moran, 1997). If the core temperature rises, the hypothalamus initiates both increased sweat production and vasodilation. The increased sweat production helps to cool the skin by evaporation, and vasodilation directs capillary blood flow near the skin surface to increase radiational and convective cooling. A drop in core temperature triggers the production of additional metabolic heat by shivering (a rapid contraction of muscle tissue) while vasoconstriction redirects capillary blood flow from the surface to conserve heat. In either case, the hypothalamic directed response ceases when the thermal balance is restored.

On occasion, the thermoregulatory system may be unable to maintain the core temperature. In both extreme heat and extreme cold stress, the body proceeds through a series of increasing more severe responses. Continued exposure may result in severe injury or even death. Frostbite occurs when the skin freezes. Thus, the body response of vasoconstriction which moves blood away from the extremities, places fingers and toes at risk of damage from frostbite. Warning signs of frostbite include a transition from feelings of cold to pain to numbness in the extremities. Damage can be minimized by promptly rewarming the skin. Frostbite is a possibility whenever the ambient temperature falls below 32°F. Should the core temperature fall below about 95°F hypothermia results. Hypothermia is a dangerous medical condition marked initially by drastic increases in shivering, progressing to lethargy, and if untreated, unconsciousness leading to death as the core temperature continues to fall. Hypothermia victims need immediate medical attention to prevent additional heat loss and restore their core temperature.

Hyperthermia is the result of rising core temperature. An increase in the core temperature to 102°F results in heat exhaustion, noted by heavy sweating, nausea, and a clammy feel to the skin. Moving immediately to a cool environment and re-hydration can usually restore thermal balance. Heat stroke occurs when the core temperature rises above 105°F. Hot, dry skin, an extremely strong pulse, and possible unconsciousness note this life-threatening condition. Immediate medical attention is required to reduce the core temperature and prevent death. Morbidity and mortality from extreme temperatures may be prevented or minimized by careful monitoring of personnel, the activity in which they are engaged, and the surrounding environment.

Climatological Measures of Thermal Stress

Numerous indices have been developed over the years to relate the effects of temperature, humidity, and wind on human comfort (Driscoll, 1985). No index is perfect because the human response is not perfect. We all perceive and respond differently to heat and cold. In general terms, a temperature stress index attempts to relate the environmental conditions to the rate of heat loss or gain by a human body. The two most common indices used today are the wind chill index and the heat index. Both are equivalent temperatures, relating ambient conditions to the temperature produced by comparison to some set of reference conditions. It is important to note that although these indices are listed as temperatures, they do not meet the physical definition of temperature. Physically, temperature refers to the average kinetic energy of the molecules composing the substance. Since the impact of wind or humidity is on the _rate_ of heat exchange, which depends to a first approximation on the temperature difference between the body and the environment, you cannot measure the wind chill index or heat index with a thermometer. A container of water will not freeze if the wind chill index is below freezing and the ambient temperature is above freezing. Thus it is a misnomer to refer to an apparent temperature as a "feels-like" temperature.

The wind chill index is the temperature of calm air in which the rate of heat loss is the same as at the ambient conditions of wind speed and air temperature. Table 1 gives values of the wind chill index over a variety of common winter temperatures and wind speeds. Note that at the lowest wind speed of 5 MPH, there is little difference between the ambient and apparent temperatures. The difference becomes much more pronounced at higher wind speeds reflecting the increasing rate of heat loss with increasing wind speed. Also note from the body of Table 1 that different combinations of wind speed and ambient temperature produce the same value of wind chill index. If the thermoregulatory system is unable to provide heat to the body surface at a rate similar to the rate of heat loss, skin temperature will drop eventually leading to frostbite or hypothermia.

The heat index is the temperature of air with a fixed moisture content (dew point temperature of 57°F) in which the rate of heat exchange with the environment is the same as at the ambient conditions of moisture and temperature. As the temperature and humidity increase the body becomes less effective at shedding heat to the environment. Table 2 provides values of the heat index for a representative range of temperature and humidity conditions. Note that as with the wind chill, identical values can be obtained for different combinations of temperature and humidity. It is also of interest to note that unlike the wind chill index, which is always cooler than the ambient temperature, very low relative humidity values produce a cooling effect where the heat index is actually lower than the ambient temperature. This is mostly due to the large increase in evaporation from the skin in very dry environments. This effect becomes less important as the ambient temperature increases. When the heat index exceeds 90°F, prolonged exposure and physical activity may lead to heat exhaustion. Heat index values over 105°F should be considered dangerous as the risk of heat stroke is very high.

Table 1. National Weather Service Wind Chill Index [°F]

Wind Speed	\multicolumn{14}{c}{Ambient Air Temperature [°F]}
	35	30	25	20	15	10	5	0	-5	-10	-15	-20	-25	-30
5	32	27	20	16	18	6	1	-5	-10	-15	-20	-26	-31	-36
10	22	16	10	4	-3	-9	-15	-21	-27	-33	-40	-46	-52	-58
15	16	9	2	-5	-11	-18	-25	-32	-38	-45	-52	-58	-65	-72
20	11	4	-3	-10	-17	-25	-32	-39	-46	-53	-60	-67	-74	-82
25	8	0	-7	-15	-22	-29	-37	-44	-52	-59	-66	-74	-81	-89
30	5	-2	-10	-18	-25	-33	-41	-48	-56	-63	-71	-79	-86	-94
35	3	-4	-12	-20	-28	-35	-43	-51	-59	-67	-74	-82	-90	-98
40	2	-6	-14	-22	-29	-37	-45	-53	-61	-69	-77	-85	-93	-101
45	1	-7	-15	-23	-31	-39	-47	-55	-62	-70	-78	-86	-94	-102

Neither the wind chill or heat index is a perfect predictor of the body response to thermal stress. Assumptions made and lack of human experiments (for obvious reasons) in developing these indices as well as changes in the efficiency of body thermoregulation as a function of age and general health make these indices useful but not absolute indicators. Other factors to be considered include protective clothing, activity level, and exposure to solar radiation or external heating. Heat index values may be as much as 10F° warmer if the body is exposed to direct solar heating. Similarly, the wind chill takes no account solar or metabolic heating. Clearly, awareness and prevention are appropriate strategies for dealing with the possible effects of excessive heat or cold in training (or operational) environments.

Table 2. National Weather Service Heat Index [°F]

Ambient Air Temperature [°F]

RH	80	85	90	95	100	105	110	115	120
10	75	80	85	90	95	100	105	111	116
20	77	82	87	93	99	105	112	120	130
30	78	84	90	96	104	113	123	135	148
40	79	86	93	101	110	123	137	151	
50	81	88	96	107	120	135	150		
60	82	90	100	114	132	149			
70	85	93	106	124	144				
80	86	97	113	136					
90	88	102	122						
100	91	108							

Note: RH = Relative Humidity (%).

Bioclimatological Assessment of Basic Training Sites

The military services conduct basic training at a variety of sites throughout the United States. These sites are subject to a variety of weather conditions during the year. Table 3 is a list of basic training sites by service with a nearby representative climatological station.

All but two of the sites are located in humid, subtropical areas. As such, they would tend to have distinct seasons and receive precipitation all months of the year. Humid, subtropical climates (Köppen: Cfa) are noted for hot summers with high humidity and occasional severe heat waves. This is principally due to their location relative to the west side of the subtropical (Bermuda) high-pressure cell, which advects moisture poleward over the region. Winters are mild with relatively few days of freezing temperatures except in the more northerly portions. Summer precipitation is usually confined to afternoon thunderstorms, but the occasional tropical storm or hurricane can bring significant rainfall.

Winter precipitation, associated with midlatitude cyclones, is mixed, with freezing rain and ice storms more common in the southerly portions and snow with occasional blizzard conditions in the north.

Table 3. US Military Basic Training Sites

Service	Basic Training Site	Climatological Site	Köppen Classification
Air Force	Lackland AFB, TX	San Antonio, TX	Cfa
Army	Fort Jackson, SC	Columbia, SC	Cfa
	Fort Knox, KY	Louisville, KY	Cfa
	Fort Leonard Wood, MO	Springfield, MO	Cfa
	Fort Sill, OK	Wichita Falls, TX	Cfa
	Fort Benning, GA	Columbus, GA	Cfa
Coast Guard	Cape May, NJ	Atlantic City, NJ	Cfa
Marine Corps	San Diego, CA	San Diego, CA	Csb
	Parris Island, SC	Charleston, SC	Cfa
Navy	Great Lakes NTC, IL	Milwaukee, WI	Dfb

The remaining two sites are located in humid continental (Dfb) and Mediterranean (Csb) climate types. Humid continental climates are known for marked seasonality with cool summers and cold, windy winters with persistent snow cover. Long spells of below-freezing temperatures resulting from arctic air outbreaks are common. Spring and Fall are transition seasons noted for extreme variability and changeability in temperature conditions. Coastal Mediterranean (Csb) climates are marked by mild, wet winters and warm, dry summers. The presence of cold ocean currents offshore cools the air sufficiently to produce thick coastal fogs and low-lying stratus clouds. The relatively stable conditions are the result of large-scale subsidence associated with the eastern margin of the subtropical (Pacific) high. A notable dry period exists from about April through September, increasing the risk of hazardous brush fires in Summer and Fall.

Table 4 provides temperature information for those climatological stations most representative of conditions at the basic training sites[2]. The months of January, April, July, and October are chosen to represent the range of temperatures encountered over the year. The maximum temperature is the highest temperature recorded during the day. The minimum temperature the lowest. Maximum temperatures tend to occur two to three hours after solar noon, while minimum temperature usually occurs just prior to sunrise. At almost every site, January minimum temperatures are near or below freezing. Thus, nighttime activity must be carefully monitored to guard against frostbite or other cold morbidity. Conversely, July maximum temperatures are near or exceed 90°F at most sites resulting in a need to monitor for heat morbidity. It is important to note that the more southerly sites have July minimum temperatures over 70°F. Heat exposure can have a cumulative effect, especially in cases when high minimum temperatures prevent the body from achieving a level of cooling during nighttime hours. Cold injury is also an important consideration in Milwaukee in January as the maximum temperature is below freezing.

Table 4. Average Maximum and Minimum Temperatures for Selected Cities (°F)

Location	January	April	July	October
San Antonio	61 38	80 58	95 75	82 59
Columbia	55 32	76 49	92 70	76 50
Louisville	40 23	67 45	87 67	69 46
Springfield	42 20	68 44	90 67	70 46
Wichita Falls	52 28	76 50	97 73	77 52
Columbus	56 35	77 52	92 72	78 54
Atlantic City	40 21	61 39	85 65	66 44
San Diego	66 49	68 56	76 66	75 61
Charleston	58 38	76 54	90 73	77 56
Milwaukee	26 12	53 36	80 62	59 42

Table 5 is a listing of average morning and afternoon relative humidity values. Relative humidity is the customary measure of atmospheric moisture but unfortunately it is not the most reliable. Relative humidity is a nonconservative property. Its values change throughout the day without a corresponding change in actual moisture content of the atmosphere. This is because relative humidity measures the percentage of saturation not the actual water vapor content. Both the water vapor content and the air temperature influence saturation. For a particular water vapor amount, the atmosphere will be become saturated at a lower temperature than a higher temperature. Thus typically, relative humidity values are highest in the cooler times of the day. From Table 5 we see that the atmosphere is at about 80% of saturation in the mornings and 50% in the afternoons.

A conservative measure of atmospheric water vapor content is the dew point temperature. The dew point is the temperature to which the air must be cooled (at constant pressure) to achieve saturation. Dew points above 65°F are usually considered uncomfortable. In the summer, all of the stations except Milwaukee and San Diego experience high dew points.

Table 5. Average Morning and Afternoon Relative Humidity for Selected Cities.

City	January	April	July	October
San Antonio	80 59	83 56	87 51	84 53
Columbia	82 54	84 43	89 54	91 51
Louisville	77 64	76 52	85 58	85 55
Springfield	78 62	79 55	87 56	82 54
Wichita Falls	80 57	80 49	78 44	83 51
Columbus	84 59	85 48	89 57	89 52
Atlantic City	78 58	77 52	84 58	88 56
San Diego	71 57	76 59	82 66	77 64
Charleston	83 56	84 49	88 62	88 56
Milwaukee	76 68	78 61	82 61	81 62

Average wind speeds (Table 6) at most of the stations tend to be moderate. Spring and winter tend to have the highest average winds for all stations. Milwaukee and Wichita Falls average winds above 10 MPH in all seasons. All stations have recorded peak 1-minute winds speeds in the 30-60 MPH range.

Table 6. Average Wind Speed (mph) for Selected Cities

Wind Speed By Month (mph)

City	January	April	July	October
San Antonio	9	10	9	8
Columbia	7	8	6	6
Louisville	10	10	7	7
Springfield	12	12	8	10
Wichita Falls	11	13	11	11
Columbus	7	7	6	6
Atlantic City	11	12	8	9
San Diego	6	8	8	6
Charleston	9	10	8	8
Milwaukee	13	13	10	12

Table 7 provides information on the number of days per year for significant weather events. Very cold days are days when the temperature does not rise above freezing. They are quite common in winter in Milwaukee, less frequent in Louisville, Springfield, and Atlantic City, and very rare events in the other cities. On such days outside activity may be impacted. Very hot days are defined as those days in which the temperature exceeds 90°F. Southern cities in Texas, South Carolina, and Georgia can expect impacts on training schedules due to summer heat. Kentucky and Missouri may also be impacted by summer heat waves. Rain days are a fairly common occurrence. Most stations will receive precipitation on 20% or more of the days in the year. Precipitation is uniformly distributed over the year except in San Diego, which has a distinct winter precipitation season, low annual rainfall, and less than 50 rain days per year.

The data presented in the Tables 4-7 represent average conditions. Any particular day or year can exhibit considerable variability and deviation from these averages (Dixon, 1997). Average conditions can be used as a planning guide but actual conditions must be monitored daily to insure the safety of all personnel. Analysis of the heat index and wind chill tables above using the climate data supplied indicate that in general, wind chill may be a significant factor in evening operations during the winter while excessive heat and humidity can impact daytime operations during the summer. The exception to this generalization is San Diego, which benefits from the moderate temperature conditions of a Mediterranean climate.

Table 7. Average Annual Number of Very Cold, Very Hot, and Precipitation Days for Selected Cities

Location	Max T < 32°F	Max T > 90°F	Precip. > 0.01"
San Antonio	1	112	82
Columbia	1	73	110
Louisville	19	31	124
Springfield	18	42	109
Wichita Falls	7	104	71
Columbus	1	76	110
Atlantic City	16	18	112
San Diego	0	4	42
Charleston	1	53	114
Milwaukee	59	9	125

Number of Days in Year

Conclusion

Basic training cannot be separated from the physical environment in which it occurs. Excessive summer heat and humidity or cold, windy winter conditions may impact training operations. Obviously, the military must conduct missions in severe environments when tasked. Basic training gives recruits exposure to physical activity in a variety of environments. The challenge to supervisors is to ensure that training is conducted with a realization of the potential dangers of severe environments and that recruits are trained properly in preventive care for themselves and others in such environments. This awareness for the health of personnel under their command must also be instilled during officer recruit training.

Notes

1. The National Weather Service Office of Meteorology maintains a web page at <tgsv5.nws.noaa.gov/om> which provides additional safety information on hot and cold weather, along with wind chill and heat index charts.

2. These data were abstracted from the 1961-1990 normals for each station. Additional information can be obtained from the National Climatic Data Center <www.ncdc.noaa.gov>.

Bibliography

Dixon, R.W. 1997. A Heat Index Climatology for the Southern United States. *National Weather Digest* 22(1): 16-21.

Driscoll, D.M. 1985. Human Health. In *The Handbook of Applied Meteorology*, ed. D. D. Houghton, pp. 778-814. New York: John Wiley & Sons.

Griffiths, J.F. 1976. *Climate and the Environment.* Boulder: Westview Press.

Morgan, M.D., and Moran, J.M. 1997: *Weather and People.* Upper Saddle River, NJ: Prentice-Hall Inc.

Training a Global Force:
SUSTAINING ARMY LAND FOR 21ST CENTURY READINESS

By

Robert B. Shaw, William W. Doe III, Eugene J. Palka,
and Thomas E. Macia

Introduction

The primary 21st-century mission of the US Army remains the same: fight and win the nation's wars. The Army must therefore be capable of prompt and sustained land combat in all types of terrain and operating environments, across the entire spectrum of conflict (Hooker, 1999).

Intensive and realistic live field training and testing, under conditions that replicate the variety of landscapes and potential threats, are fundamental to Army warfighting readiness. Technological advances in equipment and weapons, and corresponding changes in doctrine and tactics, have dramatically increased the Army's requirements for maneuver space, ranges and munitions impact areas. Concepts for Army transformation have projected a two- to threefold increase in battlespace requirements (Rubenson, et al, 1999). Moreover, proliferating regional and world threats have enlarged the number and geographical extent of potential Army deployments and operations. During 1999 the Army had a presence in 122 countries.

As today's Army rethinks its force structure and global strategy, and responds to a wide range of missions, the need for suitable land to support training and testing remains imperative. The emerging interim brigade combat team (IBCT) reflects a doctrine of tactical maneuver that will require expanded space for training. Nevertheless, the Army's continued requirement for land and the frequent impacts associated with its use invite scrutiny and competition from a variety of external sources, including environmental interest groups, recreationalists, developers, landowners and regulators. If the Army expects its land inventory to underpin its warfighting preparedness, it must demonstrate effective land stewardship and establish a clear link between its land requirements, doctrine and readiness (Bandel, et al, 1996).

Today, the Army is the largest land manager within the Department of Defense (DOD), responsible for approximately 12 million acres of federal land—almost half of the total DOD land inventory. Army installations are geographically distributed throughout the Continental United States, Hawaii and Alaska, representing a variety of landscapes and environmental conditions found throughout the world. Although the Army uses land and training areas overseas, Army lands within the 50 United States represent the major land assets for training and testing. From a readiness perspective, these lands and their associated physical attributes (such as terrain, vegetation and climate), can be viewed as operational analogs for areas where the Army may deploy to fight a major theater war or participate in a military operation other than war (MOOTW).

Battle Settings and Operational Analogs

Military history is replete with examples of how terrain, climate, weather, soil and vegetation have shaped the outcome of major campaigns and battles (Winters, 1999). The ancient Chinese warrior-philosopher, Sun Tzu, cautioned military leaders about the importance of knowing the terrain: *"The terrain is to be assessed in terms of distance, difficulty or ease of travel, dimension and safety . . . the contour of the land is an aid to an army . . . those who do battle knowing these will win, those who do battle without knowing these will lose."* (Cleary, 1991).

In recent Army campaigns, notably Bosnia and Kosovo, unfamiliar terrain and unexpected environmental conditions challenged the Army's ability to perform critical missions, such as crossing major rivers during a flood and flying helicopters at night over precipitous mountainous terrain. When US Army forces stationed in Germany deployed to the Balkans in 1995 during the initial phase of Operation Joint Endeavor, seasonal snowmelt and flooding of the Sava River impeded military bridging operations and delayed the arrival of troops to the operational area in Bosnia. In 1999 Task Force Hawk deployed Apache helicopters from Germany and the United States to staging areas near Kosovo. Observers questioned whether the pilots' training conditions had prepared them to fly missions in the mountainous Balkan terrain.

These few, but very typical examples, taken together with lessons learned in Kuwait, Somalia, Haiti and other operational deployments in the 1990s, illustrate the importance of realistic situational and geographical training. The documented successes of Army maneuver forces during Operation Desert Storm, have been attributed, in large part, to pre-conflict training at the National Training Center (NTC) and other combat training centers (CTCs).

The deserts of Southwest Asia differed distinctly from the environment in the southeastern United States and central Germany, from where major Army forces deployed in 1990 (Clancy and Franks, 1997). However, because the NTC is in California's Mojave Desert, a landscape and environment much like that of Saudi Arabia and Iraq, training there was ideal. General Frederick Franks, VII Corps Commander during the Gulf War, had trained with the 3rd Armored Cavalry Regiment at Fort Bliss, Texas, and was familiar with desert landscapes and navigating forces in such terrain (Clancy and Franks, 1997). Without the experiences of many Army officers and soldiers at the NTC and other US desert training areas, the transition to a desert operating environment may have required longer periods to adapt and train in theater.

Despite the rapid advancements in simulation-based training (virtual and constructive environments), the loss of money and land for realistic training remains a critical concern of many warfighters. Sustaining the Army's diverse land inventory throughout the United States is paramount to allaying these concerns. The Army must argue convincingly that its land base sustains readiness for future contingencies and missions across the operational spectrum.

The Army's Current Land Inventory

Army Active and Reserve Components currently manage more than 100 major Army installations, approximately one-fourth of the military installations in the United States. Over 50 of these contain troop concentrations and land sufficient to support significant training and testing activities. Notable concentrations of major active Army installations follow:

Photograph 1. Opposing force armor negotiates rough terrain during a training operation at the NTC, Fort Irwin, California.
Source: United States Army photograph.

Southeast. Fort Benning, Fort Gordon and Fort Stewart, Georgia; Fort Bragg, North Carolina; Fort Jackson, South Carolina; Fort Polk, Louisiana; and Fort Rucker, Alabama.

Southwest. Fort Bliss and Fort Hood, Texas; Fort Huachuca and Yuma Proving Ground, Arizona; Fort Sill, Oklahoma; and White Sands Missile Range, New Mexico.

West. Fort Carson and Pinon Canyon Maneuver Site, Colorado; Fort Irwin, California; Fort Lewis and Yakima Training Center, Washington.

Alaska. Fort Greely, Fort Richardson and Fort Wainwright.

Hawaii. Schofield Barracks and Pohakuloa Training Area.

The sizes of today's major Army installations vary considerably, ranging from approximately 25,000 contiguous acres to as many as two million contiguous acres. The largest Army installations with land available for training and testing are found in the Southwest and West: Fort Bliss and White Sands Missile Range are separate installations joined by a common boundary, comprising approximately 3.2 million acres; and Yuma Proving Ground is a weapon, equipment and vehicle test site, comprising approximately one million acres. The Army's largest installation dedicated to large-scale, mechanized, force-on-force exercises is Fort Irwin, covering approximately 643,000 acres.

Strategic Analysis of Army Lands

In a recent strategic-level inventory of its installation range and training land capacity, the Office of the Deputy Chief of Staff for Operations and Plans, Department of Army, selected 31 major active Army installations for analysis (Doe, et al, 1999). This analysis, the Installation Training Capacity (ITC) study, was to objectively catalog and assess the Army's existing live training assets to improve input toward future land-use decisions. The 31 installations considered represented approximately 86 percent (10.35 million acres) of the Army-controlled lands within the United States. One important criterion the ITC study considered was the operational-analog value of each installation.

To make this analog assessment, the 31 installations were superimposed on a map delineating ecological boundaries as described by Bailey's Ecoregional Classification System. Bailey's scheme is based on the concept of an ecoregion, a contiguous areal extent defined by climate and vegetation, and exhibiting a unique mix of landforms, soil, flora, fauna and ecological succession (Bailey, 1998). Previous studies have used this landscape classification methodology to compare ecological diversity on Army and other federal lands (Leslie, et a, 1996). The classification is well documented, defendable, widely accepted and global in scale. Thus, it enables comparison of Army training and testing lands throughout the United States with regional areas where Army forces may deploy.

Bailey's Ecological Classification of Army Lands

Bailey's Ecological Classification is a fourth-order, hierarchical system. The four levels of classification are domain, division, province and section. The geographic boundaries for domains (groups of ecoregions with related climates) are based upon the broad climatic zones of the earth. The four domains are polar, humid temperate, dry and humid tropical. Divisions are subunits of a domain determined by isolating areas of definite vegetative affinities within the same regional climate. There are fifteen divisions globally: icecap, tundra, subarctic, warm continental, hot continental, marine, prairie, Mediterranean, tropical/subtropical steppe, tropical/subtropical desert, temperate steppe, temperate desert, savanna and rainforest. The province level corresponds to broad vegetation regions, while sections are based on broad land-surface forms. Bailey's classification is only complete to the domain and division level at the global scale; thus, the comparison of Army lands on a global scale was limited to the two higher classification levels.

The polar domain is restricted to the northern fringes of the North American continent as well as northern Europe and Asia and the polar icecaps. It occupies approximately 26 percent of the earth's land area. Two major divisions have been defined: the tundra division (4.5%) where the average temperature of the warmest month lies between 10 degrees Celsius (C) and 0 degree C; and the subarctic division (9%), where only one month of the year has an average annual temperature above 10 degrees C. The remaining 12.5 percent of the polar domain is in icecaps. None of the 31 major active installations resides within the tundra division, but the three Alaskan installations are found within the subarctic. These installations account for approximately 1,639,000 acres, or 16 percent, of the total Army lands in the survey.

The humid temperate domain occurs at mid-latitudes (30 to 60 degrees north and south latitude) and generally consists of broad-leaved and coniferous forests. This domain covers over 15 percent of the earth's land area and is concentrated in eastern North America, Central Europe, southern China, Uruguay and adjoining parts of Argentina and Brazil, coastal southeastern Australia and New Zealand. The domain is separated into six divisions based on winter and summer temperatures. The warm continental division (1.4 percent) has very cold winters and warm summers. None of the surveyed installations resides within this division. The hot continental division (1.4 percent) has cold winters but hot summers. The installations in this division are: Fort Campbell and Fort Knox, Kentucky; Fort Drum, New York; and Fort Leonard Wood, Missouri. These installations comprise over 454,000 acres. The subtropical division (3.5 percent) is characterized by mild winters and hot summers. This division is one of the more highly represented with nine installations falling within the subtropics: Aberdeen Proving Ground, Maryland; Fort A.P. Hill, Virginia; Fort Benning, Fort Gordon and Fort Stewart; Fort Bragg; Fort Jackson; Fort Polk; and Fort

Rucker. The total area of the installations within this division is about 1,144,000 acres. The marine division (2.4 percent) is a region with mild winters and cool

Photograph 2. Snowshoed soldiers train for cold-weather operations, an Alaskan specialty, at Fort Greeley.
Source: United States Army photograph.

summers. Fort Lewis, Washington, is the only installation in this division and totals about 87,000 acres. The prairie division (1.5 percent) is generally a transitional area between humid and dry climates and could be classified as sub-humid. It generally is too dry for tree growth (except in riparian areas) but too wet to be classified as arid. Fort Riley, Kansas, is the only installation located in this division and totals 101,000 acres. The Mediterranean division (1.8 percent) is characterized by dry summers and warm winters. None of the installations studied was classified in this division. In total, the selected Army lands in the humid temperate regime account for about 17 percent of the total Army lands assessed.

The dry domain encompasses arid and semiarid areas of the mid-latitudes and covers 32 percent of the earth's land surface. Most of western and southwestern North America, northern Africa, the Middle East, Central Asia and most of interior Australia are located within this division. The dry domain can be segregated into very arid areas (deserts) or semiarid areas (steppes) which separate the arid regions from more humid climates. Four major divisions are recognized within the dry domain based on aridity and temperature. The tropical/subtropical steppe division (11 percent) is a large semiarid area that typically borders tropical deserts to the north and south. Fort Sill is the only installation studied that occurs within this division and totals 94,000 acres. The tropical/subtropical desert division (15 percent) is characterized by extremely arid conditions with high air and soil temperatures. Fort Bliss, Fort Huachuca, Fort Hood, Fort Irwin, White Sands Missile Range and Yuma Proving Ground, are in this division. Together, these installations occupy nearly five million acres. The temperate steppe division (2 percent) has a semiarid continental climate with warm summers and cold winters. Fort Carson and Pinon Canyon Maneuver Site cover a combined 373,000 acres and are the only installations in this division. The temperate desert division (4 percent) is arid with hot summers and cold winters. Dugway Proving Ground, Utah, and Yakima Training Center cover about 1,120,000 acres, and are in this division. In all, the dry domain lands in the Army active installation inventory compose about 64 percent of the total.

The humid tropical domain covers about 27 percent of the earth's land surface area and is characterized by a hot and humid climate. Every month of the year has an average temperature above 18 degrees C and there is no winter season. This area is equatorial and lies primarily between the Tropic of Cancer and the Tropic of Capricorn. The savanna division (17 percent) has a distinct wet and dry season that supports open tall grasslands with drought-tolerant trees and shrubs. No US installations represent this division. The rainforest division (10 percent) has a wet equatorial climate with no distinct dry season and occurs between the equator and 10 degrees latitude. Pohakuloa Training Area and Schofield Barracks are included within this division and cover about 165,000 acres. Overall, less than 2 percent of the Army training lands within the 31 installations are located within the humid tropical domain.

The 31 major active installations in this analysis represent approximately 60 percent of the earth's land surface area by ecoregion. The humid temperate and dry domains are extremely well represented by the lands within the US Army's control. Conversely, nearly 40 percent of the earth's land surface area is not represented by the installations studied. Notably, a large portion of the polar domain and most of the humid tropical domain are underrepresented. The polar icecaps (9.6 percent of the globe) are considered insignificant because the probability of a conflict in this region is relatively small.

However, those areas of the humid tropical domain are of great importance. The savanna division (17 percent of the global land surface) is not represented at all by the 31 installations. This division includes large areas in central Africa, central South America, the Indian Peninsula and northern Australia. Among Army training lands, the rain forest division is also underrepresented. While the two Hawaii installations are classified in the rain forest division, neither is truly indicative of a multilayered humid tropical rain forest. Schofield Barracks does have some small areas that receive significant amounts of precipitation, but the area does not have high enough temperatures to be classified as a tropical rain forest.

Photograph 3. Although the Pohakuloa Training Area and Schofield Barracks in Hawaii are classified in the rain forest division, neither is truly indicative of a multi-layered, humid, tropical rain forest.
Source: Pacific Command photograph.

Thus, the areas of Central Africa and South America, Central America, the Caribbean, Southeast Asia and Indonesia are not well correlated with any US Army training and testing land resource. The recent closing of the US Army Jungle Warfare School in Panama—45,000 acres of relatively undisturbed tropical rainforest—exacerbates the critical lack of training and testing areas within the rainforest division. The Mediterranean division (2 percent of the globe's terrestrial surface) also is not represented. Several military installations located in the western United States exhibit a Mediterranean climate but were not a part of the analysis—Camp Pendleton, California is a US Marine Corps installation and Fort Hunter-Liggitt, California, is a US Army Reserve installation.

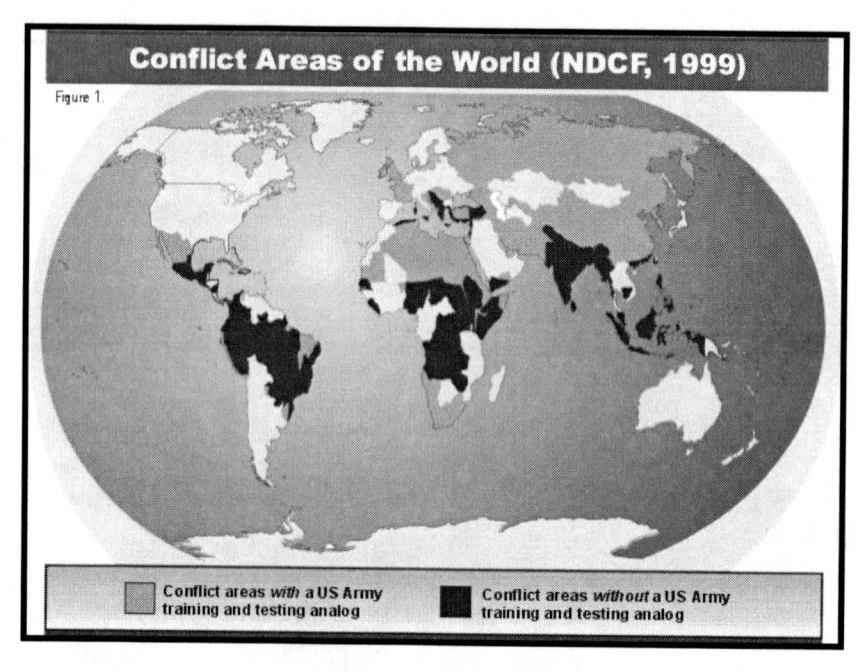

Map 1. Distribution of Areas of Conflict.

Since the mid-1980s, several regional conflicts have involved the Army. Map 1 illustrates the geographical extent and diversity of the Army's recent operating environment and identifies analogs among installations.

The Army's land inventory was adequate to prepare for conflict in temperate and dry areas, particularly those which support a desert- or continental-type climate. Conversely, the land inventory is inadequate to train personnel and test equipment in areas of conflict, which are represented by the savanna, rain forest and Mediterranean ecoregions.

In a recent report, the National Defense Council Foundation identified existing and new major conflicts areas throughout the world. The report identifies 193 nations embroiled in conflict, nearly twice the Cold War level (Raum, 1999). Comparing these conflict areas to the current Army land inventory yields obvious conclusions. Numerous Army installations reside within the temperate climates and are therefore geographically similar to major conflict areas such as Russia, Kazakhstan, Georgia, Turkey, Afghanistan, China and Korea. Areas of warm and cold deserts—such as Iraq, Iran, Sudan, Egypt and Algeria—where conflicts are occurring, are also well represented by the Army installations studied.

However, this comparison also reveals a significant lack of adequate training land to represent conflict areas in Mediterranean and tropical environments. As shown in Figure 1, the Mediterranean region is predicted to remain unstable with potential conflicts in Morocco, the Balkans, Lebanon and Israel. Similarly, the environments of conflict areas in South America (Columbia and Peru), Africa (Sierra Leone, Liberia, Nigeria, Kenya, Rwanda, Burundi, Angola and the Democratic Republic of Congo), southern Asia (India, Sri Lanka and Bangladesh), Southeast Asia (Cambodia) and Macronesia (Philippines, Indonesia and Papua New Guinea), are underrepresented in the Army land inventory. From this assessment, it is clear that the Army faces tremendous challenges to prepare for operations in a variety of potential conflict areas, with vastly different landscapes.

Nearly 50 years ago, Military Review published a two-part article that emphasized the need for professional soldiers to understand fundamental geographical regions to calculate their impacts on operations and logistics during global war (Forde, 1949). The author, Harold Forde, recognized that each environment presented different considerations for military operations, and categorized the earth's land surface into eight distinct groups: dry lands; tropical forests; Mediterranean scrub forests; mid-latitude mixed forests; grasslands; Boreal forest lands; polar lands; and mountain lands.

Forde's message was clearly based on the US Army's World War II deployments to all of the regions he identified. His insights may be even more significant in the 21st century. While current and future threats do not suggest fighting two major theater wars simultaneously, the current military strategy demands preparing for the worst case. Moreover, since the end of the Cold War, the US military has been increasingly involved in regional conflicts and MOOTW. The greater tendency to deploy soldiers and units requires that they be exceptionally trained and equipped for a wide range of short-notice missions throughout the world. While these deployments are smaller than those during World War II, the diversity of operational landscapes is much broader.

US Army history reveals a synergistic relationship between training to fight in varied operating environments and success once deployed to particular regions. Despite their deficiencies, Army training and testing lands throughout the 50 states are a precious resource fundamental to mission accomplishment. These lands represent analogs to potential areas of conflict where the Army may be deployed. The Army must therefore retain and sustain these essential land resources so that soldiers can train as they will fight.

ಶಿ ⚜

Editor's notes: This chapter was adapted from: Shaw, R.B., Doe, W.W., Palka, E.J, and Macia, T.E., 2000. "Sustaining Army Land for Readiness in the 21st Century." *Military Review,* 5: 68-77. Reprinted by permission.

Notes:

1. Due to the scale of Bailey's maps, the Pohakuloa Training Area falls within the rainforest division. At a finer resolution the installation would be classified as tropical/subtropical steppe, since it is situated within the rainshadow (leeward side) of mountains on the island).

Bibliography

Bailey, Robert G. 1998. *Ecoregions: The Ecosystem Geography of the Oceans and Continents.* New York, New York: Springer-Verlag.

Bandel, Don, et al. 1996. *Land for Combat Training: A Briefing Book,* Atlanta, Georgia: United States Army Environmental Center Report AEPI-IFP-96.

Clancy, Tom and Franks, Fred Jr. 1997. *Into the Storm: A Study in Command*. New York, New York: G.P. Putnam's Sons.

Cleary, Thomas. 1991. *Sun Tzu: The Art of War*. Boston, Massachusetts: Shambhala Publications, Inc.

Doe, William, et al. 1999. "Locations and Environments of US Army Training and Testing Lands: An Ecoregional Framework for Assessment," *Federal Facilities Environmental Journal*, Autumn: 9-26.

Doe, William. 1997. *Terrain and Ecoregion Analysis of Select US Army Installations*. Fort Collins, Colorado: Colorado State University, Center for Ecological Management of Military Lands Report TPS 97-17.

Forde, Harold. 1949. "An Introduction to Military Geography: Part I and Part II," *Military Review*, February 1949: 26-36; and March 1949: 52-62.

Hooker, Richard D. 1999. "The Role of the Army in the Common Defense: A 21st Century Perspective," Landpower Essay Series, No. 99-4, Arlington, VA: Association of the US Army Institute of Land Warfare, April.

Leslie, M., et al. 1996. *Conserving Biodiversity on Military Lands: A Handbook for Natural Resources Managers*. Arlington, Virginia: The Nature Conservancy.

Raum, Tom, 1999. "65 Conflicts Rage Around the World," *The Coloradoan*, Fort Collins, Colorado, December 30.

Rubenson, David, et al. 1999. *Does the Army Have a National Land Use Strategy?* Santa Monica, California: A: RAND Arroyo Center, Report MR-1064-A.

Winters, Harold. 1999. *Battling the Elements*. Baltimore, Maryland: Johns Hopkins University Press.

Measuring vegetative cover in a military training area.
Source: Center for Ecological Management of Military Lands (CEMML), Colorado State University, Fort Collins, Colorado photograph.

16

U.S. Army Training and Testing Lands:
An Ecoregional Framework for Assessment

By

William W. Doe III, Robert B. Shaw, Robert G. Bailey,
David S. Jones and Thomas E. Macia

Introduction

The U.S. military, represented at the federal agency level by the Department of Defense (DoD), has a unique need for land, airspace and ocean waters to provide realistic training and testing environments to meet national defense readiness goals. Of the four military Services (Army, Air Force, Navy and Marine Corps), the Army is the largest land steward with management responsibility for approximately 12 million acres of land, or almost half of the total lands under the control of the Department of Defense. These lands are used to train Army units and test Army equipment and weapons to ensure warfighting readiness. The Army's readiness posture is predicated upon a "train as we fight" doctrine that requires contiguous space and a variety of landscapes and climatic conditions to train for the eventuality of operations in all regions of the world.

In his recent Senate confirmation testimony, General Eric Shinseki, the new Army Chief of Staff, stated that his first priority would be to produce trained and ready forces to support the Theater-level requirements of the joint Service

Commanders in Chief (CINCs) for current and future missions. In order to prepare for these missions the Army trains in a variety of ways, to include mechanized and non-mechanized cross-country maneuvers, artillery and weapons firing, vehicle driving and testing, demolitions, small-arms range firing, and numerous other individual and unit-level tasks.

Land management policies and practices that sustain the integrity and vitality of Army lands are paramount because they ensure a realistic natural setting for training and testing activities. Degradation of land quality can directly effect the readiness of Army forces by denying units and their soldiers the range of environmental conditions normally encountered in military operations. Examples of this degradation include: loss of camouflage resources; reduction in cross-country trafficability due to rutting and gullying; and an increase in safety hazards to troops. Degradation of land quality can also increase rehabilitation and maintenance costs. If the land quality degrades beyond recovery and repair it may become necessary to defer its use, diminishing precious land assets available to support mission readiness.

Frequent and intensive training and testing on Army lands affects the natural environment. The physical impacts on the landscape associated with these activities have been studied and well documented, and include damage and loss of vegetation and forest resources, compaction of soil, and rutting of the surface (Shaw and Kowalski, 1996). If unmitigated, these impacts can result in increased soil erosion, atmospheric pollution and sedimentation of water features. Additionally, valuable habitat for the abundant wildlife and flora existing on Army lands, and adjacent lands, may be impacted through changes in ecosystem structure and functions.

Finally, as a federal land steward, the Army must comply with all federal and state environmental laws and directives concerning protection of its natural resources. Although its primary mission, unlike other federal natural resources management agencies, is national defense, the Army must also sustain the land and natural resources under its care. Since the early 1990's the Army, and its sister services, have followed the Department of Defense lead in implementing ecosystem management as the guiding principle for conservation and protection of these resources (Leslie et al., 1996).

The Army has instituted the Integrated Training Area Management (ITAM) program to address these concerns and mitigate the undesirable impacts of training and testing. This program, under the proponency of the Deputy Chief of Staff for Operations (DCSOPS), Department of the Army, is a $35 million annual investment in four functional areas: 1) Land Condition Trend Analysis (LCTA), 2) Land Rehabilitation and Maintenance (LRAM), 3) Environmental Awareness (EA), and 4) Training Requirements Integration (TRI) (Department of the Army, 1998). Since 1995, the ITAM program has been directly supported at installations by operational funds allocated through the Army budget.

One of the important land management concepts embedded in the ITAM program is land "carrying capacity". Derived from ecosystem science principles, "carrying capacity" is the amount of cumulative land use, or load (usually referring to a particular type of use such as livestock grazing, recreation, or military training) that a given parcel of land can accommodate in a sustainable manner. Carrying capacity is a complex, integrated variable that is a function of the inherent site characteristics (e.g., soil, slope, aspect, climate) and biological regime (e.g., flora, fauna, vegetation community, structure and composition) of the natural environment. It can be quantified by scientific observation, experimentation, and measurement, and estimated using professional judgment. Type, intensity and frequency, based upon established military doctrine can quantify load or land use (e.g., military training). For the Army, land carrying capacity is defined as the amount of training and testing which a given parcel of land can accommodate over time in a sustainable manner.

Land use at or below the carrying capacity allows the landscape to recover naturally over time. When the amount of use (load) placed on the natural system exceeds the carrying capacity, a critical threshold is reached and accelerated degradation, or ecosystem change, may occur. Human intervention, in the form of land maintenance and rehabilitation, is essential to prevent these thresholds from being crossed. These practices may include re-seeding of damaged areas, planting of trees and shrubs, structural improvements to streams, and other common land rehabilitation and maintenance solutions. The ITAM program goal is to sustain the long-term capacity on Army training and testing lands through a balance of usage, land condition trend monitoring, and land maintenance/repair.

The application of carrying capacity and sustainability principles to Army land management is a complex undertaking that incorporates training requirements and resource conservation. It requires extensive data collection and measurement to understand how key ecosystem indicators respond to disturbance at various scales and to identify critical thresholds for the natural system. Before these principles can be transferred and understood agency-wide it is useful to consider a strategic framework for assessing ecosystem condition across multiple locations and environments. A broader perspective can provide a foundation for regional and smaller scale (Army installation or training area) land management initiatives and decisions.

This approach was taken to assess the ecological diversity and potential carrying capacity of 31 major active Army installations in the United States. The objective of this assessment was to provide Army decision-makers with an integrated context for addressing current and future training and testing requirements in concert with ecosystem management goals.

The Army's Land

A Historical Perspective

The Army's role as a federal land manager is rooted in its historical mission to provide national defense for its citizens, both at home and abroad. The need for land to house troops, maintain and test equipment, and train soldiers and units has grown with the evolution of a standing Army and the periodic demands for more soldiers during times of crisis. Over the last fifty years, advances in technology have rapidly increased firepower, lethality and mobility. These trends have changed warfighting doctrine and increased the need for more dedicated land, both from a training and safety perspective.

Army lands were established as military outposts across the frontier as the nation expanded westward. These were generally small garrisons designed to house cavalry and other units who protected settlers as they traversed and settled the Great Plains and Western territories during the Indian Wars. By 1911, forty-nine of these small posts were still in use. Many of the larger lands that were initially ceded to the Army evolved from the establishment of camps and posts for mobilization in 1917, in preparation for the nation's entry into World War I. Thirty-two Army and National Guard cantonments were opened during this period, many of which today remain active Army installations. These include Forts Jackson, Lewis, Benning, Bragg and Knox (Department of the Army, 1998).

The Army expanded its land holdings dramatically in the early 1940's as the nation prepared for, and entered into World War II. The Army added forty new camps for Division training centers, ranging from 25,000 to 100,000 acres apiece. Forts Campbell, Carson, Gordon, Greeley, Hood, Irwin, Polk, Rucker, Stewart and Leonard Wood, all active Army installations today, were established in this manner. By the end of World War II, the War Department (the predecessor to the Department of Defense) had responsibility for over 46 million acres – almost twice the DoD land inventory today. Additionally, the Army acquired maneuver rights to extensive land areas in Texas, Louisiana and California for the conduct of Corps-on-Corps maneuvers and tank training. The largest maneuver area, called the Desert Training Center, consisted of 56 million acres in California and Arizona (Department of the Army, 1998).

After World War II, the Army relinquished some maneuver areas and posts, while others were expanded to fulfill new requirements for training and testing. The preponderance of land holdings remained in the West, with over seventy percent of the Army's land in the mid-1980's located within ten western States (Alaska, Arizona, California, Colorado, Nevada, New Mexico, Oregon, Utah, Washington and Wyoming) (Cawley and Lawrence, 1998).

A Modern Day Perspective

The Army's current land inventory is estimated at 12 million acres and is distributed throughout the continental United States, Hawaii and Alaska. It includes additional major training areas overseas in Korea and Germany that are utilized under Status of Forces Agreements with host nation allies. The majority of these lands are components of Army installations (more commonly called "posts"). The installation is akin to a municipality that provides the infrastructure and services for the soldiers and their families stationed there. A typical Army installation consists of the cantonment area (barracks, motor parks, support facilities, housing, etc.) and the adjoining training or testing areas where operational exercises (maneuvers, weapons firing, tactical vehicle driving, etc.) occur. Since the end of the Cold War, many forces that were previously stationed in forward areas on the European continent have been returned to the United States. However, the Army continues to face increasing worldwide missions, making installations the "power projection platforms" from which forces are deployed by sea and air to operational areas. As these operational requirements increase, the availability and variability of sustainable land on which to train becomes even more important.

There are more than one hundred Army installations under control of the Active Army, Reserves and Army National Guard, representing over one-third of the total military installations in the United States. Because of their size and the intensity of training and testing activities, 111 Army installations currently require Integrated Training Area Management (ITAM) implementation. Of that number, over fifty contain troop concentrations and land sufficient to support significant training and testing activities. Notable concentrations of major active Army installations exist in the Southeast (Fort Benning, Fort Bragg, Fort Gordon, Fort Jackson, Fort Polk, Fort Rucker, Fort Stewart), the Southwest (Fort Bliss, Fort Hood, Fort Huachuca, Fort Sill, Yuma Proving Ground, White Sands Missile Range), and the West (Fort Carson, Fort Hunter Liggett, Fort Irwin, Fort Lewis, Orchard Training Area, Yakima Training Center). Additionally, the Army has three major installations (Fort Greely, Fort Richardson, Fort Wainwright) in Alaska and two major installations (Schofield Barracks and Pohakuloa Training Area) in Hawaii.

The size of today's major Army installations varies considerably, ranging from about 25,000 contiguous acres (100 km^2) to as many as 2 million contiguous acres (8,500 km^2). The largest Army installations with land available for training and testing are found in the Southwest and Western regions of the country. These include Fort Bliss, Texas, and White Sands Missile Range, New Mexico (separate installations joined by a common boundary), comprising approximately 3.2 million acres, and Yuma Proving Ground, AZ, a weapons, equipment and vehicle test site in the desert, comprising approximately one million acres. The Army's largest installation dedicated to large-scale, mechanized, force-on-force exercises is Fort Irwin, CA, called the Army's National Training Center (NTC). Fort Irwin is located in the Mojave Desert, near the former Desert Training Center where General George S. Patton conducted the first large-scale tank maneuvers in 1941-42. Fort Irwin comprises approximately 650,000 acres of land that the Army is interested in expanding by an additional 150,000 acres for future use (Gorman, 1999).

Approximately 1.5 million acres of the current Army land inventory consists of "withdrawn lands", found primarily in New Mexico and Alaska. The majority of these are federal lands that were withdrawn for military use in 1986, for a period of fifteen years, from the Bureau of Land Management (BLM) under Congressional Act (PL 99-606). The Army's withdrawn lands are subject to renewal by Congress in 2001 and are currently undergoing Environmental Impact Statements in accordance with the requirements of the National Environmental Policy Act (Rubenson et al., 1998).

Army Land Requirements in the 21st Century

As the Army approaches the 21st century, its use of existing lands and need for additional lands is under increased scrutiny from a variety of external groups and regulators (Bandel et al., 1996). This is evidenced by the ongoing efforts to acquire additional lands in the Mojave Desert and requests to renew "withdrawn lands" in Alaska and Texas. If the Army expects to retain its current inventory, and acquire new training and testing lands as needed, it must continue to demonstrate effective land stewardship and establish a clear link between doctrine, land requirements and readiness (Bandel et al., 1996). It must make rational decisions that will defendable to a wide audience.

In a recent strategic-level analysis of its installations and land requirements, the Department of Army selected thirty-one active Army installations where major training and testing activities occur. These major installations, with their respective land areas, represent approximately 86% (10.35 million acres) of the total lands in the fifty States that are controlled by the Army (Table 1).

These lands vary, not only in geographical distribution and scale, but in their ecological settings as well. In Map 1, the locations of these 31 installations are superimposed on a map delineating the ecological boundaries described Bailey's Ecoregional Classification System. This map provides the basis for further description and analysis of Army lands from an ecological perspective.

Bailey's Ecoregional Classification System

Many efforts have been undertaken to classify and map ecological conditions within the United States at a regional scale. One of the most widely used by federal agencies is Bailey's "Ecoregions of the United States" (Bailey, 1996). Bailey's classification scheme was developed and adopted by the U.S. Forest Service for ecosystem management of its lands, and was used by the U.S. Fish and Wildlife Service for the National Wetlands Inventory. Bailey's classification has also been applied to studies concerning Department of Defense (DoD) lands. The Department of Defense Biodiversity Initiative adopted this classification scheme in comparing the biological diversity of DoD lands to those of other federal agencies (Bailey, 1996).

Table 1. List Of 31 Major Active Army Installations in the United States with Total Acreage*

INSTALLATION NAME	STATE	ACRES (IN THOUSANDS)
TOTALS		**10350**
Fort Jackson	SC	52
Fort Gordon	GA	55
Schofield Barracks	HI	56
Fort Rucker	AL	58
Fort Leonard Wood	MO	63
Fort Richardson	AK	63
Aberdeen Proving Ground	MD	73
Fort A.P. Hill	VA	76
Fort Lewis	WA	87
Fort Sill	OK	94
Fort Riley	KS	101
Fort Huachuca	AZ	103
Fort Campbell	KY	104
Fort Drum	NY	107
Pohakuloa Training Area	HI	109
Fort Carson	CO	137
Fort Bragg	NC	171
Fort Knox	KY	180
Fort Benning	GA	182
Fort Polk	LA	198
Fort Hood	TX	218
Pinon Canyon Maneuver Site	CO	236
Fort Stewart	GA	279
Yakima Training Center	WA	324
Fort Irwin	CA	643
Fort Greely	AK	661
Dugway Proving Ground	UT	799
Fort Wainwright	AK	915
Fort Bliss	TX	916
Yuma Proving Ground	AZ	1009
White Sands Missile Range	NM	2281

* NOTE: Acreage figures are total installation acres, including cantonment and other areas not directly available for training and testing. Figures are derived from several database sources and may vary depending on recent land surveys or land acquisitions.

Bailey's ecoregion classification is a fourth-order, hierarchical system that can be applied at various scales and mapping levels. The boundaries of various regions are not absolute, but rather represent broad transitional zones. The four levels of classification and their total number of categories within the United States are:

- Domain (4)
- Division (14)
- Province (52)
- Section (164)

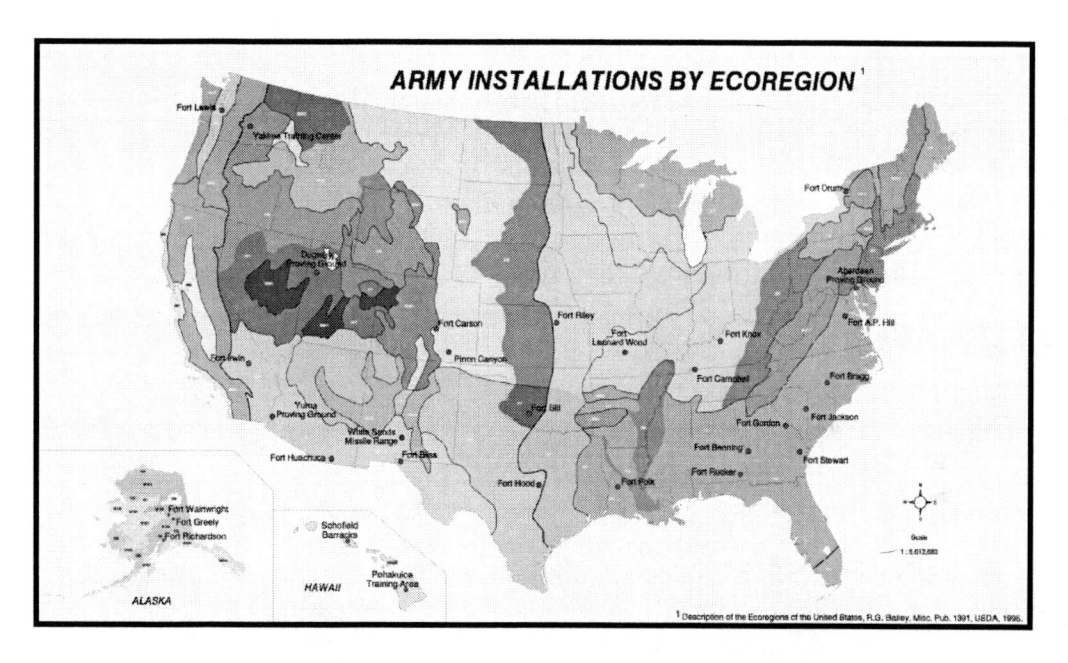

Map 1. Geographical distribution of major Army installations in the U.S., depicted on map showing ecosystem boundaries, as described by Bailey's Ecoregional Classification System.

The geographic boundaries for Domains, and within them Divisions, are based upon the broad climatic zones on the Earth. Climate subzones form the basis for subdividing Divisions into Provinces, which correspond to major plant regimes. Mountain provinces (delineated by the prefix "M" in the classification nomenclature) are those exhibiting altitudinal zonation and the climatic regime of the adjacent lowlands. Because topography exerts the major control over climate at finer scales, landforms are used as the basis for identifying sub-Province boundaries, called Sections. Eleven landscape elements are used to define each

Section. These elements include: geomorphology, lithology and stratigraphy, soil taxa, potential natural vegetation, fauna, climate, surface water characteristics, disturbance regimes, land use and cultural ecology (McNab and Avers, 1994).

Bailey's system has been applied globally (Bailey, 1995). It allows for comparison of Army training land settings in the United States with foreign areas where the Army may be deployed operationally. A recent analysis using this approach reported that the current Army land inventory provides significant training and testing lands in the Hot Continental, Sub-tropical, Temperate and Subarctic Divisions, but lacks locations in the Rainforest, Savanna and Mediterranean Divisions (Shaw and Doe, 1999).

Ecoregional Classification of Army Lands

As previously shown in Map 1, the locations of 31 major active Army installations were superimposed on a map delineating the boundaries of the 52 Bailey's Provinces within the United States. The Domain, Division, Province and Section for each installation were identified.

The classification of installations at the Domain level resulted in the following distribution:
- Humid Temperate (15)
- Humid Tropical (2)
- Dry (11)
- Polar (3)

The classification of installations at the Division level resulted in the following distribution:
- Sub-tropical (9)
- Sub-tropical Steppe (2)
- Temperate Steppe/Desert (4)
- Sub-tropical Desert (5)
- Hot Continental (4)
- Prairie (1)
- Subarctic (3)
- Rainforest (2)
- Marine (1)

At the Division classification level some useful distinctions between installation training lands become apparent. For example, there is an abundance of installations in the Sub-tropical Division, while other Divisions (Sub-tropical Steppe, Prairie, Subarctic, Rainforest and Marine) contain only one to three installations within the category (Shaw and Doe, 1999).

The Province level of classification further defines the ecological diversity of Army lands. As shown in Table 2, there are only a few installations within the same Province. This suggests that the Province-level may be the most appropriate classification for comparison and assessment of installation settings. The classification of the 31 Army installations by Province level is:

- As shown in Table 2 there are 22 unique Sections represented by the 31 installations. Three installations occupy the same Section (Forts Bragg, Jackson and Rucker in the Coastal Plains and Flatwoods, Lower Section,) while there are six pairs of installations within the same Section. All other installations occupy a unique Section within Bailey's classification scheme.
  - Great Plains Steppe and Shrub (1)
  - Southwest Plateau and Plains (1)
  - Pacific Lowland Mixed Forest (1)
  - Prairie Parkland (1)
  - Coastal Trough - Humid Taiga (1)
  - Upper Yukon Taiga-Meadow (1)
  - Alaska Range Tundra-Meadow (1)
  - Great Plains-Palouse Dry Steppe (2)
  - Intermountain Semi-desert (2)
  - American Semi-desert and Desert (2)
  - Hawaiian Islands (2)
  - Chihuahuan (3)
  - Southeastern Mixed Forest (3)
  - Eastern Broadleaf Forest (4)
  - Outer Coastal Plain - Mixed Forest (6)

Land Resiliency Concepts

Ecosystems have varying abilities in terms of resistance and resilience to both natural and human-induced disturbance. Resistance is the ability of a natural system or population to remain essentially unchanged despite disturbances or stressors (e.g., how sensitive is it?). Resilience is the ability of a system to recover after disturbance and return to its original state (Holling, 1973). These properties influence stability and persistence of plant populations, communities, and ecosystems over time. Resiliency is a function of several interrelated physical and climatic factors to include precipitation, vegetative cover type, growing season, seasonality, soil type and depth and topography. It is often expressed as a recovery period, defined as the time it takes a perturbed system to return to its former state after it has been perturbed and displaced from that state (Holling, 1973).

Conceptually, resiliency can be illustrated by the use of the "cup and ball" analogy as shown in Figure 1. Figure 1a represents a community that is both locally and globally stable, returning to its original configuration after all disturbances. This model represents the "climax" or "successional" concept commonly used in ecology. Figure 1b illustrates a community that has multiple stable points or steady states; some may be more desirable than others from a management perspective.

The depth of a trough represents the relative strength of the local stability or the energy or strength of disturbance required to force the community across a threshold into another trough (stable state) (Laycock, 1991). Many communities in North America with multiple steady states are greatly influenced by changes in fire frequency, introduction of aggressive non-native species, or other stresses such as grazing.

In either scenario, if the system is only perturbed enough to move partially up the slope it will eventually return downslope to the trough (as if under the influence of gravity). Theoretically, once the system reaches a new state it will remain in that state until additional stresses push it into another state (ever further from the original state), or significant energetic inputs (e.g., restoration activities) return it to its original state. Both cases can represent human-induced ecological change. Although the real world is much more complex than this simple analogy, these concepts are useful in explaining the interrelationships between land use and ecological response.

U.S. Army Training and Testing Lands

Table 2. Bailey's Ecoregional Classification of 31 Major Army Installations

ARMY INSTALLATION	DOMAIN	DIVISION	PROVINCE	SECTION
Aberdeen Proving Ground, MD	Humid Temperate	Sub-tropical	Outer Coastal Plain Mixed Forest	Middle Atlantic Coastal Plain
Fort A.P. Hill, VA	Humid Temperate	Sub-tropical	Outer Coastal Plain Mixed Forest	Middle Atlantic Coastal Plain
Fort Benning, GA	Humid Temperate	Sub-tropical	Southeastern Mixed Forest	Southern Appalachian Piedmont
Fort Bragg, NC	Humid Temperate	Sub-tropical	Southeastern Mixed Forest	Coastal Plains and Flatwoods, Lower
Fort Campbell, KY	Humid Temperate	Hot Continental	Eastern Broadleaf Forest	Interior Low Plateau, Highland Rim
Fort Drum, NY	Humid Temperate	Hot Continental	Eastern Broadleaf Forest	Erie and Ontario Lake Plain
Fort Gordon, GA	Humid Temperate	Sub-tropical	Southeastern Mixed Forest	Southern Appalachian Piedmont
Fort Jackson, SC	Humid Temperate	Sub-tropical	Outer Coastal Plain Mixed Forest	Coastal Plains and Flatwoods, Lower
Fort Knox, KY	Humid Temperate	Hot Continental	Eastern Broadleaf Forest	Interior Low Plateau, Highland Rim
Fort Leonard Wood, MO	Humid Temperate	Hot Continental	Eastern Broadleaf Forest	Ozark Highlands
Fort Lewis, WA	Humid Temperate	Marine	Pacific Lowland Mixed Forest	Willamette Valley and Puget Trough
Fort Polk, VA	Humid Temperate	Sub-tropical	Outer Coastal Plain Mixed Forest	Coastal Plains and Flatwoods, Western Gulf
Fort Riley, KS	Humid Temperate	Prairie	Prairie Parkland	Flint Hills
Fort Rucker, AL	Humid Temperate	Sub-tropical	Outer Coastal Plain Mixed Forest	Coastal Plains and Flatwoods, Lower
Fort Stewart, GA	Humid Temperate	Sub-tropical	Outer Coastal Plain Mixed Forest	Atlantic Coastal Flatlands
Dugway P.G.[1], UT	Dry	Temperate Desert	Intermountain Semidesert/Desert	Bonneville Basin
Fort Bliss, TX	Dry	Tropical/Sub-tropical Desert	Chihuahuan	Basin and Range
Fort Carson, CO	Dry	Temperate Steppe	Great Plains - Palouse Dry Steppe	Arkansas Tablelands
Fort Huachuca, AZ	Dry	Tropical/Sub-tropical Desert	Chihuahuan	Basin and Range
Fort Hood, TX	Dry	Tropical/Sub-tropical Desert	Southwest Plateau and Plains Dry Steppe and Shrub	Edwards Plateau
Fort Irwin, CA	Dry	Tropical/Sub-tropical Desert	American Semidesert/Desert	Mojave Desert
Fort Sill, OK	Dry	Tropical/Sub-tropical Steppe	Great Plains Steppe and Shrub	Redbed Plains
Pinyon Canyon, CO	Dry	Temperate Steppe	Great Plains-Palouse Dry Steppe	Arkansas Tablelands
Yakima Training Center, WA	Dry	Temperate Desert	Intermountain Semidesert	Columbia Basin
Yuma Proving Ground, AZ	Dry	Tropical/Sub-tropical Desert	American Semidesert/Desert	Sonoran Desert
White Sands Missile Range, NM	Dry	Tropical/Sub-tropical Desert	Chihuahuan	Basin and Range
Pohakuloa Training Area, HI	Humid Tropical	Rainforest	Hawaiian Islands	Hawaiian Islands
Schofield Barracks, HI	Humid Tropical	Rainforest	Hawaiian Islands	Hawaiian Islands
Fort Greely, AK	Polar	Subarctic	Alaska Range Humid Taiga-Tundra Meadow	Alaska Mountains
Fort Richardson, AK	Polar	Subarctic	Coastal Trough Humid Taiga	Cook Inlet Lowlands
Fort Wainwright, AK	Polar	Subarctic	Upper Yukon Taiga-Meadow	Upper Yukon Highlands

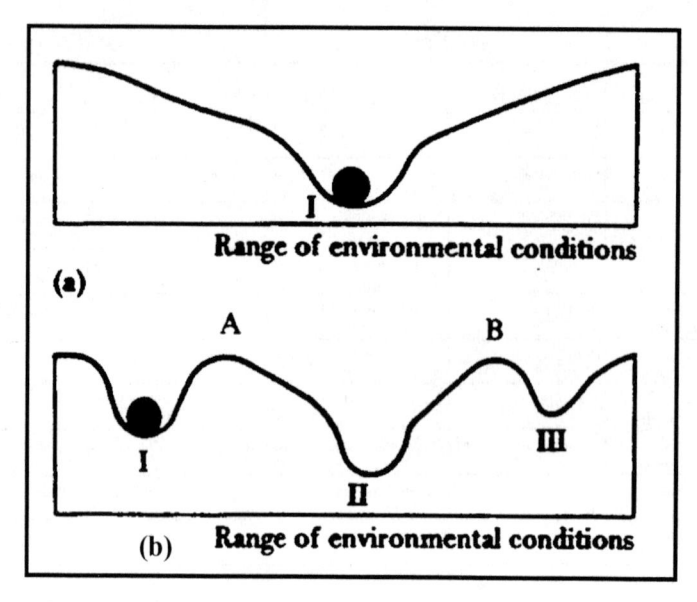

Figure 1. "Ball and Trough" Analogy for Resiliency of a Natural System.
Source: Laycock, 1991.

Resiliency of Army Training Lands

For this analysis, training land resiliency is defined as the inherent capability of the land to support intensive military training and testing activities while sustaining the existing ecological system). One example of training land resiliency can be provided from the perspective of tracked vehicle maneuvers occurring within a designated training area. Resistance defines how many times, or how frequently, a certain type of vehicle (e.g., wheeled or tracked) can travel over an area before the natural community is disturbed from its original state. Resiliency defines if and how quickly the community will restore itself (either naturally or through land rehabilitation intervention) following disturbance. Using the definitions provided earlier, four categories of potential Army land resiliency were defined as: High Resiliency; High to Moderate Resiliency; Moderate to Low Resiliency; and Low Resiliency

Bailey's Division-level classification was used to initially determine the potential land resiliency of all lands within the U.S, regardless of whether an Army installation was located within its boundaries. Each of the 14 Divisions found within the U.S. was subjectively assigned to one of the four resiliency categories. The Province-level descriptions were then used to refine the geographical boundaries of resiliency categories, where appropriate.

A panel of experts who were familiar with landscape disturbance, response, accomplished the resiliency classification and recovery, particularly with respect to impacts from Army activities. Although this initial assessment did not use quantitative monitoring data for specific ecological indicators, such an approach could be used to validate this approach. Similarly, various landscape factor overlays (e.g., soils, vegetation, topography, etc.) could be classified within a geographic information system and combined to provide a complex factor rating of resiliency for each location.

Each of the 31 installations was assigned the land resiliency rating associated with its corresponding classification in Bailey's system. In several cases, the resiliency category initially assigned to an installation's location was modified based upon more detailed knowledge of the particular landscape. For example, the Rainforest Division was initially classified in the High to Moderate Resiliency category, because the tropical environment is conducive to vegetation regrowth. However, the Pohakuloa Training Area, in Hawaii, was placed in the Moderate to Low Resiliency category, because its vegetative community, affected by local topography and rain shadow effect, is not representative of the Rainforest regime, but is more typical of the Tropical/Sub-tropical Steppe environment.

A map showing the resulting delineation and distribution of the four Land Resiliency categories, superimposed with the locations of the 31 Army installations, is shown in Map 2. Map 2 illustrates the potential land resiliency relationships between the 31 Army installations. The map indicates that seven of the key Army installations are located in regions classified with low resiliency. Conversely, only two of the installations examined – Fort Riley, Kansas and Fort Lewis, Washington, are located in areas of high resilience. The majority of installations are located within the two intermediate categories of resilience. It should be noted that a number of installations, for example those in the southeastern U.S., are located in close proximity to ecoregion boundaries. As previously discussed, these boundaries are not absolute, but rather represent transition zones between different landscapes. In those cases, the assignment of an installation to a particular resiliency category should be viewed with some caution. The limitations of scale, resolution and minimum size mapping units notwithstanding, this analysis provides a useful tool for relative comparison between installations.

Table 3. List Of 31 Major Army Installations by Land Resiliency Category

RESILIENCY CATEGORY	BAILEY'S ECOREGIONS (DIVISION-LEVEL)	INSTALLATIONS
High Resiliency	Marine	Fort Lewis
	Prairie	Fort Riley
High-Moderate Resiliency	Hot Continental	Fort Campbell
		Fort Drum
		Fort Knox
		Fort Leonard Wood
		Fort Benning
	Subtropical	Fort Bragg
		Fort Jackson
		Fort Gordon
		Aberdeen Proving Ground
		Fort A.P. Hill
	Rainforest	Schofield Barracks
Moderate-Low Resiliency	Subarctic	Fort Richardson
		Fort Wainwright
		Fort Greely
	Subtropical (Note 1)	Fort Rucker
		Fort Stewart
		Fort Polk
	Tropical/Subtropical Steppe	Pinon Canyon
		Fort Carson
		Fort Sill
		Fort Hood
	Rainforest (Note 2)	Pohakuloa Training Area
Low Resiliency	Tropical/Subtropical Desert	Fort Irwin
		Fort Bliss
		Fort Huachuca
		White Sands Missile Range
		Yuma Proving Ground
	Temperate Desert	Yakima Training Center
		Dugway Proving Ground

NOTE 1: These 3 Installations (Forts Rucker, Stewart and Polk) were placed in a lower Resiliency Category than other installations in the same Division because they have an average rainfall-runoff erosivity characteristic that is significantly higher, increasing the potential for soil loss and resulting in lower rates of recovery.

NOTE 2: Pohakuloa Training Area experiences a rain-shadow effect caused by adjacent mountain ranges which alters the vegetation characteristics of the installation from rainforest to tropical steppe. This characteristic, within the Hawaiian Islands, is not accounted for in Bailey's classification system.

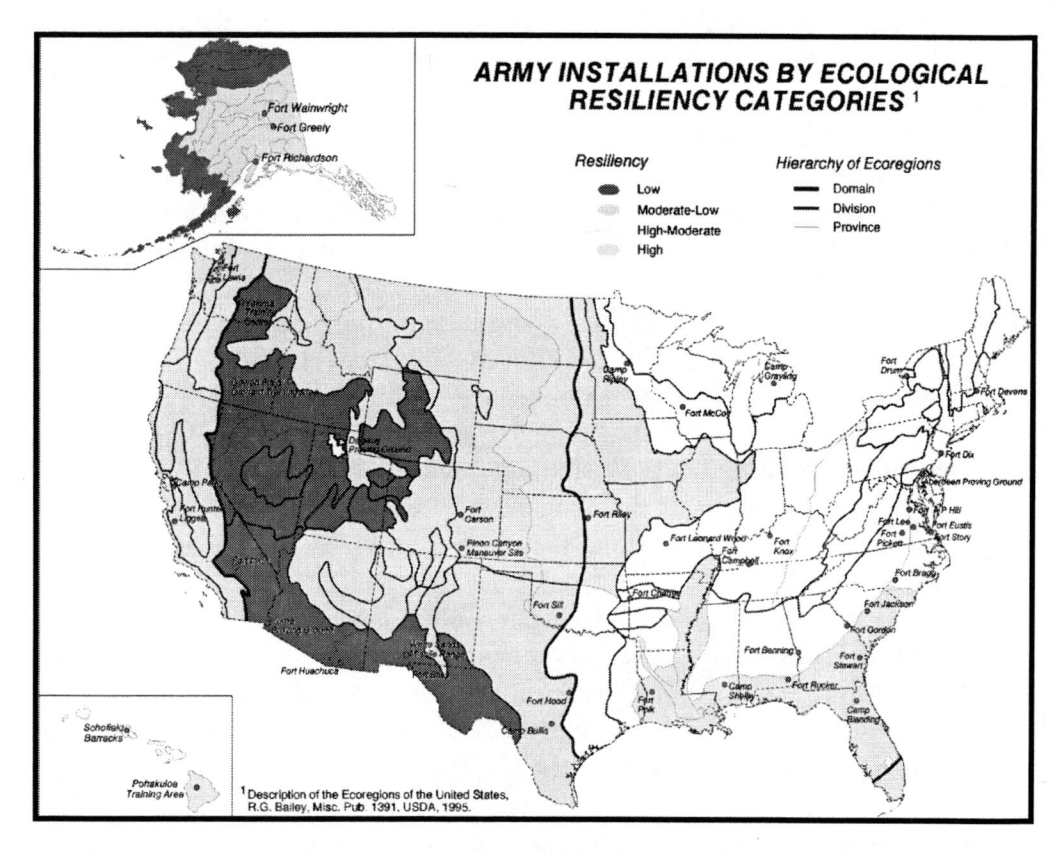

Map 2. Map Of Army Land Resiliency Categories, Showing 31 Major Army Installations.

Conclusions

The application of Bailey's Ecoregion classification system to key Army installations provides a useful framework for comparing the potential ecological resiliency of Army lands to various training and testing activities. In this initial assessment, Bailey's Province descriptions and geographical boundaries provided the most appropriate descriptions and scale for resiliency classification and comparison at the strategic planning level. Provinces also facilitate extrapolation of these concepts to the global scale that is useful because the Army deploys to and trains in foreign countries. Consideration of the ecological impacts of its operations in all locations must be considered regardless of whether the Army trains at home or abroad.

Resiliency concepts, including carrying capacity, ecological response, and thresholds, are useful in comparing Army installations based upon their inherent ability to withstand and recover from the impacts of Army training and testing activities. There are many other considerations, aside from ecological ones, to be considered by the Army in allocating training and testing activities on its lands. This analysis is not intended to suggest that key Army installations should be closed to mission-oriented training. However, this approach provides a logical basis upon which the Army can evaluate its strategic decisions concerning current and future land uses, and their potential environmental consequences. For example, High to Moderate Resiliency lands might be viewed as the "mission capable" category, meaning that training lands within this category, or with a higher resiliency, have the potential to better sustain prolonged military land use. Conversely, those training lands in the lower two Land Resiliency categories are inherently more vulnerable to the long-term effects of military training. Such information might be an important factor in making decisions about unit re-stationing and installation training density.

For example, the Army might consider several installations for re-stationing units brought back to the U.S. from foreign stationing. If the selected installations were evaluated as equal in all other decision-making factors, a higher resiliency location might weight the decision in favor of one installation, to minimize environmental consequences.

The 12 million acres of land managed by the Army is a unique federal asset that is critical to supporting the national defense mission. The long-term sustainability of these lands is in the best interests of both the Army and the public. Integrated approaches to land management decisions, which consider both mission needs and ecological consequences, will help the Army to ensure this viability.

Editor's note: Adapted from: Doe, W.W., et al. 1999. 'Locations and Environments of U.S. Army Training and Testing Lands: An Ecoregional Framework for Assessment." *Federal Facilities Environmental Journal*, Autumn. Reprinted by permission.

Bibliography

Bailey, R.G. 1995. *Ecosystem Geography*. New York, New York: Springer-Verlag.

Bailey, R. 1996. *Description of the Ecoregions of the United States, 2d ed.* Miscellaneous Publication 1391, USDA Forest Service, Washington, D.C., 108 p.

Bandel, D., C. Conrad, M. Drucker, and B. Land. 1996. *Land for Combat Training: A Briefing Book*. Report AEPI-IFP-196, U.S. Army Environmental Policy Institute, Atlanta, GA, 133 pp.

Cawley, R.M. and R. M. Lawrence. 1996. "National Security Policy and Federal Lands Policy, or, The Greening of the Pentagon," in *Post-Cold War Policy: The Social and Domestic Context*, Chicago, Illinois: Nelson-Hall Publishers, pp. 274-292.

Department of the Army. 1998. Army Regulation 350-4, Integrated Training Area Management (ITAM). Washington, D.C.: United States Government Printing Office.

Department of the Army. 1991. Training Circular 25-1: Training Land. Washington, D.C.: United States Government Printing Office, Chapter 1.

Gorman, T., 1999. "Army Plans for Bigger Base Pit Tank vs. Turtle, *The Los Angeles Times*, April 19, 1999.

Holling, C.S. 1973. *Resilience and stability of ecological systems*, Annual Review of Systems and Ecology, 4: pp.1-23.

Laycock, W. A. 1991. *Stable states and thresholds of range condition on North American rangelands: a viewpoint*, Journal of Range Management, 44 (5): pp. 427-430.

Leslie, M., G.K. Meffe, J.L. Hardesty, and D.L. Adams. 1996. *Conserving Biodiversity on Military Lands: A Handbook for Natural Resources Managers.* The Nature Conservancy, Arlington, VA, pp. 16-17.

McNab W.H. and P.E. Avers. 1994. *Ecological Subregions of the United States: Section Descriptions.* Report WO-WSA-5, USDA Forest Service, Washington, D.C.

Rubenson, D., R. Everson, R. Weissler, J. Munoz. 1998. *McGregor Renewal and the Current Air Defense Mission.* Report MR-1010-A, RAND Arroyo Center, Santa Monica, CA, 108 pp.

Shaw, R.B. and D.G. Kowalski. 1996. *U.S. Army Lands: A National Survey.* The Center for Ecological Management of Military Lands, Colorado State University, Report TPS 96-1, Fort Collins, CO, pp. 2-9.

Shaw, R.B. and W. W. Doe. 1999. "Where Does the U.S. Army Train to Fight," *Abstracts, 1999 Association of American Geographers Annual Meeting*, Military Geography Session 479, Honolulu, HI.

17

Military Land:
Multi-disciplinary Land Use Planning

By

Wiley C. Thompson

"The mission of DoD is more than aircraft, guns, and missiles. Part of the defense job is protecting the land, waters, and wildlife – the priceless natural resources that make this great nation of ours worth defending."
General Thomas D. White, Air Force Chief of Staff (1957 – 1961)

"As we enter the dawn of a new century, the men and women who wear America's uniform stand proud to serve an important role in the continuing efforts to keep our skies clear, our oceans blue and our precious soils clean. There is no greater gift we can give our children."
General John M. Shalikashvili, U.S. Army, Chairman, Joint Chiefs of Staff (1993 – 1998)

Introduction

Americans do not often equate the Department of Defense (DoD) with visions of environmental stewardship. The reputation of the DoD is more likely associated with phrases like "PCB contamination", "Super-fund clean up site", and "toxic waste". While the polluting and dumping issues have most likely received a majority of the publicity, there has also been significant environmental degradation on military training lands from disturbances caused by tracked and wheeled vehicle maneuver, the digging of fighting positions to hide vehicles (survivability), excavation of large trench-works to stop an oncoming threat (counter-mobility), the digging of individual fighting positions (foxholes) and other damage caused during routine field training. This chapter will focus on how the military is managing its training lands for sustained use and propose a model where management goes beyond sustaining training lands to an end state of pre-disturbance restoration.

Photograph 1. Vehicle disturbance in the training area at Fort Bliss, Texas.
Source: Editor's collection.

The Cold War military witnessed a build-up of large armored forces, in preparation for an armor-heavy conflict against a Warsaw Pact adversary. Consequently, these were the times when readiness was paramount and all other requirements, including environmental stewardship, appear to have been subordinate. The world political situation has changed greatly since the end of the Cold War and so have the attitudes of society and the senior military leadership regarding environmental stewardship.

The current leadership of the armed forces has put environmental stewardship and conducting realistic training with minimal disturbance to the environment at the forefront of contemporary military issues. The two documents mainly responsible for some of these changes in attitudes and practices were the National Environmental Policy Act of 1969 (NEPA) and U.S. Army Regulation 200-2 (1980). AR 200-2 is the Army's implementing regulation for NEPA. Both of these documents *"require that the Army minimize or avoid both short and long-term impacts caused by military activities"* (LCTA, 1999, p. 1). Two other environmental laws that placed requirements upon the Army are the Endangered Species Act (1973) and the Clean Water Act (1972).

Another motivator for environmental stewardship was the realization that *"Army training lands are finite while training demands and intensity due to technological changes and base realignments have increased. Therefore, it is in the Army's interest to sustain soils and vegetation resources on current training lands to meet mission requirements for realistic training and testing"* (LCTA, 1999, p. 1). Additionally, the Army seeks to acquire additional lands to train on and requests to renew "withdrawn lands" or federal lands withdrawn by the Bureau of Land Management for military use in Alaska and Texas (Doe, 1999). A history of good environmental stewardship is imperative for these requests to be granted.

Another significant event occurred in 1984 when the Secretary of the Army commissioned a panel to evaluate the natural resource management of military lands. The panel found that natural resource uses were subordinate to mission activities and that there was excessive soil erosion from tracked vehicles (LCTA, 1999). The resulting fix from this study, developed by the U.S. Army Construction Engineering Research Laboratory (CERL), was called the Integrated Training Area Management (ITAM) program. Initial pilot projects were established in 1985 at Fort Carson, Colorado and Fort Hood, Texas.

The overall goal of ITAM is that *"the Army achieve optimum, sustainable use of its training lands by implementing a uniform ITAM program that includes inventorying and monitoring land condition, integrating training requirements with land capacity, providing for land rehabilitation and maintenance, and educating land users to minimize adverse impacts"* (AR 350-4, 1998, p. 1). ITAM is divided into four functional areas: Land Condition Trend Analysis (LCTA) which is "a management procedure that provides for collecting, inventorying, monitoring, managing and analyzing" land conditions on installations, Training Requirements Integration (TRI) which is a *"decision support procedure that integrates training requirements with land management, training management, and natural and cultural resources management*", Land Rehabilitation and Maintenance (LRAM) which is a *"preventive and corrective land rehabilitation and maintenance procedure that reduces the long term impacts of training*", and Environmental Awareness (EA) which is intended to *"develop and distribute educational materials to land users"* (AR 350-4, 1998, p. 1). Currently, the Army is investing $35 million annually into the four functional areas of ITAM.

Key to the ITAM process is the concept of a land's "carrying capacity." A carrying capacity is the load, or in this case, the amount of military training that a parcel of land can accommodate in a sustainable manner (Doe, 1999). *"Carrying capacity is a complex, integrated variable that is a function of the inherent site characteristics (e.g., soil, slope, aspect, climate) and biological regime (e.g., flora, fauna, vegetation community, structure, and composition) of the natural environment"* (Doe, 1999, p. 11). The Army defines carrying capacity for its training lands as *"the amount of training and testing that a given parcel of land can accommodate over time in a sustainable manner"* (Doe, 1999, p. 11). If training activities exceeds the carrying capacity, degradation of the environment will occur. Likewise, if training activities are less than what an area can accommodate, the areas are more likely to regenerate vegetation, which can, in turn, can offer an attractive habitat for native fauna.

Agencies Involved

There are three major agencies involved with the Army's development, implementation and monitoring of the ITAM program. They are the U.S. Army Construction Engineering Research Laboratories (CERL), the United States Army Environmental Center (USAEC), and The Center for Ecological Management of Military Lands (CEMML). Today, these organizations generally work together towards a common goal: sustainable, managed use and stewardship of Army training lands.

The U.S. Army Construction Engineering Research Laboratories is part of the U.S. Army Engineer Research and Development Center (USAERDC), which is the Army Corps of Engineers' integrated research and development organization. CERL was established in 1969 and was organized for the purpose of conducting research to support sustainable military installations. Research is directed toward increasing the Army's ability to more efficiently construct, operate, and maintain its installations and ensure environmental quality and safety at a reduced life-cycle cost. An adequate infrastructure and realistic training lands are critical assets to installations, which serve as platforms to project power worldwide. As previously mentioned, ITAM was developed by CERL.

The United States Army Environmental Center (USAEC) is the other Department of the Army organization involved in the restoration of military training lands. This organization was established in 1993 by renaming the U.S. Army Toxic and Hazardous Materials Agency (est. 1975). USACE's original mandate was to restore active Army installations, but has since expanded greatly. Today the mission of the USAEC is to:

- Integrate, coordinate and oversee implementation of the Army's Environmental Programs for the Army Staff (ARSTAF)
- Provide technical services and products to Headquarters Dept of the Army (HQDA), Major Army Commands (MACOMs) and Commanders
- Promote Readiness and enhance Quality of Life

The United States Army relies on the expertise of the USAEC Conservation Branch to support Army Conservation Program goals. The USAEC Conservation Branch promotes conservation techniques that support the military mission. The teams of experts in the Conservation Branch have extensive experience and knowledge regarding mission requirements & Army operations, environmental stewardship & leadership, integrated planning & management methods, and environmental policies & regulations.

The Center for Ecological Management of Military Lands (CEMML), located at Colorado State University, Fort Collins, Colorado is also a key player ITAM program. CEMML is self-described as a group of 120 full-time biologists, data analysts, computer scientists, and resource managers who focus on the integration of sound land management practices to support the military's mission. CEMML has been performing these functions since 1985. As part of the services provided in support of LCTA, CEMML conducts training area resource inventory and monitoring. Typical data collected in the inventory would include:

- Descriptive site information
- Plant basal area, canopy cover, diversity, and frequency
- Woody plant density and basal area

- Plant productivity
- Disturbance and land uses – qualitative and quantitative
- Estimation of tactical concealment parameters
- Physical attributes – soil type, topography, weather, etc.
- Wildlife information – presence/absence for specific species and surveys for mammals, birds, and reptiles
- Cultural resource information – documenting historic and prehistoric sites to ensure compliance with laws and an appreciation for the past

(Executive Summaries, 1998)

CEMML also specializes in fire ecology, a field of knowledge that would allow for the experimentation with natural fire management practices.

The last major entitities involved in the ITAM process are the army installation managers and planners. Installation managers/planners are normally under the Deputy Chief of Staff for Plans, Training, Mobilization and Security (DPTMS). The DPTMS coordinates all of the training activities on an installation. The office approves and allocates land for training events and is therefore intimately involved with the management and restoration/rehabilitation of the training lands. The installation's environmental compliance personnel and range control personnel usually work for the DPTM.

Tools for the Military Land Use Planner

A key indicator of an installation's ability to manage training lands is vegetative health. Re-planting disturbed surfaces and elimination of invasive plant species are two problems facing planners on military installations. The restoration of natural vegetation is key to maintaining training lands. This may involve replanting areas that have been disturbed by training activities, resulting in increased erosion or the elimination of non-native species that were introduced intentionally or have encroached on their own.

To determine an installation's vegetative health, installation managers must conduct large-scale land classification to evaluate the physical condition of the training areas. Vegetation inventories are a key component of the LCTA program. Currently, most land classification surveys are conducted by inventorying plots on foot. This is time consuming and labor intensive, not to mention that with the vast holdings of some military training areas (12 million acres), it would be virtually impossible to physically inventory the entire installation. Recently, great utility has been found in the employment of remotely sensed imagery and geographic information systems (GIS) to conduct an automated land classification. These tools are used, along with the developed classification strategies to help the managers detect change in vegetation, monitor erosion, and conduct ecosystem assessment. This is key to assessing the land's capacity to support required training. As GIS capabilities improve and the spectral and spatial resolution of remote imagery gets better, the end users ability to conduct land classification will also improve.

CERL has developed a method of classifying military lands as to their ability to support training activities by integrating remotely sensed data with a GIS. Their systems incorporate the Universal Soil Loss Equation (USLE) and the Revised Universal Soil Loss Equation (RUSLE) with a GIS to create a series of training area capability overlays based upon information from soil data, remote imagery and LCTA ground-truthing (Warren, 1997). The Corps of Engineers expects their land classification system to help installations conduct inventories, improve training scheduling to avoid degraded areas, consolidate lands of similar condition and prioritize rehabilitation efforts. This ability to more efficiently manage a training area and ensure its future availability through the application of aerial imagery and GIS is critical in this time of installation closures and increased operational tempo.

In 1997, a team from the U.S. Army Environmental Center and the Topographic Engineer Center (TEC) collaborated on the "Remote Sensing User's Guide." This guide was written to help land managers apply remote sensing as a tool to aid them as they attempt to deal with the broad management objectives of vegetation, soils and soil erosion, and land disturbance detection on military installations. The guide takes these three broad management objectives and allows the user to select their ecoregion (Northern Plains, Southeast, Southern Plains, Pacific Southwest or Northwestern United States). Managers and planners can then pick from list of applicable sensors that will help them meet their specific land management objectives such as determining vegetative cover or soil erosion. They can also examine the effects of land management techniques such as prescribed burns, chemical or mechanical vegetation removal. Finally they can choose sensors that will help them detect maneuver damage or re-vegetation.

The guide lists applicable sensors for each sub area and has an enclosed sensor fact sheet that allows the user to do a side-by-side comparison of the sensor's cost and temporal and spatial resolution, to name a few. The guide also gives guidance on how and where to acquire imagery and provides sample statements of work. Once the user has selected the appropriate imagery, they can import this data into a GIS and observe how their objective indicators have changed over time. This guide, useful for both experienced and inexperienced imagery users, greatly assists installation managers and planners as they attempt to balance training requirements while minimizing environmental degradation.

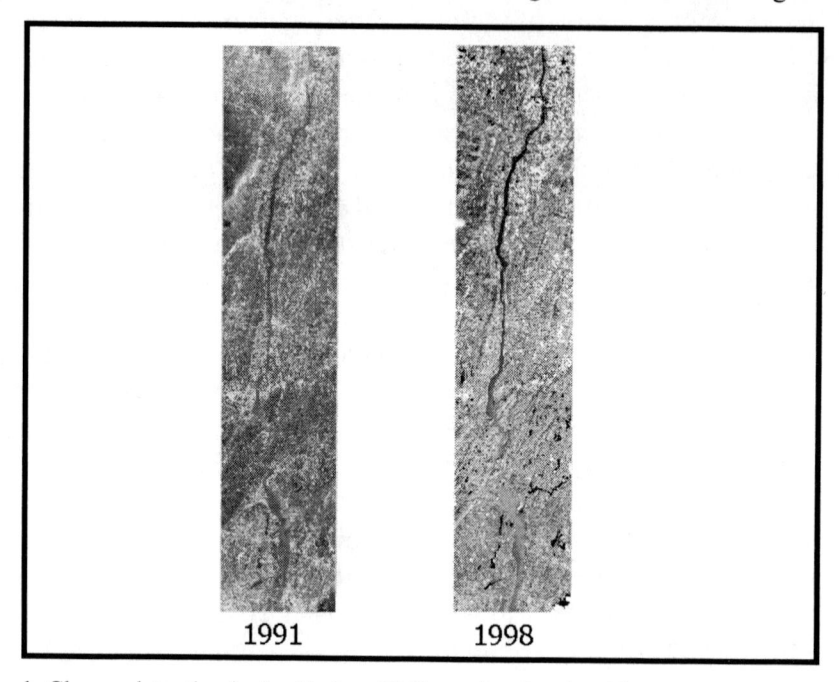

Figure 1. Change detection in the Hudson Valley using Landsat Thematic Mapper imagery. Source: USMA, West Point.

Aerial imagery and GIS have a definitive role in aiding planners and managers in managing training lands at military installations and have a key role in automating the LCTA process (see Figure 1). These tools are extremely helpful when used to classify coverages in order to determine impacts of training, ability to support future training and redesign training areas if needed. The future development of remote imagery systems with increased band availability and greater spatial resolution will help to increase the accuracy of land classifications. Additionally, as more coverages become available, activities at the local, state and federal levels can begin to share products, reducing the cost of acquisition and development.

Today, there are software systems and online applications that are available to help the planners meet their specific needs. One tool available to those involved in land re-vegetation is a program called VegSpec: Training Area Plant Species Design Software Tool. This program includes a comprehensive database of over 2,500 plant species and was developed by the Natural Resources Conservation Service, the U S. Army Construction Engineering Research Laboratories, and the U.S. Geological Service. With the knowledge of the desired land practice, soil information, climatological data, and site characteristics, VegSpec will recommend a list of native and non-native (if requested) species that will suit the users purpose(s). These purposes can range in scope from vegetative screening, restoring native plant communities, stabilizing slopes, erosion control, and creating wildlife habitat. The NRCS has divided the U.S. into about 240 Major Land Resource Areas (MLRA) each of which are delineated by geographic areas having similar patterns of soil, climate, water resources and land uses (Smith, 1999). When planning re-vegetation with a native species, the MLRAs help to account for the natural, spatial distribution of natural vegetation in these areas. One must realize however, that it is only a tool and due to the infinite numbers of possibilities posed by different environments and circumstances, one cannot solely rely upon it as a final authority for historic vegetative reconstructions.

Another software package used in the restoration of training lands is Plant Management Information System (PMIS) for Noxious and Nuisance Plant Species (see Figure 2). Researchers at the U.S. Army Engineer Waterways Experiment Station (WES) developed PMIS. This program provides training land managers with long-term, environmentally sound solutions for eliminating and controlling noxious and nuisance plant species.

At many military installations, damaging land use practices have long disturbed and destroyed native vegetation and eroded soils. The net result of these activities is the increase in surface erosion and the invasion of noxious and nuisance species of plants, which would further displace and inhibit the growth of native plant species. Cheatgrass (*Bromus tectorum L*) is highly flammable and tends to increase the frequency of fires and elevate their intensity (Westbrooks, 1998). Damage from the more frequent, hotter fires can be extremely costly. Kudzu (*Pueraria montana var. Iobata*) can also pose a problem for land use planners as it out-competes tree stands for sunlight, resulting in the stand's degradation. Some tree stands such as the longleaf pine are habitat for the red-cockaded woodpecker, an endangered species. Additionally, narrow tap rooted vines like Kudzu can actually increase erosion as it out-competes more grass, which have more binding, shallow, fibrous root systems (Westbrooks, 1998). Solutions provided by PMIS to the land use planner on how to control invasive plants such as Cheatgrass and Kudzu might include short-term control measures like chemical and mechanical control and the use of host-specific biological agents that are a more practical long-term solution.

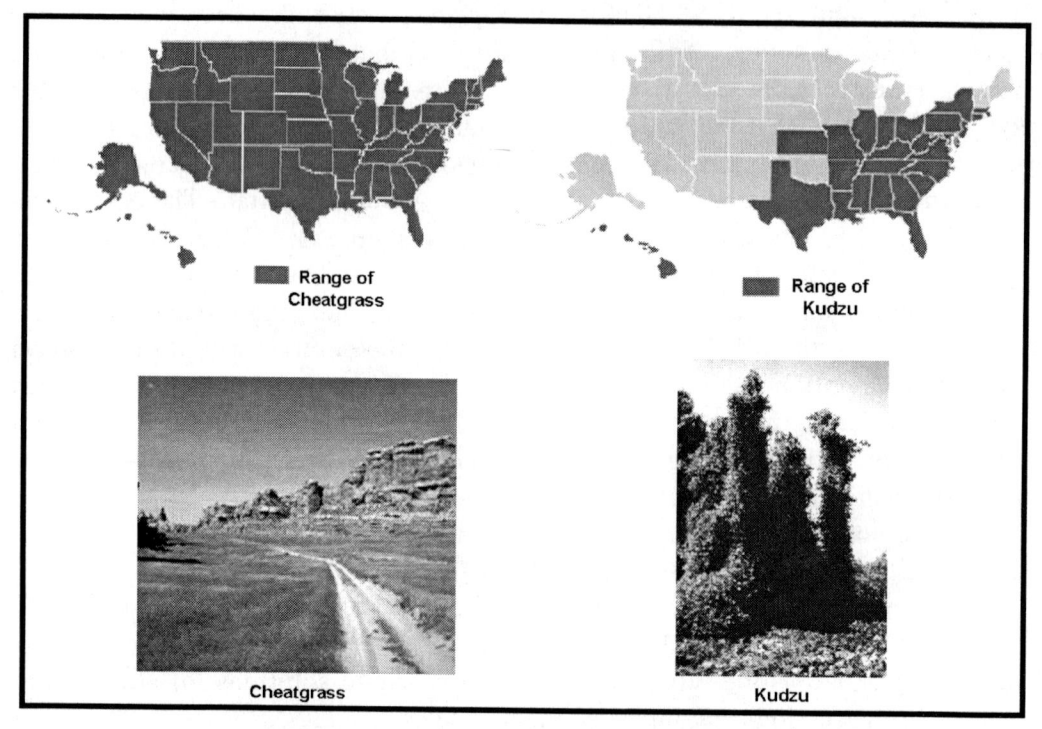

Figure 2. Range of Cheatgrass and Kudzu
Source: Photographs from Kansas Department of Agriculture and USDA.

Reestablishment of training lands to pre-disturbance regimes may not always be possible at first. Realizing that these lands must still endure the hardships of military training, there may be some compromises required. For instance, a hardier, yet compatible ground cover may have to be permanently substituted for a native plant that would not otherwise hold up well in tracked-vehicle maneuver areas. In his work as a vegetation and soils specialist for the U.S. Army Construction Engineering Research Laboratory, Dr. Muhammed

Sharif found that sometimes in order to re-establish native vegetation, planners may have to introduce a non-native species (Sharif, 1999). Dr. Sharif noted that often a hardy non-native ground cover would be planted for its ability to quickly germinate (3-4 days). These plants take hold quickly and prevent further erosion once training has stopped. Intermixed with that same seed would be some permanent, native grasses that take 4 to 6 weeks to germinate and up to a year to establish. Having stabilized the soil, the non-native species will die off after 3 or 4 months leaving native species to become the dominant ground cover. This is a form of management that Russell, author of *People and the Land Through Time: Linking Ecology and History,* describes as "*active management*" (Russell, 1997, p. 242). Active management is a human disturbance or intervention to correct the non-natural changes that were brought about by past land uses (Russell, 1997). In the above scenario, the introduction of a non-native plant as a soil stabilizer would also stop the eventual dominance of another non-native, invasive species.

There are, however, times when a non-native, hardy species is put in on a permanent basis, as it is the only species that is compatible with the intended uses in the training area. According to Smith, the introduction of a non-native that will withstand high traffic areas and will quickly re-vegetate after use makes good sense. The benefit returned from the prevention of erosion, pollution and further destruction of habitat would outweigh the omission of the native plant in this case (Sharif, 1999).

Historical Insights

Currently, there is little evidence that any installation managers or land use planners are using historical reconstruction resources such as old photographs, surveys, or accounts of early travelers as part of their training land rehabilitation. However, the techniques of the ecological historian could be easily integrated into training area management. It might be interesting to see how people like Emily Russell or William Cronon, author of *Changes in the Land: Indians, Colonists, and the Ecology of New England* would view the work that is being done and the goals of the current military land management programs? With a better knowledge of the agencies, programs and tools currently involved in the ITAM process, one can begin to look at how land use planners are managing lands, evaluate the merits of the program from a historical landscape framework and suggest ways to augment the work that is already ongoing.

One reason that a planner might want to reconstruct the past is to determine what has caused the landscape to look the way it currently does. Russell wrote her book "to help point to different aspects of current environments that bear the imprint of various past human activities, which must be considered in order to understand current process" (Russell, 1997: *xvi*). In many cases on military installations the land was devegetated by intensive military training, which lead to erosion. The military has looked at past activities to account for the degradation of the environment and is making progress in the area of halting the damage and reversing the ill effects of past activities. Likewise they are modeling the effects of current activities and are accumulating data so that they may curb, curtail or modify practices that are not obtaining the desired effect – a trained force, with minimal environmental degradation.

What data are available to the military land use planner for the historical reconstruction of military training lands? Cronon and Russell put emphasis upon using photography and historical maps (Russell, 1998), written documentation by early occupants (Cronon, 1983). Oral accounts of past human activities can also be a valuable source of information during a reconstruction. This is an area that could be rapidly expanded and used to rehabilitate the land towards a pre-disturbance environment. Many military installations have controlled the lands they are presently on dating back to the late 18th and early 19th centuries. These large tracts of land were in the past, and unlike today, readily available and often deemed to be economically worthless. Some of these installations were established in areas with sparse populations to minimize the negative effects of their often-noisy activities on their neighbors. Moreover, most military installations have long been fenced-off or have been in a limited access. Consequently, there may be little guesswork regarding the activities that have taken place in the areas over the years.

Most of the activities that have taken place on military installations have been well documented. Most installations have been well photographed for historical or survey purposes (see Photograph 2). These photographs are usually preserved in the post museum or archives. Historical resources such as photographs depict what the land looked like and what has taken place and can be used to help reconstruct the process of change on the landscape. This information can be even further enhanced or confirmed with other methods such as pollen and sediment sampling (Russell, 1998).

Photograph 2. Armor training at Camp Polk, LA and Second Army Tennessee Maneuvers, 1943. Source: U.S. Army Center of Military History.

A Model for Reconstruction

Like in many military-civil endeavors, reconstruction and restoration of training lands to pre-disturbance conditions requires a multi-disciplinary approach. Anthropologists and archeologists should be included to ensure that cultural heritage sites are protected and integrated into the training area planning. Botanists, ecologists and wildlife experts should be brought to the planning table to ensure that the proper types and amounts of habitats are preserved and if needed, reintroduced so that they result in a synergistic effect of flora-fauna interaction. This should result in a more stable, sustainable habitat. Finally, a historian/historical ecologist should be included so as to provide a check and balance on the other parties and to ensure that this reconstruction resembles, as best it can, the landscape prior to disturbance.

The first step of a theoretical model of reconstruction is to determine the activity that created the landscape that presently exists. As previously mentioned, the records of training and land uses should exist and provide a great deal of insight as to exactly what events have taken place in the past. But training records may not reveal all of the activities that have taken place over time on the installation. There may have been a time when grazing was allowed in training areas that were unused. Farming may have also been allowed in areas that were leased to local farmers. Any farming would have to require clearing native vegetation if it existed at the time. Records of these non-military uses may be found in the files of the Staff Judge Advocate's (SJA) office, as there should have been some legal, written document permitting civilian activities on federal lands. Also, the current land may have been purchased as a later addition to the installation. Again, there would be records with the SJA and most likely at the county courthouse where the sale was made. One might also try to confirm some of these discrepancies with old photographs, installation maps, surveys, or other like sources of information. Other possible sources of information might include descriptive accounts from early occupants in the area. If the installation is relatively new, within the last 40 to 50 years, there may be local residents that could provide an oral account of previous land uses.

The involvement of the team historian in this stage of the reconstruction process is critical. The historian's expertise would be needed to confirm any changes the area has undergone, be they from activities on the military installation or any other human activities prior to the installation's establishment. Additionally, while some of the other key players may have a working knowledge of history, the historian is best trained in the area of finding and evaluating available historical information.

The final step is the re-establishment of the native regime (if compatible with current or anticipated use) and the rehabilitation towards a pre-disturbance landscape. This is where the installation planners/managers get together with the ecologists, wildlife experts, agronomists and historians to decide how to re-establish the pre-disturbance vegetation. Suggested activities at this point in the process include:

- Creating a map of the installation as if fully vegetated with native plants
- Overlaying a map that shows areas of discrepancy
- Looking at the training that will take place in these areas
- Consulting VegSpec and replanting if there are no conflicting activities

- If there are conflicts,
    - Moving training to allow native vegetation to re-establish
    - Consulting VegSpec to suggest alternatives

The above is a rough outline, and is similar to current practices, but the focus is not solely on erosion control or sustained management, but more on native vegetation restoration. In the case of the model, the planning team has a better idea of what the landscape originally looked like. It's no longer just a mission of erosion control and compatible use; it is now truly an exploration into pre-disturbance restoration. The restoration of pre-disturbance or native vegetative regimes is as important to habitat restoration as it is to training realism. Degradation to native vegetation denies the sold a realistic training environment, the ability to use natural camouflage for cover and concealment, and reduction in cross-country trafficability due to increased erosion (Doe, 1999).

Although degradation of training lands has been the focus of this chapter, many DoD installations contain a broad diversity of relatively undisturbed habitats. Of these many installations, the Army alone manages over 12 million acres of federal training and testing lands (Doe, 1999). This accounts for roughly half of the land under control of the DoD and includes lands in all of the fifty states (Doe, 1999). These habitats are important ecological settings that provide refuge for many plants and animals that are in a concerned status (Doe, 1999).

In her address to the Oklahoma Association for Environmental Education, Sherri W. Goodman, Deputy Undersecretary for Defense for Environmental Security noted that DoD installations are refuge to over 200 endangered species (Goodman, 1997) (see Figure 3). On military installations throughout the southeastern United States, carefully managed use of training lands and proper fire ecology practices have helped to create viable habitat for federally listed species like the Red-Cockaded Woodpecker (Cast, 2000). At Fort Irwin, California, in the Mojave Desert, awareness and education of soldiers and leaders have allowed the Desert Tortoise to live in a habitat free of encroachment and destruction due to development. At Fort Lewis, Washington and other installations, the American Bald Eagle has found viable habitat where it can strengthen its numbers (2000).

These habitats, because they are on a military installation, have normally been protected from urban development and other destructive uses such as grazing and timber harvesting. While artillery impact areas and surface danger areas are incompatible with human existence, often having a "no-man's land" appearance, many animal populations thrive on the fringes of these lands, devoid of physical human presence. Dr Albert Bivings, of the Forces Command Environmental Office, Fort McPherson, GA, described these surface danger areas as outstanding habitats for wildlife (2000). Although some portions of training lands on installations may experience heavy vehicular traffic, some areas on military installations are seldom used, but serve as buffers with the neighboring civilian populations. With sensitive populations thriving, the restoration and maintenance of a pre-disturbance landscape could make portions of the installation's training areas seem as if they are a glimpse into the past?

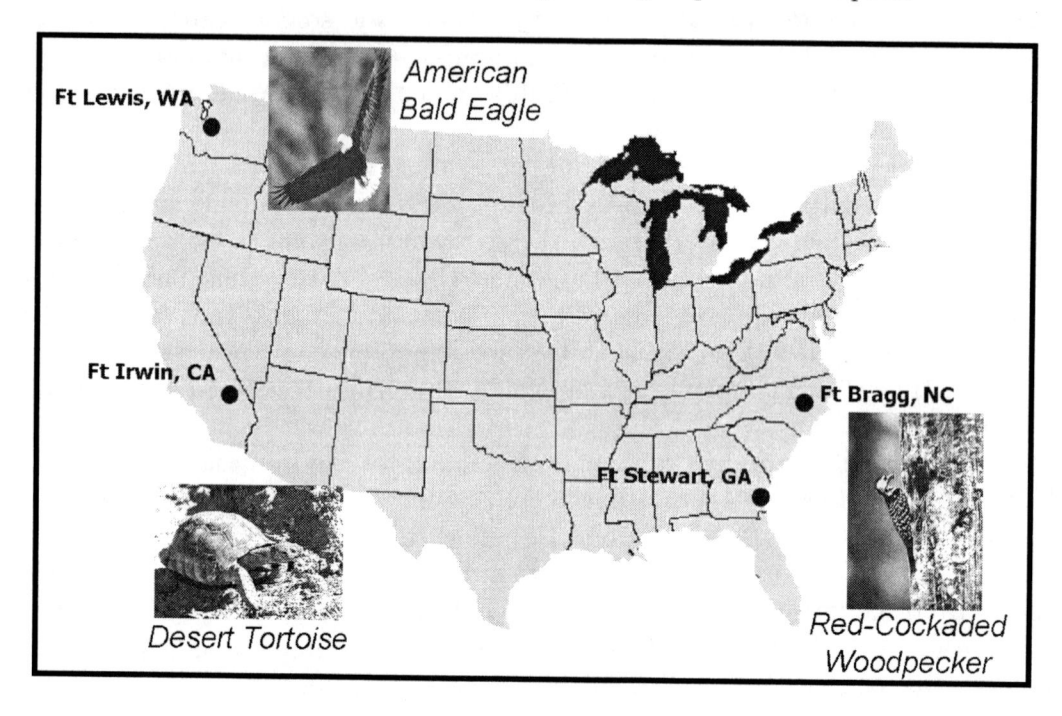

Figure 3. A few of the many endangered species thriving on military installations.
Source: Photographs from the Army Environmental Center.

Conclusion

The armed forces must maintain the highest levels of readiness at all times, especially in this rapidly changing, unstable, and occasionally volatile world. Despite the naysayers, the threats of global and regional conflicts are very real. The ever-evolving commitments to Military Operations Other Than War (MOOTW), such as Bosnia, Macedonia, Kosovo, Somalia and Haiti require just as intense field training as their combat commitment counterparts.

Nevertheless, the DoD is in a unique position to restore a significant quantity of lands back to their "native" condition. The ability to carryout this new mission is even further enhanced by the fact that these lands will be protected and the protective measures have a very high level of enforceability. The military has the manpower, resources, and enforcement potential to return these training lands or portions of them to their pre-disturbance conditions. This is an ecologically responsible thing to do and they Army would benefit greatly from the publicity of doing so.

The current land management practices of the DoD seek the same ultimate results as the ecological historian. They both examine the current landscapes as products of past activities and attempt to quantify those past activities in an effort to prevent further environmental destruction. Care of the environment is a heavy responsibility and the decisions regarding the benefits of training and their corresponding degradation of the environment are tough. But, heavy responsibilities and the need to make tough decisions are a daily part of life for the members of the armed forces. Senior military leaders now have the capability to accomplish tough, realistic training and still be Stewards of the Environment. The necessary technology, methodologies, and philosophy have arrived just in time because with increased base closures, it is imperative that we maintain our training lands as renewable resources.

Bibliography

Bright, T.A., Getlein, S., Jarrett, J., Tripp, S., and Moeller, J. 1997. *Remote Sensing User's Guide*. U.S. Army Environmental Center and the Topographic Engineering Center.

Cast, M. ed. 2000. Army Management of Training Lands Enhances Realism, Protects Wildlife. *Environmental Update* Vol. 12.

Center for Ecological Management of Military Lands. 1998. *Executive Summaries*. Fort Collins, CO: Colorado State University.

Cronon, W. 1983. *Changes in the Land: Indians, Colonists, and the Ecology of New England*. New York: Hill and Wang.

Doe, W. W., Shaw, R. B., Bailey, R. G., Jones, D. S., and Macia, T.E. 1999. Locations and Environments of U.S. Army Training and Testing Lands: An Ecoregional Framework for Assessment. *Federal Facilities Environmental Journal*. Autumn: 9-26.

Headquarters, Department of the Army. 1998. *Integrated Training Area Management (ITAM)*. Army Regulation 350-4. Washington, D.C.: U.S. Government Printing Office.

Headquarters, Department of the Army. 1999. *LCTA II Technical Reference Manual*. Washington, D.C.: U.S. Government Printing Office.

Goodman, S.W. 13 February 1998. "Environmental Issues: A Top Priority for Defense Leadership". Prepared remarks by Sherri W. Goodman, Deputy Undersecretary of Defense for Environmental Security, to the Oklahoma Association for Environmental Education, Fort Sill, Okla.

Meinig, D.W., ed. 1979. *Interpretation of Ordinary Landscapes: Geographical Essays*. New York: Oxford University Press.

Russell, E. W. B. 1997. *People and the Land Through Time: Linking Ecology and History*. New Haven, CT: Yale University Press.

Sharif, M. 17 February 1999. Telephone Interview.

Smith, D. (psmith@itc.nrcs.usda.gov). 17 February 1999. "Re: VegSpec". Email to author.

Warren, S. D. 1997. CERL Fact Sheet (LL 8). U.S. Army Construction Engineering Research Laboratories. Champaign, IL.

Westbrooks, R. 1998. Invasive Plants, Changing the Landscape of America: Fact Book. Federal Interagency Committee for the Management of Noxious and Exotic Weeds (FICMNEW), Washington, D.C.

18

Integrated Military Land Management:
The Warfighter's Terrain Perspective

By

Brandon K. Herl

Introduction

Managing military training and testing lands is a complex, ever changing, and justifiably frustrating occupation. Military land management is a broad topic and has probably best been defined in a 1996 Department of Defense public affairs videotape as:

The integrated management of natural resources and man-induced effects on Department of Defense lands owned, leased, or withdrawn from public use for the unique functions of training, testing, and garrisoning military units from all the armed services' four branches—the Army, Navy, Air Force and Marine Corps.

The following case study uses a geographical approach to help a military land manager better perform his or her duties by spatially analyzing combat training tactics to help identify locations within training areas that will experience above-average ecological disturbance.[1] First, however, a quick discussion about military land management is warranted.

Military Land Management

Traditional Methods

Natural resource managers are charged with ensuring the delicate ecological health and sustainability of areas specifically set aside for not-so-delicate combat training. These professionals must constantly seek the balance between nature and its fragile systems with the needs of soldiers to realistically train with equipment purposefully designed to wreak destruction. To further complicate matters, most tactical military maneuver land disturbance is considered unpredictable due to the complexities of unit tactics and training exercise objectives (Price et al., 1997). However, a general understanding of military tactics and a combat leader's thought processes coupled with a detailed analysis of actual military maneuvers and its associated disturbance trends can illuminate this dark corner within the military land manager's problem of spatially locating heavily disturbed areas prior to an exercise's commencement rather than after its completion.

Military training land management has its basic roots founded in the natural resources management disciplines, mainly forestry, watershed management, agriculture (both farming and ranching), and wildlife conservation (Price et al., 1997; *ITAM Strategy Guide,* 1995). As such, nearly all of the current evaluation techniques used on military lands focuses on physical and/or biological systems to assess the current ecological health of a training area.

While these techniques are sound for monitoring changes on a military reservation's maneuver area, few focus on true proactive land management. Instead, most management activity has tended to be reactive and focused on repairing or rehabiliting the land after massive disturbance, much like natural disaster remediation. The key to sound and efficient management of military lands lies first in unlocking the mysteries of the disturbance effects and the disturbance agents. While a great deal of effort has gone into detailing the effects, the focus now must shift on the disturbance agents. In this special case, the major disturbance agent itself is easily identifiable--the soldier and his/her military training. However, the nature of the disturbance agent is still very much shrouded in mystery.

Maneuvershed Concept

Natural resource managers on military installations must begin shifting their focus from a traditional management vantage point (such as a purely erosional-based watershed system) to that of managing a *maneuvershed*--a physical landscape where soldiers and their equipment operate (Doe et al., 1997). The effect of soldiers and their training must be included in all management activities. While the U.S. Army's Training and Testing Area Carrying Capacity (ATTACC) land management system and it's Integrated Training Area Management (ITAM) program are examples of strides in this direction, neither can adequately address the full spatial scale and locational needs of a land steward on the ground.

As an installation management tool, ATTACC it is excellent. It accounts for many of the training and scheduling dynamics of military land management, but it does so at a training area level. For a land steward, ATTACC provides unit size, vehicle type and density, dates, and types of military training units will conduct within certain training areas (*ATTACC Handbook,* 1999). While this information does provide a land steward with previously unknown (but sorely needed) training area use information, ATTACC still does not address many of the land steward's finer resolution data needs to help calculate actual maneuver disturbance levels and locations.

Defining a maneuvershed is two-fold process. The first process includes having a firm knowledge of a reservation's physical and biological process and its natural interrelationships. This is an area where nearly all military land stewards excel since it forms the basis for their employment. The second process in visualizing a maneuvershed is understanding combat training and tactics and recognizing its distinctive landscape footprint. The lack of common ground between these two processes can form a gap in understanding between the natural resources professional and the military professional. Bridging this gap is essential to creating sound military land management practices.

Bridging the Gap

The initial step towards bridging this gap is to form a common language between the two professional groups. This is part of the ITAM program's charter (*AR 350-4,* 1998; *DA Pam 350-4,* 1998). One of the first notions evident in the natural resources profession is the idea that some areas are better suited for an activity than others. At the most basic level, agronomists seek to plant crops in high-yield areas, rangeland managers look for desirable grazing areas for their herds, and wildlife conservationists tend to study a species and its dominant habitat or range. In any land management discipline, the concept of a *preferred area* is paramount (Walsh et al., 1998). This holds true for military land management.

One military training myth is that training exercises always totally denude entire landscapes. This myth, while false, does have some understandable qualities. If left unchecked, at a very local scale this may be true. (Military arms and vehicles are not designed to be gentle—it's not a design priority.) However, in the conduct of warfare, most traditional military operations center on destroying an enemy force and/or controlling a critical location in battle (*FM 100-5,* 1993). This idea, in-and-of itself, hints at the notion that not all portions of an area hold an equal martial importance.

Battles will usually be fought in spots that have some domineering influence over an area or definitively control access to and flow within an area. These places are often called *key* or *critical terrain*. Other highly-prized areas in a maneuver space is land that can be easily traversed and links parent units with subordinate units or provides the best access into an enemy-controlled area. These areas are respectively called *lines of communication* and *avenues of approach*. All of these areas vary in size and shape, but all have important tactical advantages and form preferred marital terrain.

The second gap-bridging step comes from understanding a tactical leader's combat decision-making process and the maneuver tactics that result from those decisions. Understanding the former adds the perspective of a combat leader and answers the "how and why" combat units maneuver certain places. A heavy unit (armor or mechanized infantry) combat arms leader at the company-level or below is constantly focused upon a few criteria when mentally preparing mission orders. Those parameters are:

1. Accomplish the mission quickly and easily.
2. No "approved" or "perfect" solution ever exists, but "better solutions" do.
3. A good solution always addresses these two aspects of combat:
    a. Maximize enemy destruction, and
    b. Minimize friendly casualties.
4. End the current mission postured for the most likely future mission(s).

These simple rules guide nearly every front-line decision made by a leader at the tactical level.[2] They may not be fancy, but they are extremely effective. They also give a clue as to how focused a leader is in wartime. Peacetime training exercises are just as important. Natural resource managers must remember that while training exercises may be some steps removed from war, they are viewed in a military person's mind as just as important *or more important than* actual combat.

Understanding how this process materialize on the ground helps answer "where" units fight and adds a spatial dimension to combat. Here is where a natural resources manager must have a basic understanding of military tactics. Unfortunately, this area is often poorly documented and the least studied aspect of military land management (Price et al., 1997). Consequently, little common ground exists between land stewards and military trainers and a huge stumbling block between these two professional cultures results.

Creating a Tactical Template

Military trainers will often cite that tactical training is never the same for any exercise; too many dynamics go into human decision-making and too many variables exist in a combat scenario. Creating a "real world" tactical template based upon individual vehicle maneuver is therefore a time-consuming, error-prone process. However, it is possible to create an idealized or generalized template given some basic rules that account for leader tactical thinking, sound tactical doctrine, and documented maneuver examples.

Ideally, tracking individual vehicles during several major exercises over a single land parcel would be the best source data to create a map illustrating important tactical areas. Surveying post-exercise vehicle footprints and ruts or GPS-tracking each individual vehicle during an exercise could provide this data. While instrumenting every M1A1 tank and M2/3A2 fighting vehicle on an installation for a repeated number of exercises sounds enticing, it would introduce a huge training distracter that might interfere with unit training. There is, however, a source of similar data that could be used as a proxy data set to gather trends on actual tactical maneuver.

The Combat Maneuver Training Center (CMTC) at Hohenfels, Germany tracks individual vehicle movement during a unit's training rotation via a specialized computer system (*CTC Handbook,* 1998). Each instrumented vehicle emits a telemetered burst that broadcasts it current GPS position, ammunition load, and current status ("alive" or "dead") every few seconds.

The data is received and monitored at a central location manned by an observer/controller from the CMTC's operations group. This information is used to track and monitor a battle and enforce exercise "rules of engagement." Designed to provide realistic battles. A time-condensed replay is used later to facilitate unit After Action Reviews (AARs) so they can critique their own performance. Tapping into this resource provides a means to create a tactical template based upon individual vehicle movements.

The Hohenfels Maneuver Data Set

To create a dataset of an actual tactical battle, contact was established with the Center for Army Lessons Learned and the CMTC Battle Analysis Training System operators. Data depicting sixteen different battles from five different 1999 rotations was collected. Each battle had five to eleven time-based "tactical snapshots." The snapshots were the basic data used to create a vehicle location overview image for a rotation.

For each snapshot, each friendly vehicle location was referenced from its snapshot location to a georeferenced CMTC color infrared image mosaic to create a series of ArcView GIS shapefiles. Every friendly vehicle was entered as a point data source without reference to its type or status (as indicated by the snapshot graphic). One shapefile was created for each tactical snapshot. Each individual snapshot shapefile was later merged into a new shapefile that consolidated all the vehicle positions for each battle in a rotation. By doing this, an illusion of a rotation's vehicular battlefield movement was created in accordance with other field-based battle tracking systems (see Figure 1).

A weakness in this dataset was that the tactical snapshots were not taken at uniform time intervals. Additionally, the snapshots did not always have a uniform scale. Some images covered the entire battlefield at a scale of 1:50,000 while others represented smaller, more detailed portions of a battle at a 1:25,000 (or larger) scale. The time-condensed unit AAR battle replays were also provided on videotape as an additional data source. While extremely helpful as background information and insight into the thought processes behind tactical decisions, the tapes did not provide enough detail to be a source of "hard" data to enhance the tactical snapshots.

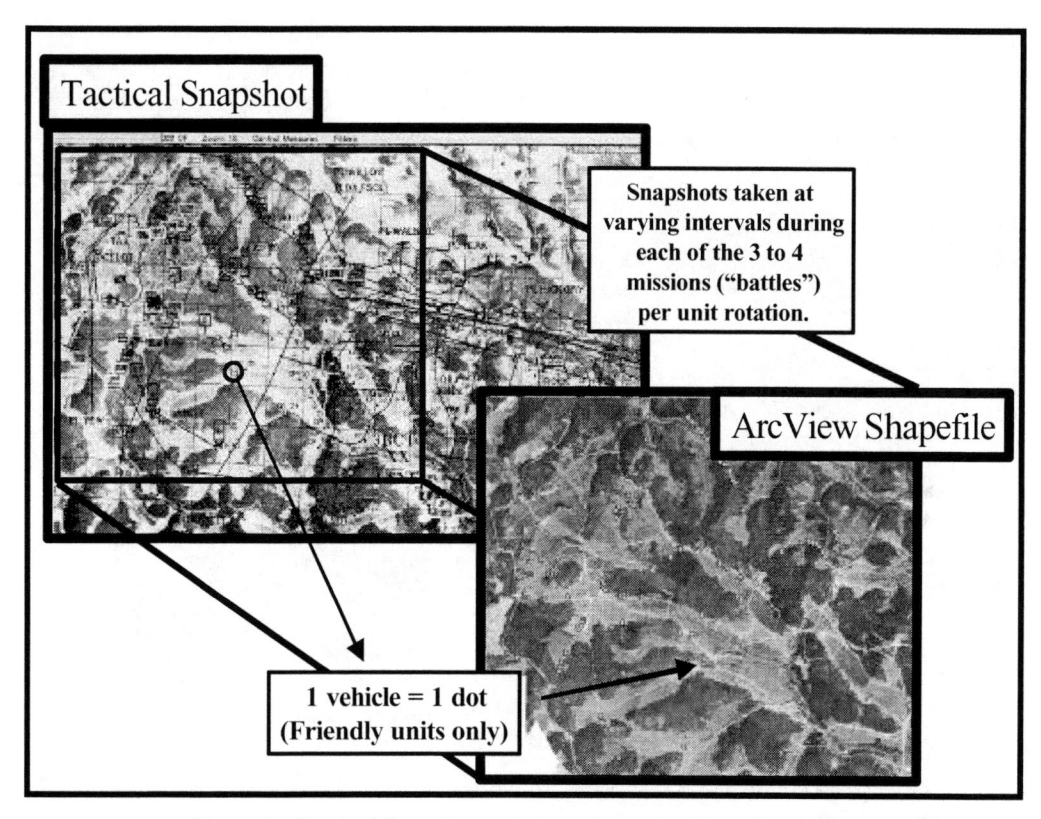

Figure 1. Tactical Snapshot and Associated ArcView Shapefile.
Source: Herl, 2000.

Despite these limitations, merging each battle's shapefile into a rotational roll-up created a detailed overview of one rotation. In an attempt to account for errors when creating shapefiles from the original data, two buffer rings were generated around every vehicle data point within the rotational roll-up shapefile. One ring was generated at a distance 25m from the vehicle data point to compensate for vehicle data point positioning error. A second outer ring with a 50m radius was also generated to give an idea of where the vehicle would most likely move within the next few seconds after a snapshot. In effect, the second ring helped to somewhat offset the time inconsistencies between snapshots (see Figure 2).

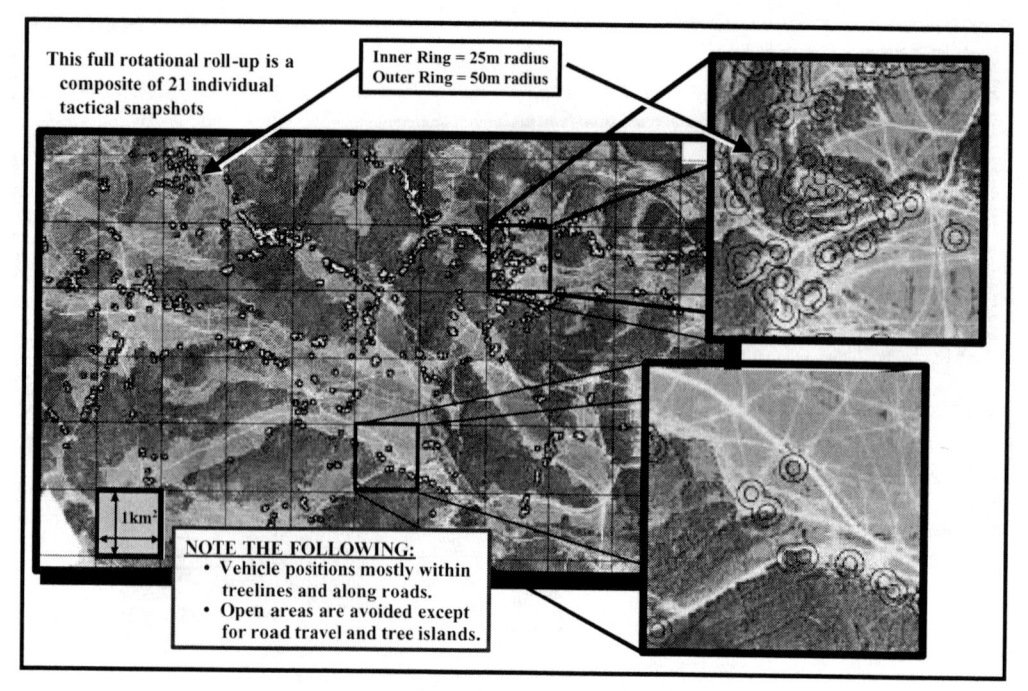

Figure 2. Complete CMTC Hohenfels Rotation.
Source: Herl, 2000.

Visible trends appeared within and near definable terrain locations. Table 1 generalizes these preferred, non-preferred area trends, and their landscape characteristics. As such, the generalizations in this table seem to be a viable approach towards spatially identifying maneuver impact locations.

The next step in this study was to apply these generalizations to a maneuver area. The vehicle for validating this step was a newly created training event called the CMTC-Live Fire Exercise held at Grafenwoehr, Germany.

Case Study: Grafenwoehr's CMTC-Live Fire Area

Prince Luitpold, the Regent of Bavaria, formally created the Grafenwoehr Training Area in 1907. It became an ideal site for artillery unit exercises and was designed to garrison and train members of the III Bavarian Corps. Later, infantry and cavalry commanders considered the area suitable for their training as well. The training area was officially opened for active training on 30 June 1910 when the first artillery round was fired. Since then, Grafenwoehr has seen constant, intensive, war and peacetime use along with numerous facility upgrades (Kneidl and Meiler, 1990).

Table 1. Martial Preferred and Non-Preferred Areas (CMTC-Hohenfels Analysis)

PREFERRED AREA TYPES		
	1	On or near established or improved surface roads
	2	Within forested area fringes nearby large open areas
	3	Nearby or within tree islands or taller vegetation clumping
	4	Major road junctions, especially those near easily identifiable choke points
	5	Reverse-slope of hills perpendicular to travel direction
	6	Low-ground and defiles that "shield" (provide cover) from enemy fire
	7	Around combat service support (CSS) assets marked on maneuver graphics
	8	Around tactical field headquarters locations marked on maneuver graphics
NON-PREFERRED AREA TYPES		
	W	Wide-open, mostly flat areas
	X	Heavily wooded areas and forests other than forest-clearing fringe areas
	Y	Far-side treelines of open areas
	Z	Large forest clearing "cul-de-sacs" perpendicular to general travel direction

The latest upgrades established the Training Area as it is today. In April of 1982, many of Grafenwoehr's ranges were redesigned to accommodate the (then new) M1-series Abrams main battle tank and M2/3-series Bradley infantry/cavalry fighting vehicle. The "Graf-82" range modernization program created the current Range 201 and Range 301 complexes (Burckhardt, 1994).

Under the new concept, the two fixed ranges, 201 and 301 (and the area between them), will become a live-fire "maneuver box" where armored vehicles roam and fire freely over the terrain instead of remaining confined to existing fixed course roads and firing lanes. In Southern Germany's characteristic rolling-hill terrain and moist climate, this training practice may increase land maneuver damage (*Preliminary Watershed Reconnaissance,* 1998). Predicting where the most intense activities will take place has both ecological and financial consequences.

The 7th Army Training Command (7th ATC) authorized the range conversion largely in part due to the increased training benefits a live-fire maneuver range holds over a fixed firing range. The 7th ATC also decided to slow full implementation of the conversion and conduct a modified environmental impact statement (M-EIS) in order to fulfill the obligations and intent of U.S. national laws and Department of Defense regulations. The 7th ATC Integrated Training Area Management (ITAM) coordinator convinced Army commanders to fund an initial two-year environmental monitoring contract awarded to the Center for Ecological Management of Military Lands (CEMML), a Colorado State University research and service organization that specializes in monitoring military lands (*Environmental Monitoring CMTC-LF,* 1998). All this was done despite the fact that German laws do not have similar National Environmental Protection Act-like requirements that apply to all military kassernes (bases). This setting provided the ideal location to combine a land impact assessment study with the creation of a tactical advantage map outlining the exercise's high tactical value areas.

The Tactical Advantage Map and High Tactical Value Areas

A tactical advantage map identifies and locates high tactical value areas an attacker and/or defender would consider most important to mission success. It is a fusion of the martial terms that describe both key/critical terrain and avenues of approach, but it is at a detail level much greater than that provided by a more traditional, larger area terrain analysis method such as the Modified Combined Obstacle Overlay--MCOO (*FM 5-33,* 1990; *FM 34-130,* 1994).

Translating this cognitive concept into real, "on-ground" locations is one of the most basic (albeit one of the most illusive) tasks fundamental to the art of warfare. To do this well, a military leader should first physically see the place of battle and understand the nature of the ground. Early military history speaks about how successful generals personally chose the ground upon which they fought.

Advances and innovations in weaponry and battle tactics have forced leaders to now reconnoiter future battlefields with maps, photographs, satellite imagery, and computer model representations. Of these, computer-generated terrain models can be extremely beneficial when attempting to identify areas of controlling ground or low-lying maneuver routes for low-level echelon exploitation.

For Grafenwoehr, an on-site reconnaissance was initially performed in June 1999. Insights from this visit were used as the basis for determining if the MCOO could feasibly predict tactical training pattern disturbance development. The intent behind that trip was not to locate and identify the tactically advantageous terrain within the CMTC Live-Fire Area. Discussions with the 7th ATC ITAM Coordinator in Grafenwoehr and the 264th Base Support Battalion's Environmental Section at Hohenfels led to acquisition of aerial imagery and a high-resolution digital elevation model with a 4m grid cell elevation posting of Grafenwoehr. With this new data, a three-dimensional model of the CMTC Live-Fire Area was generated using the commercial software package ERDAS *Imagine* and its "image drape" options. Another public domain (military-oriented) terrain visualization software package, Terrabase II, provided easily usable, military-specific functions that added a second area of analysis perspective. By combining the knowledge gained on the previous June 1999 trip with the computer generated 3-D representations of the Live-Fire area from these software packages, a workable facsimile of Grafenwoehr was created vastly assisting with high tactical value area identification.

High tactical value (HTV) areas are those areas that an armor or mechanized unit leader would consider valuable or desirable ground in combat. HTV areas provide maximum protection to vehicles and crews but still allow them to fight and destroy an enemy. They also provide the best defensive positions that have a commanding view of the terrain AND are the best attack routes into an enemy position. Since these areas are desired terrain, anticipated maneuver disturbance should likewise be higher than average.

Low tactical value areas are those that have little value in attacking or defending an enemy (Table 1). In general, these areas are places a vehicle commander or unit leader avoids unless necessary. When occupying low tactical value areas, a combat arms leader would be in a position where he could not engage an enemy or not be engaged by an enemy. It could also be a place where exposure to enemy fire is extremely high and friendly casualty potential is maximized. Since these areas are non-desired terrain, anticipated maneuver disturbance would be non-existent or light. Note that these low tactical value areas could still witness disturbance, but its severity would be less than that in an HTV area.

Plotting High Tactical Value (HTV) Areas

The map image was created in ArcView. The background is a georectified aerial photograph mosaic image of the Live-Fire Area. High tactical value areas represent likely tank or mechanized infantry vehicle platoon battle positions, choke points, or the best tactical routes (see Figure 3).

The ellipse-like polygons are battle positions (BPs) and have roughly a 400m long axis and a 200-300m short axis--roughly corresponding to a platoon battle position's area it covers on the ground, sometimes referred to as a "tactical footprint" (*FM 7-7J*, 1993; *FM 17-15*, 1998). These positions usually set on dominating high ground usually with over one kilometer of unrestricted visibility. Better positions were near cover (trees) and/or had a reverse slope hide position.

Major choke points are represented as irregular-shaped polygons located near areas of restricted maneuver. The major choke points in the CMTC Live-Fire area are near or on Range 201 within the PA9707 and PA9808 grid squares where narrow roads pass through densely spaced trees.

The lines connecting the platoon battle positions represent the most likely low-lying ground routes vehicles would take to the platoon battle positions in a general west-to-east attack zone or an east-to-west defense sector. A single line was traced on the image to represent the general travel route and a 25m buffer was added to each side of the line to create a 50m wide vehicle movement.

These HTV areas were initially created *without regard for* off-limits maneuver areas in order to account for a fully unrestricted maneuver exercise. After the HTV areas were created, two layers representing environmental and cultural off-limits areas were created based upon an AutoCAD file print-out of the exercise maneuver plan agreed upon by both Grafenwoehr's Range Control Operations Office and Environmental Sections.

Without the artificial off-limits areas, the projected maneuver damage areas seemed to be regularly dispersed due to the lay of the land and routes provided by roads. The damage areas would be densest near the exercise start point in the west and gradually widen as the exercise boundaries open-up towards the east. This eastern separation is exaggerated due mainly to the dense forests separating Ranges 301 and 201. The lack of well-defined road networks and defendable terrain also precludes heavy mounted maneuvering within these areas. The wide open, flat areas needed to establish target arrays in both ranges helps discourage massed movements and its associated concentrated maneuver damage.

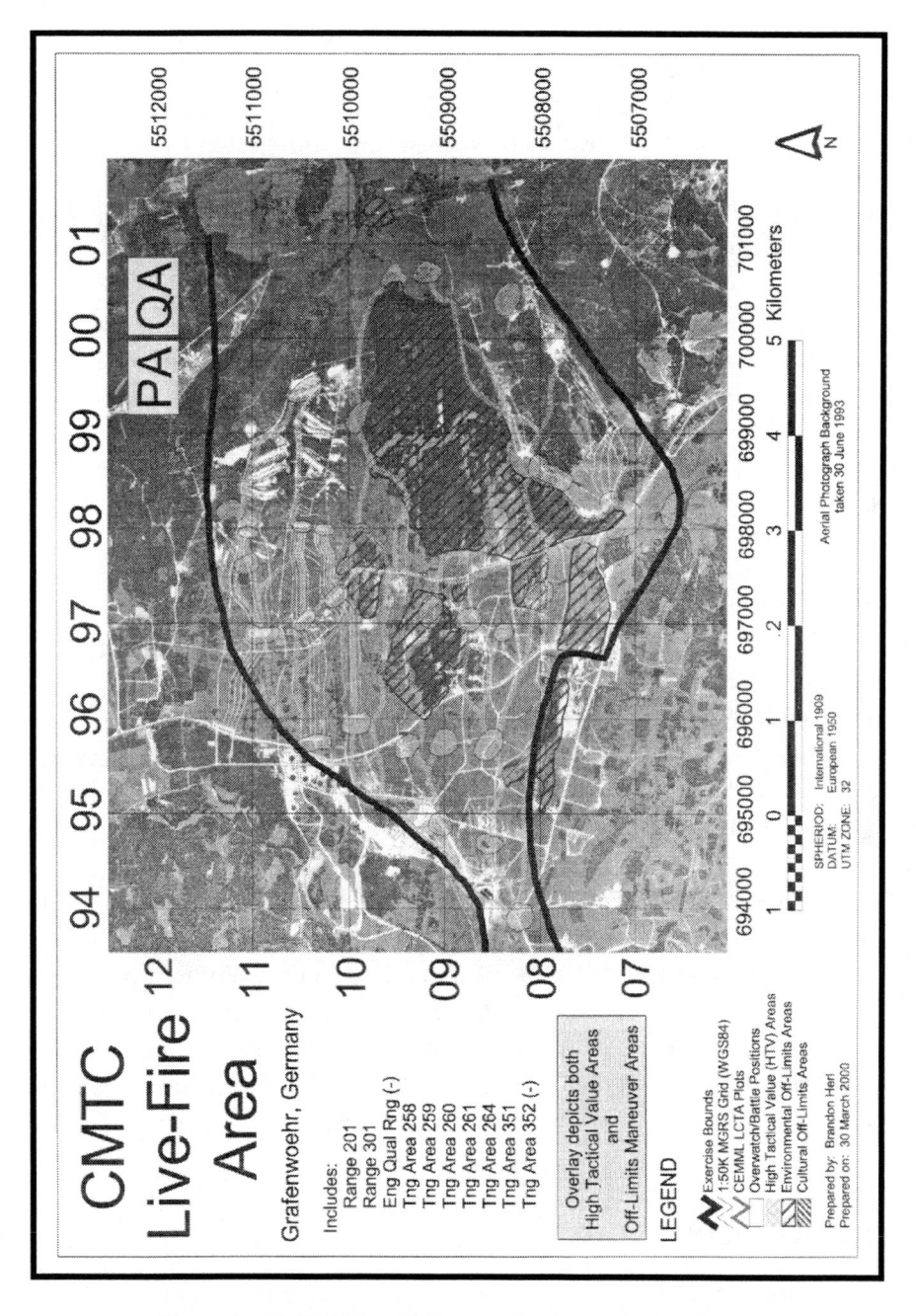

Figure 3. CMTC Live-Fire Area, Grafenwoehr Training Area.
Source: Herl, 2000.

With the inclusion of the off-limits areas, roughly seven of twenty-nine possible battle positions and four-to-five of ten major east-west movement corridors are either eliminated, interrupted, or both. The net effect of this action has two possible outcomes: 1) Reduction in training quality; training distractors[3]; and 2) Intensifying probable maneuver damage in on-limits areas.

The purpose of creating the natural area off-limits areas was to help mitigate wildlife disturbances and soil erosion losses within the anticipated heavily used CMTC Live-Fire maneuver box. Currently, all the off-limits areas within the exercise's bounds comprise over 5.3 km^2 (or 22%) of the live-fire's total land area. A very close look at the trade-off between disturbance mitigation and more concentrated maneuver damage assessments may warrant further investigation.

Preliminary Indications

To run an initial accuracy check on the HTV area model, some land-use trends were needed. For this information, disturbance measurements from a set of special use Land Condition Trend Analysis (LCTA) 100m transect plots was used. Data collected on these special use plots coincided with the previously mentioned environmental monitoring contract between the 7th ATC and the Center for Ecological Management of Military Lands. The LCTA dataset used had three separate monitoring periods, two in 1998 and one in 1999. These monitoring periods started nearly simultaneously with the first CMTC Live-Fire exercise event.[4]

A GIS data layer consisting of the LCTA special-use plot transects was created and overlaid on the high tactical value (HTV) area map. Transects that fell within the high tactical advantage areas were analyzed based upon their military disturbance values. Transects that did not fall within the templated HTV areas were used as comparison points. In theory, LCTA plots falling within the HTV areas should witness a higher overall level of military disturbance. Conversely, plots outside the HTV areas should have an overall lower military disturbance rate.

Method & Assessment

Out of thirty CEMML special-use LCTA plots, three (#317, #327, and #328) were discarded since they did not fall within the live-fire area's maneuver boundaries. The military-related disturbance values of the remaining plots were generalized into two categories of above average and below average disturbance (see Figure 4).

PLOT ID No.	SURVEY 98-1 M+P+R+T	SURVEY 98-1 N+O	SURVEY 98-2 M+P+R+T	SURVEY 98-2 N+O	SURVEY 99-1 M+P+R+T	SURVEY 99-1 N+O	ALL SURVEYS AVG M+P+R+T	ALL SURVEYS AVG N+O
301	45	55	69	31	64	36	59.3	40.7
302	74	26	68	32	70	30	70.7	29.3
303	75	25	46	54	77	23	66.0	34.0
304	59	41	38	62	70	30	55.7	44.3
305	76	24	63	37	72	28	70.3	29.7
306	100	0	45	55	88	12	77.7	22.3
307	12	88	0	100	19	81	10.3	89.7
308	6	94	0	100	0	100	2.0	98.0
309	0	100	0	100	0	100	0.0	100.0
310	0	100	0	100	17	83	5.7	94.3
311	48	52	8	92	16	84	24.0	76.0
312	0	100	0	100	3	97	1.0	99.0
313	0	100	10	90	2	98	4.0	96.0
314	7	93	46	54	78	22	43.7	56.3
315	51	49	7	93	37	63	31.7	68.3
316	87	13	56	44	83	17	75.3	24.7
317	OUTSIDE OF CMTC LIVE-FIRE AREA – NOT INCLUDED IN CALCULATIONS							
318	0	100	0	100	0	100	0.0	100.0
319	0	100	0	100	0	100	0.0	100.0
320	0	100	6	94	12	88	6.0	94.0
321	0	100	0	100	2	98	0.7	99.3
322	2	98	0	100	15	85	5.7	94.3
323	71	29	61	39	47	53	59.7	40.3
324	49	51	7	93	16	84	24.0	76.0
325	44	56	46	54	64	36	51.3	48.7
326	21	79	37	63	53	47	37.0	63.0
327 328	OUTSIDE OF CMTC LIVE-FIRE AREA – NOT INCLUDED IN CALCULATIONS							
329	30	70	24	76	43	57	32.3	67.7
330	0	100	0	100	0	99	0.0	99.7

LCTA PLOT MILITARY DISTURBANCE STATISTICS

	Survey 98-1	98-2	99-1
Mean	31.7	23.6	35.1
Median	21.0	8.0	19.0
Std Error	6.4	5.0	6.1
Median	21.0	8.0	19.9
Std Dev	33.2	25.8	31.8
Range	100.0	69.0	88.0
Min	0.0	0.0	0.0
Max	100.0	69.0	88.0

All Surveys – Averaged –

Mean	30.1
Median	20.4
Std Error	5.5
Median	24.0
Std Dev	28.4
Range	77.7
Min	0.0
Max	77.7

Disturbance Key
M = Military (non-maneuver)   N = No Disturbances
P = Vehicle Pass   P = Other (non-military) Disturbances
R = Vehicle Road
T = Vehicle Trail

LCTA PLOT MILITARY DISTURBANCE CATEGORIZATION

LEGEND
- Plot Outside Live-Fire Area (Not Included)
- Below Average Use (NO Military Disturbance)
- Below Average Use (0-30% Military Disturbance)
- Above Average Use (30-50% Military Disturbance)
- Above Average Use (>50% Military Disturbance)
- ⊠ LCTA Plot within High Tactical Value (HTV) Area

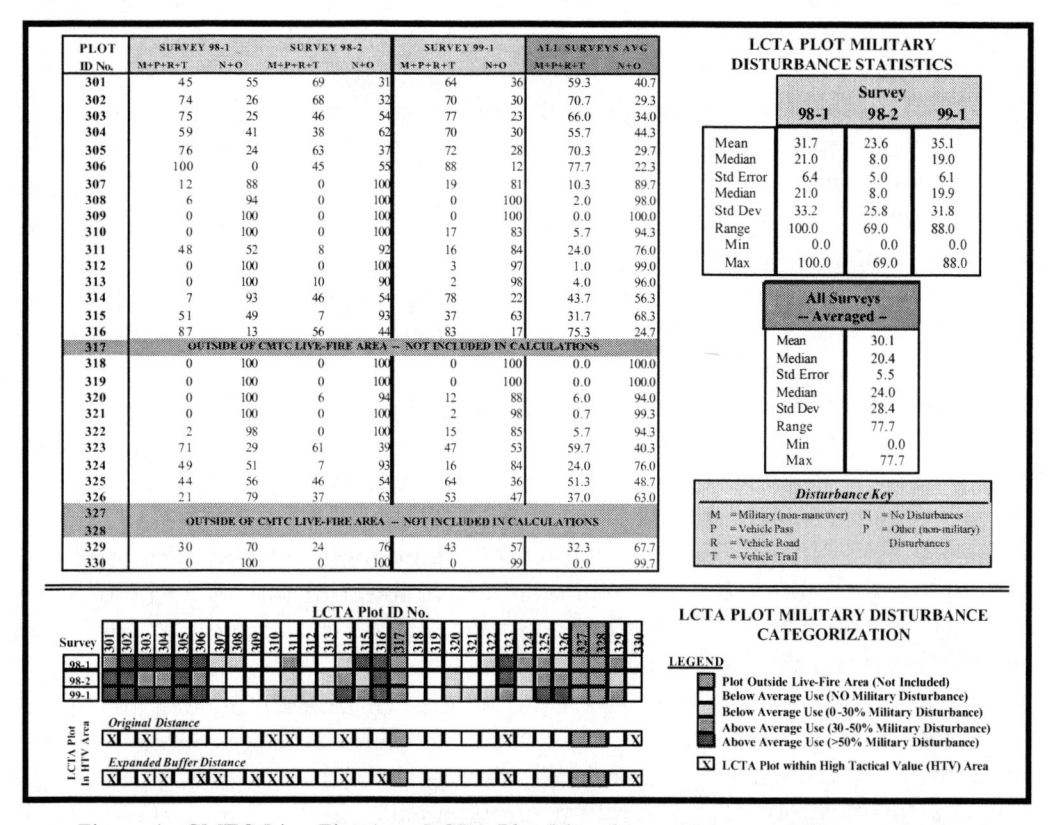

Figure 4. CMTC Live-Fire Area LCTA Plot Disturbance Values and Characterization. Source: Herl, 2000.

Each survey's military disturbance average values ranged from 23.6% to 35.1%. To form a baseline disturbance measurement, a summary statistical test was performed on all three surveys' combined military disturbance values. A resultant average military-related disturbance value of 30.1% was generated. As such, LCTA plots witnessing over 30% or more military disturbance on any survey were considered "above average use." Plots with less than 30% military disturbance were labeled "below average use."

To account for the wide range of military disturbance values (anywhere from 0% to 100% in some instances), each above and below average use category was broken into two tiers. The lowest-end tier of below average use was established for plots that had no military disturbance. The above-average, upper-end military disturbance tier tracked plots that witnessed disturbance values of over 50%.[5] Each plot's military disturbance value was then compared to the usage key and labeled as below average use or as above average use (and also placed into one of each categories upper or lower tiers).

LCTA plots that intersected high tactical value areas were then identified. Seven of the twenty-seven available LCTA plots (<25%) fell within HTV areas that compromised approximately 3.3 km^2 of the live-fire's total available maneuver area (18%). An initial accuracy assessment percentage rating was generated for the plots that fell within the HTV areas. This accuracy percent was generated by dividing the number of HTV plots having above average use by the total number of HTV area plots within the sample pool. A below average use accuracy rate was generated in a like manner. The number of below average use plots within the HTV area was divided by the total number of HTV area plots.

Of the seven LCTA plots that fell within the HTV area, five had above average use and two had below average use. The above average use accuracy rate was roughly 71% and the below average accuracy was 29%. Both values were consistent with the theoretical idea that high tactical value terrain would show higher usage while other terrain would have lower disturbance values.

To double-check the validity of the HTV area accuracy rates, the remaining twenty LCTA plots that fell *outside* the HTV areas were run through the same accuracy assessment test. Of the twenty remaining plots, ten had above average use and ten had below average use resulting in a 50% accuracy rate for both above and below average use plots. This did not hold to the HTV terrain theory. (The above average use rate of plots outside the HTV area should have been low while the below average use rate should have been high.)

Plotting HTV Areas—Doctrinal Revisitation

With the mixed results of the original HTV area delineation, a deeper look at the data was warranted. Part of the results seemed to come from skewed sample sizes between the number of plots falling inside and outside of the HTV area delineation (seven compared to twenty). Another portion of the sample size error seemed rooted in the differences between predominantly open ground within Range 301 and its adjacent training areas as compared to the relatively closed terrain surrounding Range 201.

In accordance with military maneuver tactics, when open areas are encountered, tactical formations open-up correspondingly. In other words, if doctrinal distance between vehicles under normal terrain conditions is 100m (such as on Range 201), in more open terrain, intervehicular distances would increase accordingly (such as on Range 301).

To account for this spreading-out effect, the same HTV areas were uniformly expanded across the CMTC Live-Fire Area. The original overwatch/battle position template was increased by 100m in all directions from its original 400m width and 200-300m depth. The new position now had dimensions of roughly 600m in width by 400-500m in depth, making the increasing in surface area up to 40% larger. Choke points also grew outward by 100m from their original perimeters. Movement paths were also expanded. Each linear movement path was given an extra 25m on either side of the path's centerline. This expansion created 100m wide paths (vice 50m wide paths in the original HTV template) effectively doubling each path's surface area. Overall, the expanded buffer distance now compromised over 7.2km^2 (nearly 40%) of the available maneuver area within the CMTC Live-Fire boundaries.

These uniform expanded buffer distance template adjustments helped account for unit spreading in the predominantly open terrain comprising the majority of the CMTC Live-Fire maneuver area. It also (unexpectedly) nearly doubled the sample pool of LCTA plots that fell within the new buffered distance high terrain value areas. The number of plots now falling within the HTV areas increased from seven to twelve while the number of plots outside the HTV areas decreased from twenty to fifteen. An accuracy assessment having nearly identical sample pool sizes was now possible.

The same accuracy assessment procedures were used for this expanded buffer HTV area template as used on the original HTV area template. The numbers of above average use plots falling within HTV areas was divided by the total number of HTV area sample pool plots. Below average use accuracy rates were also calculated. The increased sample pool decreased the accuracy rate of above average use plots from 71% to 67% and increased the below average use rate from 29% to 33%; however, the results were still consistent with the HTV area prediction.

Of more interest, however, was the accuracy rate change for plots falling outside the HTV areas. The accuracy rate for above average use plots falling outside the HTV areas dropped 10%. (The original template rate was 50% and the buffered distance template rate was 40%). Likewise, the below average use plot accuracy rate increased from its original value of 50% to a buffered distance value of 60%. The buffered distance template findings *were consistent* with the HTV theory.

Despite a 4% drop in the buffered distances above average accuracy ratings for HTV areas, the percent accuracy difference between both above and below average use plots increased. The difference in accuracy rates for above average HTV area plots and non-HTV area plots grew by 6% from 21% to 27%. In other words, while the accuracy rate may have dropped slightly, the accuracy rate gap between inside and outside HTV area plots increased (see Figure 5).

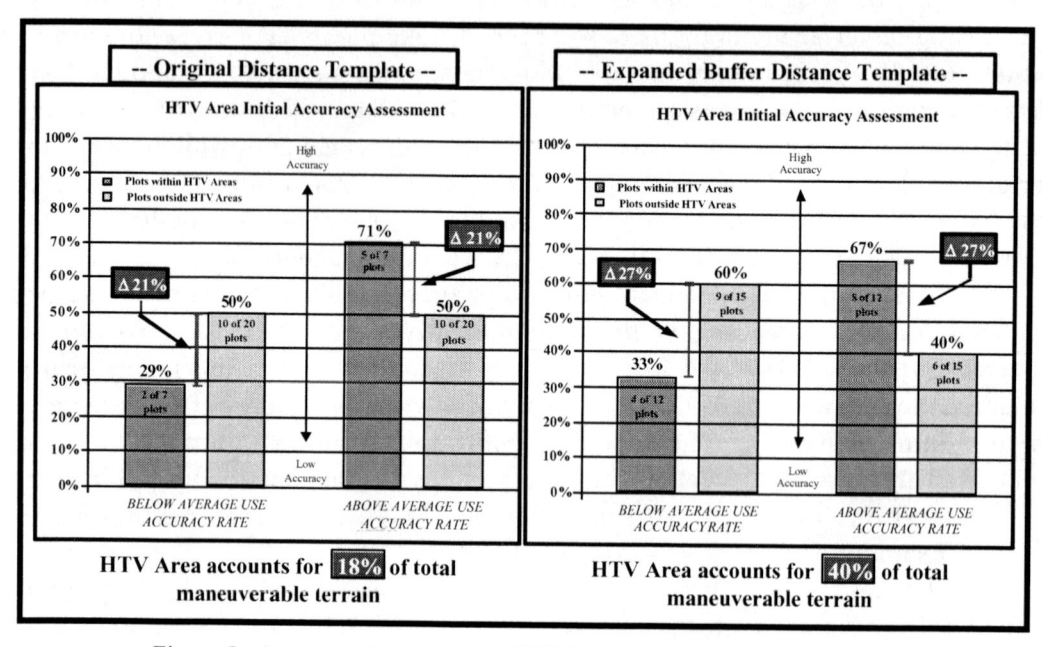

Figure 5. Accuracy Assessment of HTV Areas and LCTA Disturbance.
Source: Herl, 2000.

Discussion

This chapter started by discussing a common assumption made about military lands -- maneuver damage is not predictable. While the accuracy assessment technique used in this portion study and its associated small statistical sample pools cannot fully validate that maneuver damage locations are completely predictable, this study does indicate that it is possible to narrow-down the areas that most heavy damage will occur in.

Despite low accuracy ratings generated for plots falling within both HTV area distance templates, a distinct separation between disturbance trends is evident. This low accuracy rating is tempered by the fact that the original distance HTV area's above average use 71% accuracy rating was based upon an analysis of land area that compromised less than 1/5th of the CMTC Live-Fire's total available maneuver area. With the expanded buffer distance's above average use accuracy rating of 67%, 2/5ths of the live-fire's total available maneuver space HTV area--still less than half of all available maneuver lands.

A corollary to this study is that military training land management decisions cannot be made from an uninformed point of view. A land steward lacking the knowledge of how a maneuver unit tactically utilizes a training area (or selected portions thereof) will more than likely make unwise (and unpopular) land management decisions.

There are three probable reasons behind the low accuracy for plots lying outside HTV areas. The first is that the LCTA data supporting the accuracy ratings was not designed to perform this type of an analysis. To truly validate this method, a separate study design and sampling method would be needed.

A second possible reason could be attributed to the HTV area identification method itself. The method is based upon identifying combat maneuver unit high tactical value areas. Disturbances within other areas could easily be the result of disturbance caused by units supporting the combat maneuver elements (self-propelled mortars, field artillery, logistical support vehicles, or command and control headquarters). Furthermore, this method assumes that the unit leaders and soldiers participating in a live-fire exercise are moderately trained and have a working knowledge of tactical maneuver. Untrained units or leaders who lack martial terrain sense maneuvering over this area would not conform to this model since the model is based upon executing sound tactical doctrine.

The third possible reason for higher disturbance values in the western half of the CMTC Live-Fire area is that much of this land is still an open general training area. Disturbances west of the 97-Easting and south of the 09-Northing could be the result of training not associated with any CMTC Live-Fire training. A combination of two or all of these reasons could be a possibility.

Summary

This entire analysis was originally designed to "bridge the gap" between trainers and ecologists in both language and systems. It was purposefully structured to fill a known void in the military lands management study literature by providing a user-based perspective on how a warfighter views and utilizes training lands. This study also justifies conducting further research into refining the methods used to spatially identify disturbance areas prior to land degradation.

Unfortunately, this method is an experience-based model set in typical Western European terrain. The tactical trends evident here will not apply everywhere, especially in places where climatic conditions and terrain are drastically different. Directly applying the lessons from this study to a widely different region would most likely result in unsettling inconsistencies or complete failure.

However, just as terrain changes from place to place in the world, tactics to fight on other terrain evolve to best utilize the surrounding environment. This adaptive technique was somewhat demonstrated in this study when the HTV area template was adjusted to account for subtle terrain differences between Range 301 and Range 201. By applying knowledge of how maneuver units doctrinally maneuver in different regions, the principles and concepts underscoring this specific method are validated.

Notes

1. This paper's information was originally collected and compiled as the author's Colorado State University graduate degree professional paper titled *Predicting Tactical Training Patterns on Military Lands* as a derivative from an earlier research report generated for the U.S. Army Research Office (Shaw et al., 2000).

2. No U.S. Army field manual singularly states all of these items in this particular way. This list is actually a compilation from numerous sources (written, oral, and learned) that is a function of all training officers and noncommissioned officers receive throughout their military careers and professional field experiences.

3. Discussions with the CMTC Live-Fire design team previously indicated that this restriction might hamper training by limiting viable tactical options to combat leaders. Some of the video-taped after action reviews (AARs) from 2-63d Armor's February 1999 CMTC Live-Fire Rotation mentioned items about maneuver restrictions. Personal discussions with the commander of C Company attested to this training distracter. The company commander stated that while his major concerns were for the safety of his soldiers and the

training benefit of the exercise, ensuring that all his soldiers understood the locations of all the off-limits areas was both difficult and frustrating.

4. The first CMTC Live-Fire rotation was a "proof-of-principle" exercise held in March 1998 that demonstrated the feasibility of conducting live-fire maneuvers at Grafenwoehr. CEMML established their special-use LCTA plots after this exercise in June 1998. As of June 1999, three full rotations and three LCTA data collection surveys had been held.

5. With a median disturbance value of 24.0% and a mean disturbance value of 30.1% with a standard deviation of 28.4%, a high-use area determination point was not easily generated. Plotting $+1\sigma$ from both the median and mean gave a range of high-use benchmark values from 52.4% to 58.5%. For simplicity, areas witnessing military disturbance values greater than 50% were selected as "highly-used" areas. The background for this disturbance data came from a series of different reports prepared by CEMML during the 1998 to 2000 monitoring period.

Bibliography

Burckhardt, P. 1994. *The Major Training Areas: Grafenwoehr/Vilseck, Hohenfels, Wildflecken.* Amberg, Germany: Druckhaus Oberpfalz.

Doe, W., Julien, P., and Ogden, F. 1997. Maneuversheds and Watersheds: Modeling the Hydrologic Effects of Mechanized Training on Military Lands. *American Water Resources Association Conference Proceedings, 29 June-03 July 1997.*

Herl, B. 2000. *Predicting Tactical Training Patterns on Military Lands.* M.S. Degree (Forest Sciences) professional paper, Fort Collins, CO: Colorado State University.

Kneidl, D. and Meiler, O. 1990. *Truppenubungsplatz Grafenwoehr (Training Area Grafenwoehr).* Pressath, Germany: Buchhandlung Eckhard Bodner.

Price, D., Anderson, A., McLendon, T., and Guertin, P. 1997. *The U.S. Army's Land-Based Carrying Capacity. U.S. Army Construction and Engineering*

Research Laboratories (USACERL) Technical Note 97/142. Champaign, IL: USACERL.

Shaw, R, Doe, W., and Herl, B. 2000. Using Military Terrain Analysis Techniques and the Modified Combined Obstacle Overlay to Predict Training Area Land Use Trends at the Combat Maneuver Training Center (CMTC) Live-Fire Range in Grafenwoehr, Germany. *Short Term Innovative Research (STIR) Report P-40109-EV-II.* Fort Collins, CO: Center for Ecological Management of Military Lands.

Walsh, S., Entwisle, B., and Rindfuss, R. 1998. Landscape Characterization through Remote Sensing, GIS, and Population Surveys. In *GIS Solutions in Natural Resources Management*, ed. S. Morain, pp. 251-265. New Mexico: OnWord Press.

_____. 1998. *Army Regulation 350-4, Integrated Training Area Management.* Washington, DC: Headquarters, Department of the Army.

_____. 1990. *Field Manual 5-33, Terrain Analysis.* Washington, DC: Headquarters, Department of the Army.

_____. 1999. *Field Manual 7-7J, The Mechanized Infantry Platoon and Squad (Bradley).* Washington, DC: Headquarters, Department of the Army.

_____. 1998. *Field Manual 17-15, The Tank Platoon.* Washington, DC: Headquarters, Department of the Army

_____. 1994. *Field Manual 34-130, Intelligence Preparation of the Battlefield.* Washington, DC: Headquarters, Department of the Army

_____. 1993. *Field Manual 100-5, Operations.* Washington, DC: Headquarters, Department of the Army.

_____. 1998. *Department of the Army Pamphlet 350-4, Integrated Training Area Management (Coordinating Draft).* Aberdeen Proving Grounds, MD: U.S. Army Environmental Center.

_____. 1998. *Combat Training Center Handbook.* Washington, DC: Headquarters, Department of the Army.

_____. 1995. *ITAM Strategy Guide.* Aberdeen Proving Grounds, MD: U.S. Army Environmental Center.

_____. 1999. *U.S. Army Training and Testing Area Carrying Capacity (ATTACC) Handbook for Installations--Version 1.1.* Aberdeen Proving Grounds, MD: U.S. Army Environmental Center.

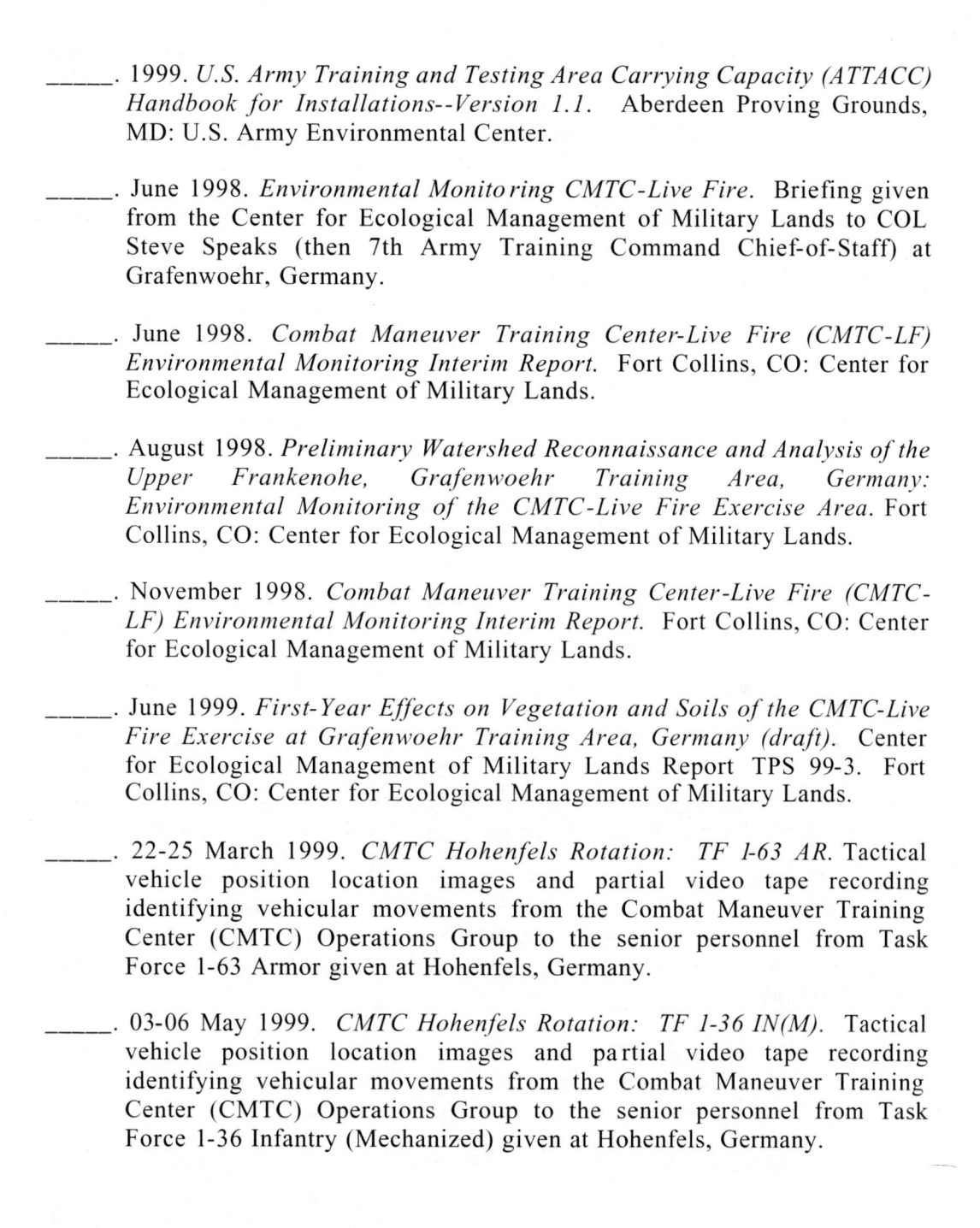

_____. June 1998. *Environmental Monitoring CMTC-Live Fire.* Briefing given from the Center for Ecological Management of Military Lands to COL Steve Speaks (then 7th Army Training Command Chief-of-Staff) at Grafenwoehr, Germany.

_____. June 1998. *Combat Maneuver Training Center-Live Fire (CMTC-LF) Environmental Monitoring Interim Report.* Fort Collins, CO: Center for Ecological Management of Military Lands.

_____. August 1998. *Preliminary Watershed Reconnaissance and Analysis of the Upper Frankenohe, Grafenwoehr Training Area, Germany: Environmental Monitoring of the CMTC-Live Fire Exercise Area.* Fort Collins, CO: Center for Ecological Management of Military Lands.

_____. November 1998. *Combat Maneuver Training Center-Live Fire (CMTC-LF) Environmental Monitoring Interim Report.* Fort Collins, CO: Center for Ecological Management of Military Lands.

_____. June 1999. *First-Year Effects on Vegetation and Soils of the CMTC-Live Fire Exercise at Grafenwoehr Training Area, Germany (draft).* Center for Ecological Management of Military Lands Report TPS 99-3. Fort Collins, CO: Center for Ecological Management of Military Lands.

_____. 22-25 March 1999. *CMTC Hohenfels Rotation: TF 1-63 AR.* Tactical vehicle position location images and partial video tape recording identifying vehicular movements from the Combat Maneuver Training Center (CMTC) Operations Group to the senior personnel from Task Force 1-63 Armor given at Hohenfels, Germany.

_____. 03-06 May 1999. *CMTC Hohenfels Rotation: TF 1-36 IN(M).* Tactical vehicle position location images and partial video tape recording identifying vehicular movements from the Combat Maneuver Training Center (CMTC) Operations Group to the senior personnel from Task Force 1-36 Infantry (Mechanized) given at Hohenfels, Germany.

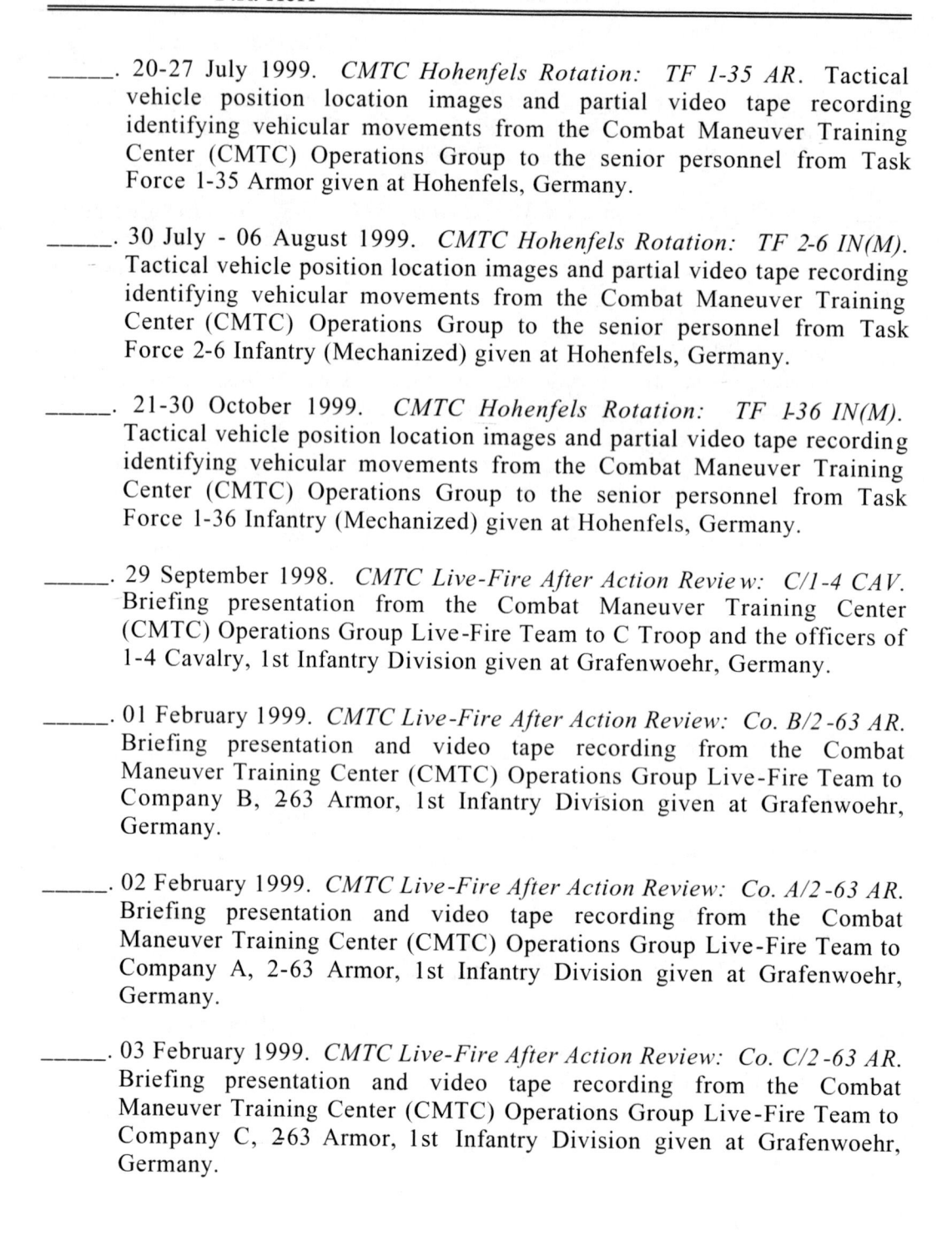

_____. 20-27 July 1999. *CMTC Hohenfels Rotation: TF 1-35 AR.* Tactical vehicle position location images and partial video tape recording identifying vehicular movements from the Combat Maneuver Training Center (CMTC) Operations Group to the senior personnel from Task Force 1-35 Armor given at Hohenfels, Germany.

_____. 30 July - 06 August 1999. *CMTC Hohenfels Rotation: TF 2-6 IN(M).* Tactical vehicle position location images and partial video tape recording identifying vehicular movements from the Combat Maneuver Training Center (CMTC) Operations Group to the senior personnel from Task Force 2-6 Infantry (Mechanized) given at Hohenfels, Germany.

_____. 21-30 October 1999. *CMTC Hohenfels Rotation: TF 1-36 IN(M).* Tactical vehicle position location images and partial video tape recording identifying vehicular movements from the Combat Maneuver Training Center (CMTC) Operations Group to the senior personnel from Task Force 1-36 Infantry (Mechanized) given at Hohenfels, Germany.

_____. 29 September 1998. *CMTC Live-Fire After Action Review: C/1-4 CAV.* Briefing presentation from the Combat Maneuver Training Center (CMTC) Operations Group Live-Fire Team to C Troop and the officers of 1-4 Cavalry, 1st Infantry Division given at Grafenwoehr, Germany.

_____. 01 February 1999. *CMTC Live-Fire After Action Review: Co. B/2-63 AR.* Briefing presentation and video tape recording from the Combat Maneuver Training Center (CMTC) Operations Group Live-Fire Team to Company B, 2-63 Armor, 1st Infantry Division given at Grafenwoehr, Germany.

_____. 02 February 1999. *CMTC Live-Fire After Action Review: Co. A/2-63 AR.* Briefing presentation and video tape recording from the Combat Maneuver Training Center (CMTC) Operations Group Live-Fire Team to Company A, 2-63 Armor, 1st Infantry Division given at Grafenwoehr, Germany.

_____. 03 February 1999. *CMTC Live-Fire After Action Review: Co. C/2-63 AR.* Briefing presentation and video tape recording from the Combat Maneuver Training Center (CMTC) Operations Group Live-Fire Team to Company C, 2-63 Armor, 1st Infantry Division given at Grafenwoehr, Germany.

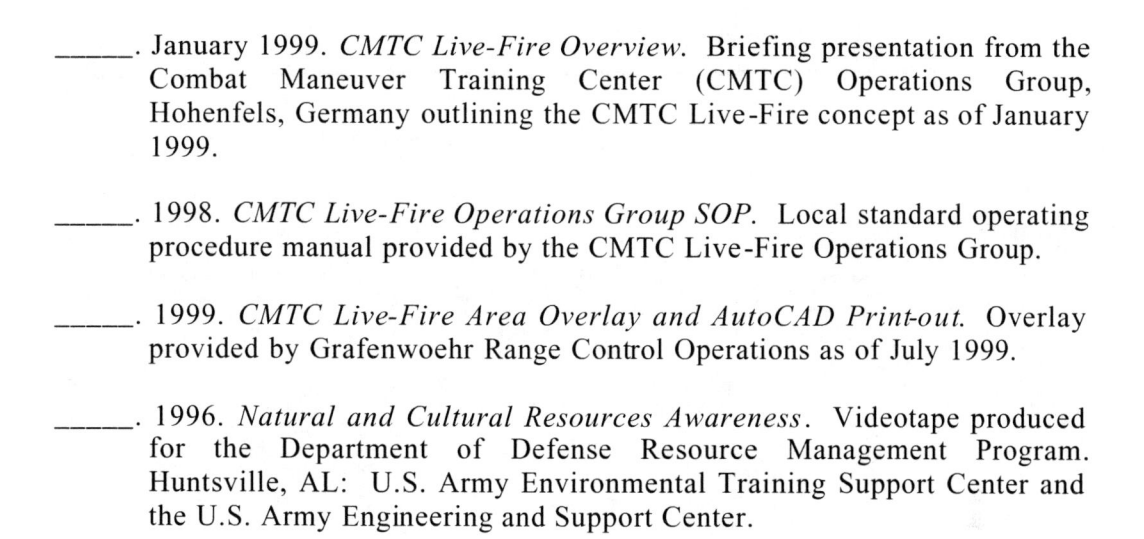

_____. January 1999. *CMTC Live-Fire Overview.* Briefing presentation from the Combat Maneuver Training Center (CMTC) Operations Group, Hohenfels, Germany outlining the CMTC Live-Fire concept as of January 1999.

_____. 1998. *CMTC Live-Fire Operations Group SOP.* Local standard operating procedure manual provided by the CMTC Live-Fire Operations Group.

_____. 1999. *CMTC Live-Fire Area Overlay and AutoCAD Print-out.* Overlay provided by Grafenwoehr Range Control Operations as of July 1999.

_____. 1996. *Natural and Cultural Resources Awareness.* Videotape produced for the Department of Defense Resource Management Program. Huntsville, AL: U.S. Army Environmental Training Support Center and the U.S. Army Engineering and Support Center.

M1A1 Abrams tanks occupy a firing position.
Source: Center for Ecological Management of Military Lands (CEMML), Colorado State University, Fort Collins, Colorado.

Conclusion

Although it is tempting to date the use of military geography to 1479 BC and appropriate to recognize its well-established history in Europe, in the United States the subfield has evolved from infancy during the course of the twentieth century. Within academia, military geography gained definition and distinction during the course of the World Wars and the Korean Conflict, only to have its legitimacy questioned during the Vietnam era. After a period of stagnation during the 1970s and early 1980s, military geography has rebounded and accelerated to new heights during the 1990s. Within U.S. Military circles, however, the subfield has maintained its utility and vitality throughout the twentieth century. Military geography has long been recognized as an imperative for adequately equipping, training, and deploying forces to places around the world. Moreover, U.S. military doctrine has long recognized that the sub-field is an integral part of all military plans and operations, regardless of the context.

On any given day, the U.S. Military is represented in nearly half the countries the world, actively engaged in training, peacekeeping, humanitarian assistance missions, environmental security, and other military operations other than war (Binnendijk 1998). And, this trend is likely to continue for the foreseeable future. While a specific military unit is not actively engaged in a "real world operation," it is nevertheless charged with training and preparing to respond to crises across the full range of military operations from humanitarian assistance to fighting and winning major theater wars. The problems encountered within each scenario are varied, complicated, and often unexpected. Many of the problems, however, are innately geographic. The tools, techniques, and knowledge of the geographer have proven to be relevant and unquestionably useful in solving military problems in peacetime or war.

It has been our intent to reconfirm the utility of the longstanding tradition of approaching military geography from either an applied or historical perspectives, during wartime or in hindsight. After all, the primary mission of the U.S. Military is to fight and win the nation's wars. That mission is the *raison d'être* for any army and has remained virtually unchanged for throughout history.

Nevertheless, this book has introduced two additional contexts outside of the wartime arena. The MOOTW and Peacetime contexts offer fertile ground for military geographic inquiry. These contexts are not new to the U.S. Military, but military geography practitioners have been previously unexposed to them. The dynamic nature of military affairs poses a considerable, if not, continuous challenge to the subfield because the Military's doctrine is always in the process of evolving in response to national objectives, new technology, the variable nature of the threat, and the world's ever-changing balance of power. Thus, military geographers must stay appraised of recent developments in both geography and military science. To these ends, it seems prudent to embrace the opportunities afforded by the expanded scope of the subfield and forge ahead into the two new contexts.

Our goal has been to produce a forward-looking book on military geography that is useful to the military professional, yet more importantly, serves to stimulate interest within academia. Thus, we have taken the middle of the road approach in some instances in an attempt to satisfy the needs, yet balance the interests of military personnel, college students, and professional and academic geographers. Our intent was to generate a product that could serve as the basis, if not point of departure for a course on military geography at universities and in military schools alike. We believe strongly in the relevance of geography to military and political affairs across the spectrum from peacetime to war.

껳 ꩜

Glossary of Common Military Terms

— A —

ACTIVE DEFENSE: A multi-layered defense in depth via multiple engagements. The defending unit is typically mobile and does not fight from one position.

AERIAL PORT OF DEBARKATION: An airfield for sustained air movement at which personnel and material are discharged from aircraft.

AERIAL PORT OF EMBARKATION: A airfield for sustained air movement at which personnel and material board or are loaded aboard aircraft to initiate an aerial movement.

AIR STRIKE: An attack on specific objectives by fighter, bomber, or attack aircraft on an offensive mission.

AIR SUPPORT: All forms of support given by air forces on land or sea.

AIR SUPREMACY: The degree of air superiority wherein the opposing air force is incapable of effective interference.

AMPHIBIOUS OPERATION: An attack launched from the sea by naval and landing forces embarked in ships involving a landing on a hostile shore.

ASSAULT: To make a short, violent, well-coordinated attack against an objective.

ASSAULT FORCE: In an amphibious, airborne, ground or air assault operation, those units charged with the seizure of an objective or lodgment area.

ATTACK: An offensive action characterized by movement supported by fire.

ATTRITION: The reduction of the effectiveness of a force caused by loss of personnel and equipment or supplies.

AVENUE OF APPROACH: An air or ground route of an attacking force leading to its objective or key terrain in its path.

AXIS OF ADVANCE: A line of advance for the purpose of control; often a road or a group of roads; or a designated series of locations, extending in the direction of the enemy.

B

BALLISTIC MISSILE: Any missile, which does not rely on its aerodynamic surfaces to produce lift and consequently follows a ballistic trajectory when thrust is terminated.

BASE: A locality from which operations are projected or supported.

BATTALION: An independent tactical unit of approximately 500-1,000 soldiers; composed of 4-5 companies. A battalion typically includes it's own fire and logistical support apparatus.

BATTLE: A series of related tactical engagements that last longer than an engagement, involve larger forces, and could affect the course of the campaign. They typically occur when division, corps or army commanders fight for a significant objective.

BEACHHEAD: A designated area on a hostile shore that, when seized and held, ensures the continuos landing of troops and material, and provides maneuver space requisite for subsequent operations ashore.

BREACH: A tactical task where any means available are employed to break through or secure passage through an enemy defense, obstacle, minefield, or fortification.

C

CAMPAIGN: A series of related military operations aimed at accomplishing a strategic or operational objective within a given time and space.

CANALIZE: To restrict operations or movment to a narrow zone by the use of terrain, obstacles or fire.

CAPTURE: To take into custody a hostile force, equipment, or personnel as the result of military operations -- to take control of terrain.

CASUALTY: Any person who is lost to the organization by having been declared dead, missing, ill, wounded, or injured.

CEASE-FIRE: A command or status given to any unit or individual firing a weapon to stop engaging the target.

CHAIN OF COMMAND: The succession of commanding officers from superior to subordinate through which command is exercised.

CHOKE POINT: A geographic location on land or water that restricts the movement of forces and can be natural or man-made.

COLLATERAL DAMAGE: Unintended and undesirable civilian personnel injuries or material damage adjacent to a target produced by the effects of friendly weapons.

COMBAT SUPPORT: Units and soldiers that provide critical combat functions in conjunction with combat arms units.

COMPANY: A small tactical organization of 75-150 combat soldiers usually organized into 3-4 platoons. Led by a captain and capable of limited combat tasks; but cannot fight independently for long periods.

COMPARTMENT: Areas bounded on at least two sides by terrain features affected by drainage and relief, such as woods, ridges, or ravines, that limit observation or observed fire into an area from points outside the area; and effect avenues of approach.

COVERING FORCE: A covering force operates apart from the main body to develop the situation early; and deceives, disorganizes, and destroys enemy forces. Unlike screening or guard forces, a covering force is tactically self-contained -- that is, it is organized with sufficient combat support (CS) and combat service support (CSS) forces to operate independently of the main body.

COUNTERATTACK: Attack by all or part of a defending force against an enemy attacking force, for such specified purposes as regaining lost ground, or cutting off or destroying enemy advance units.

COUNTERINSURGENCY: Those military, paramilitary, political, economic, psychological and civic actions taken by a government to defeat an insurgency.

COUNTERTERRORISM: Offensive measures taken to prevent, deter, and respond to terrorism.

CORPS: An administrative headquarters to which divisions and support units are attached. A typical corps may consist of two or three divisions and the attendant combat support and combat service support units.

CROSS-COMPARTMENT: A terrain compartment, its long axis generally perpendicular to the direction of movement of a force.

CULTURAL RESOURCE: Monuments, nationality, identifiable or distinctive buildings and structures, archives and libraries, ancient artifacts and structures, archeologically important sites, historically important sites, mosques, churches, cathedrals, temples or other sacred structures, sacred sites or areas, museums, and works of art.

D

D-DAY: An unnamed day on which a particular operation commences. The execution date of the operation.

DEBARKATION: The unloading of troops, equipment or supplies from a ship or aircraft.

DECISIVE ENGAGEMENT: In land and naval warfare, an engagement in which a unit is considered fully committed and cannot maneuver or extricate itself. The action must be fought to a conclusion and the unit will either win or lose.

DEFEAT: May or may not entail the destruction of any part of the enemy army; rather, the objective is to disrupt or nullify his plan and/or subdue his will to fight so that he is unwilling or unable to pursue his course of action.

DEFEND: A combat operation designed to defeat an attacker and prevent him from achieving his objectives.

DEFILE: A narrow gorge of pass that tends to prevent easy movement of troops.

DELAYING OPERATION: An operation usually conducted when the commander needs time to concentrate or withdraw forces, to establish defenses in greater depth, to economize in an area, or to complete offensive actions elsewhere. In the delay, the destruction of the enemy force is secondary to slowing his advance to gain time.

DELIBERATE ATTACK: An attack planned and carefully coordinated with all concerned elements based on thorough reconnaissance, evaluation of all available intelligence and relative combat strength, analysis of various courses of action, and other factors affecting the situation. It is generally conducted against a well-organized defense when a hasty attack is not possible or has failed.

DELIBERATE CROSSING: A crossing of an inland water obstacle that requires extensive planning and detailed preparation.

DEMONSTRATION: An attack or show of force on a front where a decision is not sought. Made with the aim of deceiving the enemy. It is similar to a feint except that enemy contact is not sought.

DEPLOYMENT: (1) The movement of forces within an area of operations. (2) The positioning of forces in preparation for battle. (3) The relocation of forces and material to a designated area or theater of operations. (4) All activities from a home station as part of a movement to a continental or intercontinental locations.

DESTROY: A tactical task to physically render the enemy force combat ineffective.

DESTRUCTION: Renders enemy force combat ineffective unless reconstituted. Puts a target out of action permanently.

DISRUPTION: To counter attacker's initiative and prevent him from concentrating overwhelming combat power against a part of the defense. The defender must disrupt the synchronization of enemy's operation.

DIVERSION: The act of drawing the attention of forces of an enemy from the point of the principal operation.

DIVISION: A large tactical unit consisting of 10-11 battalions organized into three or four brigade-sized forces. A division is capable of sustained, long-term operations.

DOCTRINE: Fundamental principles by which the military force or elements thereof guide their actions in support of national objectives.

E

ECHELON: A sub-division of a headquarters or separate level of command.

ECONOMY OF FORCE: The allocation of minimal combat capacity or strength to secondary efforts.

EMBARKATION: The process of putting personnel or material into ships or aircraft.

ENCIRCLEMENT: The loss of freedom of maneuver to one force resulting from an enemy force's control of all routes of egress or reinforcement.

ENGAGEMENT: A small tactical conflict usually between maneuver forces.

ENVELOPMENT: Basic form of maneuver or doctrine seeking to apply strength against weakness. Envelopments avoid enemy's front where his forces are most protected and fires most easily concentrated. Successful envelopment requires discovery or creation of an assailable flank. It places a premium on agility since its success depends on reaching enemy's vulnerable rear before he can shift forces and fires. An offensive maneuver where main attacking force passes around or over enemy's principal defensive positions to secure objectives in his rear.

EXPLOITATION: An offensive operation that usually follows a successful attack to take advantage of weakened or collapsed enemy defenses. Its purpose is to prevent reconstitution of enemy defenses, to prevent enemy withdrawal, and secure deep objectives.

F

FEINT: An offensive operation intended to draw the enemy's attention away from the area of the main attack, which induces him to move reserves or shift fire support in reaction to the feint. Feints must appear real; therefore, some contact with enemy is required. Usually a limited-objective attack ranging in size from a raid to a supporting attack is conducted.

FIX: To hold an enemy force in place through an attack or demonstration.

FORCE PROJECTION: The movement of military forces from the Continental United States or theater in response to requirements of war or stability and support operations.

FRONT: The line of contact between opposing forces.

FRONTAL ATTACK: An offensive maneuver in which the main action is directed against the front of the enemy forces, and over the most direct approaches.

G

GUERRILLA WARFARE: Military or paramilitary operations conducted in enemy-held or hostile territory by irregular, predominantly indigenous forces.

H

HASTY ATTACK: An offensive operation without extensive preparations. It is conducted with resources immediately available in order to maintain momentum or take advantage of enemy situation.

HASTY CROSSING: The crossing of an inland water obstacle using the means at hand and made with pausing operations for elaborate preparations.

H-HOUR: The specific hour on D-Day in which a particular operation commences.

HOST NATION: A country which receives the forces and/or supplies of allied or coalition nations to be located on, to operate in, or to transit through its territory.

HUMANITARIAN ASSISTANCE: Programs conducted to relieve or reduce the results of natural or man-made disasters or other endemic conditions such as human pain, disease, suffering, hunger or privation that might present a serious threat to life or that can result in great damage to or loss of property.

I

IN EXTREMIS: A situation of such exceptional urgency that immediate action must be taken to minimize the imminent loss of life or catastrophic degradation of the political or military situation.

INFILTRATION: Covert movement of the attacking force through enemy lines to favorable positions in their rear. Successful infiltration requires above all, avoidance of detection. Since that requirement limits the size and strength of the infiltrating force, infiltration can rarely defeat the defense by itself, but rather, is normally used in conjunction with some other form of maneuver.

INITIATIVE: The ability to set or change the terms of battle; implies an offensive spirit.

INSURGENCY: An organized movement aimed at the overthrow of a constituted government using subversion and armed conflict.

INTERNAL SECURITY: The state of law and order prevailing within a nation.

INTERVISIBILITY: The condition of being able to see one point from another. This condition may be interrupted by weather, fog, smoke, or terrain.

IRREGULAR FORCES: Armed individuals or groups who are not members of the regular armed forces, police or other internal security forces.

ISOLATE: The tactical task given to a unit to seal off an enemy force from its source of support, to deny enemy freedom of movement, and prevent enemy units from having contact with other enemy units.

J

JOINT FORCE: A general term applied to force composed of significant elements of two or more Military Departments, operating under the command of a single joint force commander.

K

KEY TERRAIN: Any locality or area, the seizure or retention of which affords a marked advantage to either combatant.

L

LANDING BEACH: That portion of a shoreline usually required for the landing of a battalion landing team. However, it may also be that portion of a shoreline constituting a tactical locality over which a larger force may be landed.

LIAISON: That contact or intercommunication maintained between elements of military forces to ensure mutual understanding and unity of purpose and action.

LINE OF CONTACT: A general trace delineating the location where two opposing forces are engaged.

LINE OF DEPARTURE: A coordination line or feature to coordinate the departure of friendly forces as they move forward to contact the enemy.

LINES OF COMMUNICATIONS: All routes, land, water and air, which connect an operating military force with its base of operations and along which military supplies and military forces move.

M

MAIN ATTACK: The principal attack or effort into which the commander places the bulk of offensive capability at his disposal. An attack directed against a chief objective of the campaign or battle.

MANIFEST: A document specifying in detail the passengers or items carried for a specific destination.

MEDICAL THREAT: The composite of all on going or potential enemy actions and environmental conditions that reduce the performance and effectiveness of the soldiers. Effectiveness can be reduced through wounds, injuries or diseases.

MOBILITY CORRIDORS: Areas where a force will be canalized because of terrain constrictions. The mobility corridor is usually free of obstacles and allows a military force to move freely along the axis of the corridor.

MOBILE DEFENSE: Focus on destruction of an attacking force by permitting enemy advance into a position, which exposes him, to counterattack and envelopment by mobile reserves.

MOVEMENT TO CONTACT: An offensive operation designed to gain initial ground contact with the enemy or to regain lost contact.

N

NATION ASSISTANCE: Civil and/or military assistance rendered to a nation by foreign forces within that nation's territory during peacetime, crises or emergencies, or war based on agreements mutually concluded between nations.

NEUTRALIZE: (1) To render ineffective or unusable. (2) To render enemy personnel or material incapable of interfering with a particular operation. (3) To render safe mines, bombs, missiles and boobytraps. (4) To make harmless anything contaminated with chemical agents.

NONBATTLE INJURY: A person who becomes a casualty due to circumstances not directly attributable to hostile action or terrorist activity.

NONCOMBATANT: An individual, in an area of combat operations, who is not armed and is not participating in any activity in support of any factions or forces involved in combat.

O

OBJECTIVE: The clearly defined, decisive, definable aims, which every military operation should be directed towards.

OPERATION: A military action or the carrying out of a strategic, tactical, service, training or administrative military mission.

OPERATIONAL CONTROL : Authority delegated to a commander to direct forces assigned so that he may accomplish specific missions or tasks that are usually limited by function, time, or location.

ORDER OF BATTLE: The identification, strength, command structure, and disposition of the personnel, units and equipment of any military force.

P

PARAMILITARY FORCES: Forces or groups that are distinct from regular armed forces of a country, but resembling them in organization, equipment, training and mission.

PEACE BUILDING: Post-conflict actions, predominantly diplomatic and economic, that strengthen and rebuild governmental infrastructure and institutions in order to avoid a relapse into conflict.

PEACE ENFORCEMENT: Application of military force, or the threat of its use, normally pursuant to international authorization, to compel compliance with resolutions or sanctions designed to maintain or restore peace and order.

PEACEKEEPING: Military operations undertaken with the consent of all major parties in a dispute, designed to monitor and facilitate implementation of an agreement and support diplomatic efforts to reach long-term political settlement.

PEACEMAKING: The process of diplomacy, mediation, negotiation, or other forms of peaceful settlements that arranges the end of a dispute.

PENETRATION: Used when enemy flanks are not assailable and when time does not permit some other form of maneuver. Attempts to rupture enemy defenses on a narrow front and thereby create both assailable flanks and access to enemy's rear. Penetrations typically comprise three stages: 1) initial rupture of enemy positions; 2) roll-up of flanks on either side of the gap; and 3) exploitation to secure deep objectives. Because the penetration itself is vulnerable to flank attack, especially in its early stages, penetrating forces must move rapidly and follow-on forces must close behind to secure and widen the shoulders.

PHASE: A particular part of an operation – usually defined by a period of time or by a geographic objective.

PURSUIT: An attempt to annihilate fleeing enemy forces requiring unrelenting pressure to prevent their reconstitution or evasion. An offensive operation against a retreating enemy force. It follows a successful attack or exploitation and is ordered when the enemy cannot conduct an organized defense and attempts to disengage. Its object is to maintain relentless pressure on the enemy and completely destroy him.

R

RAID: An operation, usually small-scale, involving a swift penetration of hostile territory to secure information, confuse the enemy, or to destroy installations. It ends with the planned withdrawal of the raiding force.

RECONNAISSANCE IN FORCE: A limited-objective operation conducted by at least a battalion task force to obtain information, locate and test enemy dispositions, strengths and reactions. Although a reconnaissance in force is executed primarily to gather information, the force conducting the operation must seize any opportunity to exploit tactical success. If the enemy situation must be developed along a broad front, reconnaissance in force may consist of strong probing actions to determine enemy situation at selected points.

RECONSTITUTION: Focussed action to restore ineffective units to a specified level of combat effectiveness.

REORGANIZATION: Action taken to shift resources within an attritted unit to increase its level of combat effectiveness. Consists of measures such as internal redistribution of equipment and personnel and formation of composite units.

RIVER CROSSING: An operation required before ground combat power can be projected and sustained across a water obstacle.

RIVERINE OPERATIONS: Operations conducted by forces organized to cope with and exploit the unique characteristics of a riverine area and inland waterways.

RULES OF ENGAGEMENT: Directives issued by a competent military authority which delineate the circumstances and limitations under which U.S. forces will initiate and/or continue combat engagement with other forces encountered.

S

SALIENT: Protrusion or bulge in the trace of the front lines.

SECURE: To gain possession of a position or terrain feature, with or without force, and deploy in such a manner to prevent its destruction or loss to enemy action.

SECURITY ASSISTANCE: Group of programs authorized by the Foreign Assistance Act of 1961. The provision of military training, equipment and other defense-related services by grant, loan, credit, or cash sales in the furtherance of national policy.

SEIZE: Clear and gain control of a designated area.

STATUS OF FORCES AGREEMENT: An agreement, which defines the legal position of a visiting military force, deployed in the territory of a friendly state.

STRATEGY: The art and science of developing and using political and military force as needed during peace and war to achieve national objectives.

SUPPORT FORCE: Those forces charged with providing intense direct overwatching fires to assault and breaching forces.

SUPPRESSION: Direct and indirect fires, electronic countermeasures (ECM), or smoke brought to bear on enemy personnel, weapons, or equipment to prevent effective fire on friendly forces.

T

TACTICS: The employment of units in combat.

TERRAIN ANALYSIS: The collection, analysis and evaluation, and interpretation of geographic information on the natural and man-made features of the terrain, combined with other relevant factors, to predict the effect of the terrain on military operations.

TRAFFICABILITY: capability of the terrain to bear traffic.

TURNING MOVMENT; A variation of the envelopment in which the attacking force passes around the enemy's principal defensive positions to secure objectives deep in his rear to force the enemy to abandon his position or divert major forces to meet the threat.

U

UNCONVENTIONAL WARFARE: The broad spectrum of military and paramilitary operations normally of long duration, predominantly conducted by indigenous forces, which are organized and equipped by an external source.

W

WEAPONS OF MASS DESTRUCTION: Weapons that are capable of a high order of destruction and/or being used in such a manner as to destroy large numbers of people.

WITHDRAWAL: A planned operation in which a force in contact disengages from an enemy force.

A U.S. KFOR soldier, and his interpreter talk to a Kosovar Serb man after giving him a copy of "Dialog", a KFOR publication printed in Serbo-Croatian and Albanian, on May 4, 2000, in the Novo Brdo Obstina, Kosovo. KFOR is the NATO-led, international military force in Kosovo on the peacekeeping mission known as Operation Joint Guardian.
Source: DoD photograph by Sgt. Drew Lockwood, U.S. Army.

About The Authors:

Eugene J. Palka received a B.S. from the United States Military Academy at West Point in 1978. He earned an M.A. in geography from Ohio University and a Ph.D. from the University of North Carolina at Chapel Hill. He is a Colonel in the U.S. Army and an Associate Professor of Geography. He is currently the Director of the Geography Program at the U.S. Military Academy at West Point, New York. He has previously authored two books, a bibliography of military geography (in four volumes), several book chapters, two instructor's manuals to accompany college level geography textbooks, and more than thirty articles on various topics in cultural, historical, and military geography.

Francis A. Galgano, Jr. received a B.S. from the Virginia Military Institute in Lexington, Virginia in 1980. He earned his Masters Degree in geography and a Ph.D. from the University of Maryland at College Park. He is a Lieutenant Colonel in the U.S. Army and an Assistant Professor of Geography at the United States Military Academy, West Point, New York. Presently, he oversees the Physical Geography Program and the Academy's core course in Physical Geography. He has authored numerous articles on various topics in military and physical geography, as well as a physical geography study guide.

Jon C. Malinowski, Department of Geography and Environmental Engineering, United States Military Academy, West Point, New York. Dr. Malinowski is an Assistant Professor of Geography and oversees the Human Geography Program at West Point. He received a B.S. in Foreign Service from Georgetown University and earned an M.A. in geography and a Ph.D. from the University of North Carolina at Chapel Hill.

Ewan W. Anderson, Centre for Middle Eastern and Islamic Studies, University of Durham, Durham, England. Dr. Anderson is Professor of Geopolitics at the University of Durham and Distinguished Research Fellow at the Center for International Trade and Security, University of Georgia, U.S. He holds doctorates in both geography and politics. Professor Anderson has also published eleven books and more than 200 articles.

Julian V. Minghi, Department of Geography, University of South Carolina, Columbia, South Carolina. Dr. Minghi is a political geographer and Professor of Geography at the University of South Carolina. He holds a B.A. from Durham University, U.K. and an M.A. and a Ph.D. from the University of Washington.

Richard W. Dixon, Department of Geography, Southwest Texas State University, San Marcos, Texas. Dr. Dixon is an Assistant Professor of Geography at Southwest Texas State. He received a B.S. in physical oceanography from Rutgers University and earned a M.A. in geography from Southwest Texas State University. He has a doctorate from Texas A&M University.

Robert B. Shaw, Director, Center for Ecological Management of Military Lands, Colorado State University, Fort Collins, Colorado. Currently, Dr. Shaw is the director of the Center for Ecological Management of Military Lands (CEMML) and a professor of forest sciences at Colorado State University. He received a B.S. from Southwest Texas State University and an M.S. and Ph.D. from Texas A&M University.

William W. Doe, III, Center for Ecological Management of Military Lands, Colorado State University, Fort Collins, CO. Dr. Doe is a retired Army officer and is an associate director with the CEMML. He received a B.S. from the U.S. Military Academy, an M.S. from the University of New Hampshire and a Ph.D. from Colorado State University.

Mark W. Corson, Department of Geology/Geography, Northwest Missouri State University, Marysville, Missouri. Dr Corson received a BA. in government from the University of San Francisco in 1983 and earned an M.A. in geography and a Ph.D. from the University of South Carolina. He is currently an Assistant Professor of Geography at Northwest Missouri State University.

Robert G. Bailey, USDA-USFS Inventorying & Monitoring Institute, Fort Collins, Colorado. Dr. Bailey is a geographer with the USDA-USFS Inventorying and Monitoring Institute, Fort Collins, CO and leader of the Institute's Ecoregion unit.

Kurt A. Schroeder, Department of Social Science/Geography, Plymouth State College, Plymouth, New Hampshire. Dr. Schroeder is an Associate Professor of Geography and received a B.A. in geography from the University of Minnesota. He earned an M.A. and a Ph.D. from Pennsylvania State University.

Thomas E. Macia, Project Officer, Integrated Training Area Management (ITAM), Headquarters, Department of the Army, Office of the Deputy Chief of Staff for Operations, Washington, D.C. Thomas E. Macia, U.S. Army, Retired, is a geographer and currently assigned to CEMML as a project officer.

Joseph P. Henderson, Department of Geography and Environmental Engineering, United States Military Academy, West Point, New York. Major Joe Henderson received a B.S. in geography from the United States Military Academy at West Point in 1987. He earned an M.S. in geomorphology from the University of Tennessee in 1997. He served as an Assistant Professor of Geography at West Point from 1997 to 2000.

Andrew W. Lohman, Department of Geography and Environmental Engineering, United States Military Academy, West Point, New York. Major Lohman is an Instructor of geography at the United States Military Academy at West Point. He received a B.S. in geography from West Point and earned an M.A. from the University of South Carolina in 1999.

Wiley C. Thompson, Department of Geography and Environmental Engineering, United States Military Academy, West Point, New York. Major Thompson is a geography Instructor at West Point, where he received a B.S. in geography in 1989. He earned an M.S. in physical geography from Oregon State University in 1999.

Brandon K. Herl, Department of Geography and Environmental Engineering, United States Military Academy, West Point, New York. Captain Herl is an Armor officer and geography Instructor at West Point. He received a B.S. in geography at the United States Military Academy in 1990, and earned an M.S. at Colorado State University in 2000.